CIVIL ENGINEERING CONTRACTS
Volume 1: The Law, Administration, Safety

ELLIS HORWOOD SERIES IN CIVIL ENGINEERING

Series Editors

Structures: Professor H.R. EVANS, Department of Civil Engineering, University College, Cardiff

Hydraulic Engineering and Hydrology: Dr R.H.J. SELLIN, Department of Civil Engineering, University of Bristol

Geotechnics: Professor D. MUIR WOOD, Cormack Professor of Civil Engineering, University of Glasgow

CIVIL ENGINEERING CONTRACTS
Volume 1: The Law, Administration, Safety

KARL WILLIAMS
Department of Civil Engineering, University of Glasgow

ELLIS HORWOOD
NEW YORK   LONDON   TORONTO   SYDNEY   TOKYO   SINGAPORE

First published in 1992 by
ELLIS HORWOOD LIMITED
Market Cross House, Cooper Street,
Chichester, West Sussex, PO19 1EB, England

A division of
Simon & Schuster International Group
A Paramount Communications Company

Printed and bound in Great Britain
by Hartnolls, Bodmin

British Library Cataloguing in Publication Data

A catalogue record for this book is available from the British Library

ISBN 0–13–132986–3

Library of Congress Cataloging-in-Publication Data

Available from the publisher

TABLE OF CONTENTS

PREFACE

This book is primarily aimed at students on civil engineering degree and diploma courses, and at graduate civil engineers on the so-called 'route to chartered status'.

The principal aim of the book is to provide basic information regarding the formation and administration of civil engineering contracts, the established procedures associated therewith and the legal and contractual responsibilities of the personnel involved in such. Whilst there are a number of excellent texts which provide more detailed coverage of certain topics dealt with herein, it is intended that this book provide what is, in effect, a platform from which individual topics may be studied in greater depth if necessary.

The contents of the book have been divided into three sections, the first of which deals with the framework of law within which the construction industry operates. Following a general introduction to the law, its workings and the liabilities imposed on individuals and organisations thereby, detailed coverage is given of the two principal branches of civil law which affect the industry, namely: the law of Tort (or Delict in Scotland), which covers general liability, and the law of Contract which covers arranged liability.

By way of illustrating particular legal principles, and of indicating how the law views certain aspects of the construction industry's *modus operandi*, this section of the book contains a liberal sprinkling of examples of legal disputes and the resultant findings of the courts. Whilst each example has been taken from real life, the majority of such are not formally identified, with reference to the law reports or other sources from which they have been abstracted, unless such identification is relevant. Whilst such a practice is positively *abhorred* by lawyers and those of a legal bent, it should be emphasised that this book has not been written for lawyers and is, most certainly, *not* intended for use by those who might wish to quote legal 'chapter and verse' to a Site Agent or Resident Engineer at the first hint of disagreement. That apart, however, many examples have been subjected to a certain amount of 'editorialisation' to ensure clarity, brevity (neither of which are conspicuous features of the average law report!) and an absence of irrelevant detail. Consequently, formal identification is considered inadvisable on account of it being potentially misleading.

It is the author's experience that the vast majority of books, which have been written for the 'engineering' or 'construction' market, and which deal with the relationship between the UK construction industry and the law, studiously ignore the fact that Scotland has a separate legal system to that of England. Those that *do* recognise the existence of a separate legal system north of the border, however, tend to restrict any mention thereof to the use of one or two trite phrases along the lines of '...there may be slight differences for contracts operating under Scots Law – for further information, consult a Scots lawyer ...'. By way of remedying that deficit, the 'legal' section of this book makes specific reference to the major differences between Scots and English law in so far as those differences are relevant to the particular topic under consideration.

The second section deals with the formation and administration of civil engineering contracts and covers such matters as: the functions, duties and responsibilities of the parties involved, the various types of contract which may be used, the associated procedural aspects and the necessary documentation.

Recognising the nature and form of the vast majority of civil engineering contracts carried out in the United Kingdom, this section includes detailed coverage of the Bill of Quantities – with particular reference to its compilation in accordance with the Civil Engineering Standard Method of Measurement – and the I.C.E. Conditions of Contract. In both cases, and by way of enabling the reader to gain experience in the use of the aforementioned standard documents, an 'exercises' section is included in both cases.

The third and final section of the book deals with the important topic of construction safety.

It is the opinion of this author that – perhaps owing to the fact that it is rarely, if ever, given more than cursory coverage other than in a text specifically and exclusively dedicated thereto – the perception exists amongst those preparing to enter, or only recently having entered, the construction industry that construction safety, somehow, is a topic which requires to be considered separately from any other aspect of contract management and administration. As anybody who has been involved with the construction industry for any length of time will no doubt confirm, this is a false perception.

The 'safety' section, therefore, is included with the specific intent of correcting this perception by encouraging the reader to appreciate that construction safety *is* part and parcel of contract management and administration and, furthermore, for reasons which the section demonstrates, is inextricably intertwined with the law.

In simple terms, the content of this final section of the book has been designed to provide the reader with answers to the following questions: "What sort of accidents can occur on my site?", "Why might they occur?", "How can I reduce their occurrence?" and "What are the potential consequences, to me and my company, of my failure to attempt to do so?".

Despite the obvious 'civils' bias of this book, members of allied disciplines, such as building, architecture and quantity surveying, may find its contents of interest and value. It should also be mentioned that, although a number of topics dealt with in this book specifically relate to construction contracts carried out in the United Kingdom, many of the general principles discussed will be as valid overseas as they are in the UK.

Finally, and by way of mollifying those who have a tendency to get 'hot under the collar' about such matters, the author wishes to emphasise that his exclusive use of male pronouns, when referring to the parties involved in the construction industry, should not be taken as anything other than a reflection of his desire to avoid the repetitive (and, arguably, slightly distracting) usage of phrases such as: 'he/she' and 'his/her'. Thus the reader may take it that 'he' implies 'he/she' etc. etc.

Karl Williams
Glasgow
July 1992

ACKNOWLEDGEMENTS

I am indebted to the Institution of Civil Engineers for permission to reproduce material from:

The Conditions of Contract and Forms of Tender, Agreement and Bond for use in connection with Works of Civil Engineering Construction (5th Edition);

The Civil Engineering Standard Method of Measurement (2nd Edition);

I would like also to express my thanks to all those who have given me assistance in the production of this book, especially to those of my erstwhile colleagues in the industry who, in a variety of ways (not all of them *positive*!), unknowingly provided me, over the years, with the incentive and the wherewithal to produce this book.

My particular thanks must go to Alexander Black (Department of Private Law, University of Glasgow) for his patient and valuable assistance in the preparation of the 'legal' chapters of the book, to John Dunbar (Partner – Crouch Hogg Waterman, Glasgow) for his helpful suggestions as to the content of the book and, lastly, to David Muir Wood (Cormack Professor, Department of Civil Engineering, University of Glasgow) for giving me the idea in the first place and for helping me to get the project off the ground.

Finally, a debt of gratitude is owed to Sylvia, Joe and Robbie, without whose unstinting support and encouragement over the last year or so this book would probably *never* have been finished.

PART 1

THE LAW

Chapter 1

THE LAW AND LEGAL LIABILITY

To a professional lawyer having a vested interest in extolling its virtues, law is possibly regarded as the single most important influence on society. It reflects society's progress and development and provides the cement which binds together and stabilizes society's complex framework, preventing its collapse into the dark abyss of chaos and disorder.

In the minds of certain other members of society, however, the word 'law' conjures up a somewhat different picture − a picture, perhaps engendered more by hearsay than hard evidence, which portrays the litigation process as an inordinately expensive exercise culminating in a lengthy judicial hearing which carries no guarantee of justice being done at the end of the day. Litigation, to such people, is an exercise designed principally to fill the pockets of lawyers, and is something which should be avoided in all instances save those in which there is no alternative!

Although these points of view are somewhat extreme, there is a certain element of truth in both: whilst it is generally accepted that the law suffers from many shortcomings, and can be shown to be an ass in a number of instances, it must be recognised that an ordered society cannot survive in the absence of a firm foundation of rules and regulations to govern its day−to−day existence.

1.1 The Function of Law

In an ordered society, the function of the law is twofold: it protects the rights of the individual whilst, at the same time, imposing upon him certain obligations and responsibilities with respect to his interaction with the rest of society.

For example, the individual has a legally protected right to free speech but is restricted in the extent of such by a legally enforceable obligation to avoid slander. In such a case, the imposed obligation is of a *civil* nature and can be enforced by the possibility of his having to pay damages to the slandered party. If, however, the individual's perceived right to free speech extends to the public divulgence of information deemed to be detrimental to national security, he will incur the wrath of the State and, if found guilty of such a *criminal* act, be punished by a fine or incarceration.

For the purpose of providing detailed academic coverage, it is convenient to 'compartmentalise' the various facets of the law, as suggested by the above distinction between civil and criminal law, and deal with them individually under the appropriate headings. In that respect, this text is no exception. However, for the person seeking to determine the extent of his legal liabilities, the law does not come in such conveniently delineated packages.

In reality, the law is a complex and fluid entity which is rooted deep in history, and which has been developed and adapted over the years to suit the ever-changing nature and demands of society. In this light, and in order to provide a better understanding of the structure of the law before discussing certain of its aspects in greater detail, it is helpful to consider the various sources from which it emanates.

1.2 The Sources of Law

The Law, in its entirety, stems from a plethora of sources too numerous to merit individual and detailed consideration in anything other than a treatise on legal history. However, it is generally accepted that the greater portion of law originates from four principal sources: *Legislation*, *Common Law*, *Equity* and, to a lesser extent, *Custom*.

1.2.1 Legislation

Legislation consists of that body of law which is promulgated with the authority of Parliament and the Monarch, and which appears in the formally enacted documents known as *Statutes* or *Acts of Parliament*.

It is the supreme form of law in the land since, under the doctrine of *Parliamentary Sovereignty*, an Act of Parliament can only be overruled by a subsequent Act of Parliament. Whilst it is possible for the judiciary to temper the effects of certain enacted legislation by exercising *statutory interpretation*, thereby creating a legal precedent (see section 1.2.2), it must be recognised that Parliament always has the power to counteract such a decision with new legislation. In this respect, it is worthwhile noting that Parliamentary Sovereignty obliges the judiciary to give effect to the will of Parliament as expressed by Statute. Thus the scope for statutory interpretation may be limited by clear express wording of Statutes.

Although it is only Parliament which has the power to legislate, it may, in certain instances and for the purpose of convenience, delegate some of its legislative powers to other bodies, such as Ministers of the Crown, government departments and local authorities. Legislation prepared in this manner is known as *Delegated* or *Subordinate* legislation.

Noteworthy examples of such legislation in the context of the construction industry are the Construction Regulations (see section 11.2.1) compiled between 1961 and 1966 by the Minister of Labour (nowadays the Secretary of State for Employment) under authority delegated by the Factories Act 1961.

Again, it is worthwhile to note that such delegated authority may be withdrawn at any time, thus reinforcing Parliament's legislative supremacy.

1.2.2 The Common Law

In the years preceding the Norman Conquest, 'justice' was largely dispensed on an *ad-hoc* basis and in a form almost entirely reliant upon the whim of those who dispensed it. As might be imagined, it lacked a certain amount of consistency to say the very least!

Post 1066, however, the situation markedly improved with the advent of commissioners appointed by the King to travel the length and breadth of England and dispense justice in his name. Since the new system was effectively 'centralised', in that the various commissioners were all ultimately responsible to the same authority, similar principles began to be applied to similar cases, giving rise, for the first time, to a system of law common to the whole country – the Common Law – which, until the advent of parliamentary democracy, constituted the basis of English Law.

In the years following the birth of the parliamentary system of government, the legislative powers of the judiciary became increasingly overshadowed by the legislative powers of Parliament to the extent that, nowadays, the Common Law is of secondary importance to the legislation enacted by Parliament.

The essential feature of Common Law, and that which distinguishes it from other systems of law both in the United Kingdom and in many other countries, is that it is based entirely on a system of evolving precedent with no written principles from which the precedent stems. Broadly speaking, this means that when a judge reaches a decision on a point of law arising from a case brought before him, that decision is *binding* upon judges dealing with a *similar point of law* in subsequent cases of a *similar factual nature*. In other words, the first judge has set a *precedent* which other judges must follow even if they personally dislike it or consider it to be based on defective reasoning – hence the fact that the Common Law system is alternatively referred to as *Case Law* or *Case Precedent*.

It is worth noting that, before such a so-called *judicial precedent* is regarded as binding, it must comply with certain rules. If it emanates from any court outside the English court hierarchy (see section 1.4), the precedent is regarded merely as *persuasive* and need not be followed if the judge in question has reservations as to its accuracy or the cogency of the reasoning on which it was based. Additionally, a court is only bound by a precedent set by a *superior* court. For example, the High Court would be bound by a decision made in the Court of Appeal but the reverse would not apply.

A further qualification needs to be added at this point. Before a court announces its decision on a case brought before it, the judge is obliged to issue a concluding statement giving his reasons for reaching the particular decision. Very often, these reasons (*ratio decidendi* – literally 'ground of the decision') are accompanied by additional statements concerning matters which may have arisen in the course of the case but are not central to the issue in hand. Since these statements (*obiter dicta* – literally 'remarks by the way') are not strictly relevant to the original decision, they are not regarded as binding in future cases.

1.2.3 Equity

In medieval times, largely resulting from the rigidity of its substantive and procedural formalities, the Common Law became somewhat monolithic and unwieldy in its *modus operandi* and, as a consequence, very often failed to provide justice in deserving cases. In such instances, the only remaining course of action open to the disgruntled parties was to address a petition to the Crown in the hope of obtaining a just settlement. In time, the matter came to be reviewed by the Lord Chancellor, on behalf of the Crown, who would attempt to remedy the situation to the satisfaction of all concerned.

In due course the Lord Chancellor established his own court, the Court of Chancery, and there dispensed justice and provided remedies for those who found no remedy in Common Law. Such remedies were formalised in the eighteenth century and, since they were initially a set of rules based on the concepts of fairness and justice (as opposed to legal remedies), became known as *Equitable Rules*.

Although the system, subsequently formalised and called *Equity*, was initially applied only in the Court of Chancery (not to be confused with the present Chancery Division of the High Court), the Judicature Acts of 1873–5 decreed that it should be applied in the law courts along with the common law and, in the event of conflict with such, should prevail.

It is worth emphasising at this point that Equity is a *discretionary* form of relief which does not exist as of right, and which acts *in personam* by addressing the consciences and equities of the parties rather than particular matters of fact arising from a case.

1.2.4 Custom

Despite the obvious role played by custom in the development of the law, its importance nowadays as a source of law is minimal in comparison to those previously dealt with.

Nevertheless, there are still occasions on which custom will make its presence felt. In the context of the construction industry, the majority of such instances usually involve disputes over land usage, and are generally 'negative' in character in that Party A is seeking to prevent Party B building something in a particular location on the grounds that the structure will impinge on Party A's *customary right of way* or affect his *customary right of common*.

In the majority of cases of the type cited, the invoking of 'customary rights' is undoubtedly justified in that there is a genuine concern for the continuance of such. However, it is not unreasonable to suggest that, in some instances, such an approach is employed as a convenient 'means to an end'.

With this in mind, possibly, the courts have instituted certain rules which must be complied with before the custom in question is considered binding:

(a) The custom must have been in existence since time immemorial. In practice, it is sufficient to prove its existence over a substantial period of time, although the exact period is unspecified. However, if there is sufficient evidence to show that the custom could not have been exercised *at least once* since the arbitrary year of 1189, the rule is contravened.

(b) The custom must have been exercised continuously, and without interruption, over the given period.

(c) The custom must have been exercised peacefully, openly and without the need for prior permission at any time. In the latter case, the practice would be considered as having operated 'under licence' and could not, therefore, be construed as a custom *per se*.

(d) The custom must be reasonable, compliant with established law and definitive in its scope.

The sources of Scots Law are much the same in character as those of English Law, giving rise to extensive similarities between the two systems.

For example, since Parliament in Westminster became the supreme legislative body of the United Kingdom in 1707, the Statutes emanating from such apply equally to England and Scotland. Additionally, although their historical origins are different, English and Scots case law systems both utilise the doctrine of Judicial Precedent and, resulting from the fact that the House of Lords is the court of final appeal for civil law cases in both countries, have become intertwined over the years.

It is noteworthy that Equity, as a separate body of 'judge–made' law, does not exist in Scotland as it does in England. However, unlike English Law, Scots Law proceeds with a basic set of remedies and is prepared to grant relief if a litigant can prove that he has a right to such. Scots Law has never distinguished between courts and principles of law and equity in that it has always been a mixed body of legal rules and equitable principles. However, in addition to its ordinary equitable juridiction, the Court of Session (see section 1.4.2) has a Nobile Officium *which is an extraordinary power to do justice where ordinary procedures would provide no remedy.*

A noteworthy difference between the systems of Scots and English Law arises from the significantly greater influence of Roman Civil Law *on the former than on the latter. The essence of English Law is* empiricism, *that is to say the* inductive *derivation of a general legal principle, from examination of previous cases, and subsequent application of that principle to the case in question. As a result of the influence of Roman Civil Law, however, Scots Law relies to a greater extent on the* deductive *application of documented legal principles to individual cases. Practitioners of Scots Law consider that this results in a more rational system of law and, in the context of European unity, has the additional advantage of making the decision–making process of Scots Law more akin to the European systems of law than is the case with English Law.*

Having considered the general composition of the law in terms of its principal sources, noting the specific differences between Statute Law, Case Law and Equity, it is now necessary to consider the law in terms of its constitutive parts – the so–called *divisions* into which law is categorised for practical and administrative purposes.

1.3 The Divisions of Law

The first major division of law is that between *Public Law* and *Private Law*.

Public Law relates to the State on one hand, and the community on the other. It contains Criminal Law (see below), contravention of which is punishable by the State on behalf of the community, and a subdivision known as *Constitutional Law* which relates to the functions and organisation of the State itself.

Conversely, Private Law concerns the rights and obligations imposed by individuals on one another and is not generally a matter for State intervention.

The division between Public and Private Law is not only relevant to national law (the law of the land) but also to *International Law*.

Private International Law, also known as *Conflict of Laws*, concerns cases in which the laws of more than one country apply and may well conflict with each other, necessitating a decision as to which system of laws is to prevail. On the other hand, *Public International Law* concerns the relationships between countries and involves the laws of war, international recognition and the like.

As a consequence of the fact that it applies a system of law quite separate from English Law, Scotland is regarded by the English legal system as a foreign country. Indeed, the 1707 Act of Union between the two countries saw fit to constitutionally preserve the separate Scottish legal system. Thus, in a case involving both countries, it may be necessary to apply the principles of Conflict of Laws *to determine the system of law which prevails.*

It is worth noting in the context of International Law that, since 1972, the United Kingdom has been subject to European Community Law which consists mainly of delegated legislation (see section 1.2.1) drawn up under the terms of the Treaty of Rome. Community Law restricts the legislative powers of the British Parliament, takes precedence over national laws and prevails in the event of conflict with such.

The second major distinction drawn is that between *Criminal Law* and *Civil Law*.

Broadly speaking, the more serious obligations owed to society are the province of Criminal Law and are enforced by the criminal courts on behalf of the Crown (i.e. the State). In the event of contravention of Criminal Law, the Crown (represented in court by the *prosecutor*) brings a *prosecution* against the alleged offender (the *accused*) who, if proved guilty, is punished with a fine, a partial restriction of activities (e.g. probation) or imprisonment.

The less serious obligations are contained within Civil Law and are enforced by the civil courts on behalf of the individual. In the event of an alleged contravention of Civil Law, the injured party (the *plaintiff*) brings a *suit* (as opposed to a *prosecution* in the context of a criminal action) against the party from whom he is seeking redress (the *defendant*). In the event of the suit being upheld by the court, the defendant is *generally* (but not *always*) required to pay damages to the plaintiff in recompense for the injury suffered.

It is worth noting that, whilst the main function of Criminal Law is punishment, the main function of Civil Law is compensation.

Consequently, the damages resulting from a civil action must not contain a punitive element and must be quantified purely on the basis of the actual or estimated loss incurred by the plaintiff as a result of the actions of the defendant.

There are a number of accepted sub–divisions of Civil Law, each of which relates to a particular set of circumstances under which legal obligations arise. However, the sub–divisions generally accepted to be of greatest relevance to the practising civil engineer in the course of his duties are the Law of Contract (covering obligations associated with arranged liability) and the Law of Tort (covering obligations of a more general nature), both of which are dealt with in depth in subsequent chapters.

Before moving on, it is necessary to draw attention to an important and fundamental difference between civil and criminal law in respect of the acceptability of facts asserted to support an action in such.

Every fact in a dispute which is necessary to establish a claim, whether the dispute is civil or criminal in character, must be proved to the judge by admissable evidence, with the so–called 'burden of proof' generally resting with the person asserting the fact. In a civil action the proof need not be absolute, in the sense that it is completely incontravertible, but it must comply with a standard based on a test of probability. Thus, the facts in a civil dispute must be proved 'on the balance of probabilities'. Since the available sanctions under criminal law are generally more severe than those available under civil law, the required standard of proof in a criminal action is understandably higher than that in a civil action, and the facts in such require proving 'beyond all *reasonable* doubt'.

Although the above divisions of law are generally accepted to be the most important, a distinction which is worth mentioning is that between *Substantive Law* and *Procedural Law*.

Substantive Law is the actual law which is applied in the courts, whilst Procedural Law (rather unsurprisingly) concerns the procedures to be undertaken in seeking redress for a breach of Substantive Law. Amongst other things, Procedural Law defines the court in which redress may be sought, the form that the proceedings may take and the nature of the evidence required to support the claim.

These, then, are the principal divisions of the law and an awareness of such is a fundamental step on the road to developing an understanding of the structure and composition of the law. It is also of value to note that, in the context of seeking legal advice in response to a particular problem, such divisions also tend to define the areas of expertise of professional lawyers.

In this context, however, it should be remembered that such an expert is retained in an *advisory* capacity only. Whilst he can provide valuable advice in respect of the method of approaching the problem from a legal standpoint, and can also advise on the chances of eventual success, the final arbiter on the matter will probably be a court of law.

1.4 Courts of Law

Broadly speaking, and with one notable exception, courts of law may be regarded as belonging to either of two mutually exclusive categories: *Criminal Courts* and *Civil Courts*. The sole exception to this rule is the House of Lords which, being the supreme court in the land, deals with matters both Criminal and Civil.

1.4.1 The Criminal Courts

In the event of an alleged breach of the criminal law, the matter invariably enters the court system via a *Magistrates' Court* presided over by a Justice of the Peace. Such people, although generally well-respected members of the community, hold no legal qualification and, being essentially laymen, are advised on legal matters by the Clerk of the Court.

As a consequence of this, together with the fact that no jury is involved, Magistrates' Courts have jurisdiction over cases involving only comparatively minor breaches of the criminal law.

The more serious breaches of criminal law, the so-called *indictable* offences, are referred to the *Crown Court* to be tried before a judge and jury.

Following conviction and sentencing in the event of guilt being proved 'beyond all reasonable doubt', the offender may lodge an appeal before the *Court of Appeal (Criminal Division)* but, unless the appeal involves a point of law, requires the permission of the convicting court to do so.

It is worthwhile to note that the Court of Appeal has no brief to entertain appeals by the prosecution against what might be considered to be 'lenient' sentencing (although the possibility of introducing such a facility is, at the time of writing, the subject of political and legal debate), and it cannot consider appeals against aquittal.

Following an unsuccessful appeal to the Court of Appeal, a final appeal may be made to the *House of Lords*, the supreme court in the land. However, this requires leave from the Court of Appeal to do so, and such leave will only be granted if the appeal involves a point of law of major public importance.

It is of interest to note that the House of Lords, unlike the Court of Appeal, *may* consider a prosecution appeal against aquittal.

Criminal jurisdiction in Scotland is divided into summary *and* solemn *jurisdiction, the former concerning comparatively minor offences not requiring trial by jury and the latter concerning the more serious offences requiring trial by jury on indictment before a judge or a sheriff. In the latter case, it is interesting to note that Scots Criminal Law requires a jury of fifteen, as opposed to the English requirement of twelve.*

In the majority of cases, the so—called 'court of first instance' in Scotland is the Sheriff *Court. There are Sheriff Courts in all the major towns and each has extensive inherent criminal jurisdiction over matters arising within its sheriffdom.*

Cases not serious enough to warrant their being brought before the Sheriff Court are dealt with by the District Court *which is presided over by a Stipendiary Magistrate or Justice of the Peace. The most serious criminal cases, such as murder, are initially brought before the* High Court of Justiciary *which sits as a 'trial court' in Edinburgh and 'circuit towns' such as Glasgow, Perth, Aberdeen and Dundee.*

All criminal appeals, from whatever court, are heard in the High Court of Justiciary *which sits as an 'appellate court' only in Edinburgh. This constitutes the 'end of the line' since there is no right of further appeal to the House of Lords as would be the case in England.*

1.4.2 The Civil Courts

Civil actions will usually enter the civil court system *via* a *County Court* unless large sums of money are at stake, or difficult points of law are in dispute.

Each County Court serves a local district and deals largely with disputes arising in that district. However, for judicial and administrative purposes, County Courts in a particular area are grouped together in a so-called 'circuit' and are presided over in turn by the Circuit Judge.

In matters of Contract Law or Tort, the juridiction of a County Court is (at the time of writing) limited to cases involving £750 or less. In other matters, such as disputes over trusts, mortgages or the administration of deceased persons' estates, jurisdiction is limited to £5,000 or less.

More serious cases, involving disputes outwith the limits of juridiction of a County Court, are brought before the *High Court of Justice* which has three divisions:

(a) *The Queen's Bench Division* – which deals with most common law cases such as Tort and Contract;

(b) *The Chancery Division* – which deals with cases concerning company or partnership disputes, trusts, bankruptcies, the administration of estates and the redemption or foreclosure of mortgages;

(c) *The Family Division* – which deals with matrimonial cases, wardships, guardianships and property disputes between husband and wife;

In the context of the construction industry, the majority of civil disputes will probably be contractual or tortious in character, will probably involve sums of money in excess of £750 and will, therefore, probably be dealt with by the Queen's Bench Division of the High Court.

However, in instances where the dispute centres around matters of great technicality, it may be deemed more appropriate to bring the case before judges in the *Official Referee's Court* who specialise in trying technical cases.

Generally, either party in a case brought before any of the aforementioned courts has automatic right of appeal to the *Court of Appeal (Civil Division)* which will re-hear the case on its merits by considering the evidence given at the original trial. It is then open to the court to order a re-trial or to uphold or reverse the decision of the original court.

As with the criminal justice system, a final appeal may be lodged before the *House of Lords*, but leave to appeal must be sought and will only be given in extreme cases involving a point of law.

In Scotland, since it has civil and criminal jurisdiction, the 'court of first instance' in most civil law cases is the Sheriff Court. *The exceptions to this rule are: cases involving sums of money less than £1500 (at the time of writing), which are dealt with by a branch of the Sheriff Court known as the* Small Claims Court, *and cases outwith the jurisdiction of the Sheriff Court which enter the system* via *the* Outer House *of the* Court of Session *in Edinburgh.*

In civil business, the Sheriff Court is both a 'court of first instance' and an appeal court, in that cases first heard by a Sheriff may be appealed to a Sheriff Principal who is a senior member of the bar and is, in effect, the senior Sheriff in any given sheriffdom. The provision for appeal from a single judge to another single judge is anomalous but it does provide for an appeal to be heard locally without undue expense, and appears to give general satisfaction.

The right of appeal concerning all civil law cases, including those which have been previously heard on appeal by a Sheriff Principal, lies to the Inner House *of the* Court of Session, *the so-called 'appeal arm' thereof, and thence to the* House of Lords.

From the standpoint of persons employed within the construction industry (or any other industry, for that matter), it is worthwhile noting the existence of the *National Industrial Relations Court*. This court was established under the Industrial Relations Act 1971 with the specific brief of hearing complaints related to unfair industrial practices, and considering appeals from Industrial Tribunals on points of law. As with other civil courts, appeal lies from this court to the appropriate Court of Appeal.

Although the above classification rightly implies mutual exclusivity, in that a criminal court cannot deal with a civil matter and *vice versa*, it would be wrong to infer from such that any given case will be *entirely* criminal or *entirely* civil in character. There are many instances in which a case might involve a contravention of Civil *and* Criminal Law. However, in such an event, the civil and criminal aspects of the case would be dealt with separately in the appropriate courts.

In the context of the construction industry, a noteworthy example of such an instance would concern a breach of the Construction

Regulations (see section 11.2.1) which resulted in an accident in which a third party was injured. Since the Construction Regulations form part of the Criminal Law, the breach of such would be dealt with by a Criminal Court. However, a possible claim for damages on behalf of the injured party would be regarded as a matter of Civil Law and would therefore be processed in a Civil Court.

1.5 Legal Liability

From the standpoint of the practising engineer, legal liability may be considered in three different contexts: the liability of the individual for his own actions, the liability of the individual in his capacity as an employee of an organisation and the liability of that organisation as a whole.

In all three contexts, the distinction must be made between *civil* and *criminal* liability since, as shown earlier, such a distinction will determine the law which applies, the court system which will deal with the matter and the price to be paid at the end of the day.

1.5.1 Personal Liability

In describing personal liability, it is sufficient to state that no single person operates in isolation from the rest of society and, as a consequence, all individuals are subject to the rule of law and may be held liable for their own actions.

If an individual is acting on behalf of the organisation by which he is employed, and is specifically authorised to do so, it is most likely that liability will be assumed by that organisation. However, as will be seen in due course, there are certain specific instances in which it may not be possible to shelter under the 'corporate umbrella'.

1.5.2 Corporate Liability

In order to deal with corporate liability, it is necessary to make the distinction between a *Partnership* and a *Limited Company* since the extent of legal liabilities of such is distinctly different.

According to the Partnership Act 1890, a *partnership* is described as 'the relation which subsists between persons carrying on a business in common with a view to profit'. The key phrase here is 'in common' since it implies *equal* sharing of profits and losses and, equally importantly, implies the authority of any given partner to bind *all* the others by his actions.

It may well be that, within a registered partnership, the extent of each partner's authority to enter into contracts, and bind the others thereby, has been defined by mutual agreement. However, in the event that a given partner has entered into a contract outwith his internally defined area of authority, the law will take no account of the internal arrangement and will regard each partner as equally bound by the terms and conditions of the contract in question.

This legal point is particularly important if, as a result of the said contract, the partnership incurs substantial debts which it cannot realise out of disposable capital. In such a case, the assets of the partnership will be liquidated and the partners left to share *equally* the balance remaining after payment of all debts. If such debts exceed the total value of the assets, each partner is held personally liable for an *equal* share of the outstanding balance. In other words, liability in this context is *unlimited*.

The situation is somewhat different with a *limited company* since, as the name would suggest, the liabilities of members of such is strictly limited.

This is due to the fact that, when such a company is formed, it is invested with what the law is pleased to call *corporate personality*. In essence, this means that the law recognises the company as a body distinct from its members – it is, in fact, a new 'legal person'. As a result, the members of the company can lend money to it, borrow from it and, in the event of its being wound up, can only be held financially liable to the extent of the shares they hold in it.

It is worth noting in this context that, if members' financial investment in the company consists of loans in return for *debentures* (certificates issued by a company as proof of a loan to it, and giving secured preference over ordinary creditors), such members stand to lose nothing at all in the event of the company being wound up, providing the liquidated assets of the company are sufficient to cover the sums involved after winding–up costs and preferential debts (taxes, wages, salaries etc.) have been met.

One final point should be made. The question of liability in the context of *corporate personality* extends beyond matters purely financial. Just like any other 'legal person', a company can sue for damages or bring an action to recover debt, it can be summonsed and found guilty of a crime and it can enter into a contract or defend itself against a claim for negligence.

It should be noted, however, that the concept of companies being treated like individual persons in the eyes of the law can only extend so far. For example, a company cannot get married or divorced, it cannot commit an assault or be personally assaulted (although the possibility does have a certain attractive quality in certain instances!) and, although it can be fined, it cannot be sent to prison.

Having considered liability in a personal and a corporate context, it should be remembered that the practising engineer will undoubtedly, at some stage in his career, be in a position whereby he exercises control over the activities of his subordinates. In such a position, he would be well advised to be aware of the legal concept of *Vicarious Liability*.

1.5.3 Vicarious Liability

The principle of vicarious liability has its origins in Tort where it is well established that a master can be held liable for all acts of his servants performed in the course of their employment. Although such a principle has no *general* application in criminal law, it should be appreciated that certain statutes make *specific* provision for vicarious criminal liability.

By virtue of its 'corporate personality', a company can be held vicariously liable for the actions of its employees both in a civil and, where statutorily provided for, in a criminal context.

However, it must not be inferred from this statement that the employees of a company have *carte blanche* to disregard the law and then, claiming that they were acting on 'company business', run for shelter under a 'corporate umbrella'. For purposes of imposing vicarious liability, the law recognises that, in any company, there are certain personnel who control and direct its activities. In an attempt to encourage such persons to avoid criminal actions themselves, and to make them think twice about sanctioning such actions by those under their control, certain statutes (such as the Health and Safety at Work Act 1974 — see section 11.3) provide for *individual* criminal liability in certain circumstances (see sections 2.2 and 2.3).

CHAPTER 2

TORTIOUS LIABILITY

As stated in Chapter 1, the obligations imposed by civil law can be broadly classified as belonging to either of two distinct categories, depending upon the circumstances under which they arise.

On the one hand there are those obligations which arise from legally enforceable arrangements drawn up between individuals or organisations for their own convenience, such obligations being classified as *contractual* and subject to the *Law of Contract*. On the other hand, there are those obligations which do *not* arise from arranged liability, but arise instead from the role of such individuals and organisations as members of society. Such obligations are regarded as *tortious* in nature and are covered by the *Law of Tort*.

In view of the *modus operandi* of the construction industry, it is very probable that the majority of disputes requiring the attention of the practising civil engineer will be *contractual* in character.

However, it bears repeating that no single person or organisation operates in isolation from the rest of society. This is particularly so in the case of the construction industry where the construction process will almost certainly intrude upon the lives of those members of the public in its immediate environs. In the event of such intrusions being beyond the bounds of *reasonable* acceptability, an action in *tort* could be brought in which the civil engineer, by virtue of his position of responsibility within a construction company, could find himself playing a starring role!

Thus, in terms of relevance to the practising civil engineer, the obligations and responsibilities imposed by the Law of Tort should be regarded as no less important than those imposed by Contract Law.

2.1 The Law of Tort

The Law of Tort is that body of civil law which imposes liability on members of society in respect of certain specified actions, known as *torts*, which may impinge upon the rights and interests of other members of society.

Although there are numerous torts, including *Libel* and *Slander*, the principal ones of relevance to the construction industry are generally recognised to be: *Negligence, Nuisance* and *Trespass*. However, before moving on to cover these in greater depth, together with the concept of *Strict Liability*, it is worthwhile to consider the question of *tortious liability* in general terms.

As with all the recognised divisions of civil law, the remedy for a *breach of tort* is compensatory rather than punitive in character, in that the offending party is required to pay damages to the wronged party in compensation for any loss incurred as a consequence of the breach. Although it is not unknown, on occasions, for a punitive element to creep into an award of damages – as, for example, in a libel case where the amount of damages is decided by the jury – such elements may be removed by the Court of Appeal following an appeal by the defendant, unless the court wishes to confirm judicial disapproval of the defendant's conduct by upholding the award of *exemplary* damages. On the other hand, and notwithstanding the success of an action in law, it is not unknown for a court to show its dissapproval of the *plaintiff's* conduct by awarding *contemptuous* damages of equivalent value to the 'smallest coin of the realm' (currently 1p in the United Kingdom). In such a case, the court may also deprive the plaintiff of his costs.

Until comparatively recently, it was considered that liability in tort should not extend to a purely financial loss. Thus, for damages to be payable, the loss had to be 'physical' in the first instance but could have financial ramifications. However, this state of affairs was changed in 1982 by a House of Lords decision which held that, *depending on the circumstances*, there was no reason why a purely financial loss should not be recoverable in tort.

A further, and not altogether surprising, point should be mentioned in the context of recompense, namely that no damages can be sued for unless the offending action constitutes a breach of a *recognised* tort irrespective of whether or not the action was motivated by malice. It is worth noting, however, that the existence of a malicious motive can transform an otherwise innocent act into a breach of tort where *malice* constitutes an integral part of the tort — for example: *malicious damage* or *malicious prosecution*.

By and large, the defendant in an action in tort will be the person who has committed the breach of tort. However, this need not always be the case since the Law of Tort recognises the principle of *vicarious liability* (see section 1.5.3) in which one person can be held liable for the actions of another. In this respect, and with particular reference to employer/employee relationships, the general rule in tort is that an employer is held liable for *all* the actions of an employee providing they are carried out *within the scope of that employee's authority or conditions of employment*. The employer may even be liable if the employee commits a breach of tort in the process of doing something he has been expressly forbidden to do, *provided it is within the normal scope of his employment*.

> *A firm of contractors employed a driver for the specific purpose of transporting its employees to and from a site at which they were employed. On one occasion, the driver carried another firm's employee who was injured as a result of the driver's negligence.*
> Held: *the employer was not liable for the negligent act of his employee since the driver was acting outside his conditions of employment.*

It should be noted, however, that this general rule of vicarious liability applies *only* in circumstances where there is an employer/employee relationship in the strict sense of the 'master/servant' relationship, referred to in Chapter 1, where the employer exercises strict control over the employee's conduct in the performance of his work. It does not apply in circumstances where no such control exists — as, for example, in the case of a so-called 'independent contractor' being employed to perform a specific service. By virtue of the contractor's independence, the employer exercises control *only* over the *standard* of the service and *not* over the *conduct* of the parties engaged in its performance. Consequently he cannot be held vicariously liable in the event of a possible breach of tort by the contractor.

There are exceptions to this general principle, however. In cases where the employer has either *implicitly* or *explicitly* authorised the actions of the independent contractor, he will be liable for any resultant breaches of tort *as if he had committed them himself.*

> *A company employed an independent contractor to carry out excavations in connection with services running underneath a street. A passer-by fell over rubble left by the contractor.*
> Held: *the company was liable on the ground that it had authorised the nuisance.*

Additionally, the employer will be held similarly liable if the nature of the required service is sufficiently dangerous to preclude delegation of his 'duty of care' to an independent contractor. This will be the case if, for example, the service is required to be performed on or adjacent to a public highway.

> *An independent contractor was employed to lay telephone wires along a street. In the course of the operation, the contractor injured a passer-by whilst negligently using a blowlamp to solder the ducts carrying the wires.*
> Held: *the employer was liable since the work was being carried out on a public highway, thus precluding delegation of his duty of care.*

It follows that, if the nature of the service is not sufficiently dangerous *under normal circumstances* to warrant the taking of specific safety precautions by the employer, the employer cannot be held vicariously liable for the actions of the independent contractor.

> *An independent contractor was employed to fit casement windows. Whilst in the process of doing such, a workman employed by the independent contractor negligently placed a hammer on the window sill where he was working. A gust of wind blew open the window and the hammer fell and struck a passer-by.*
> Held: *the employer was not vicariously liable since the workman had chosen a negligent method where the normal method would have created no reasonably foreseeable danger.*

A final point is worth noting in respect of the time limit within which an action in tort must be brought. The Limitation Acts 1939 and 1963 impose a limitation period for breaches of tort of 6 years except in instances where physical injury has been suffered by the plaintiff, in which case the limitation period is 3 years. In other words, an action in tort will only be entertained by the courts if proceedings commence within the appropriate period following the date of the breach.

In cases where the tortious act results in physical damage to property, or injury to a person, it is quite clear that the limitation period runs from the date at which the damage or injury was sustained. There are cases, however, where the tortious act and the resulting damage or injury are separated by a substantial period of time. For example, defects in a building that are the direct result of the builder's negligence may not come to light until a considerable time after the building has been completed. In such a case, the limitation period is deemed to run from the date at which the defect is discovered or, in the interests of fairness to both parties, the date by which the plaintiff ought *reasonably* to have discovered it. It is worth noting that, in cases where a delay of the type cited possibly results from *fraudulent* concealment of a defect, an extension to the limitation period may be granted by the courts.

Although actions brought within the limitation period are perfectly valid despite delay, it should be appreciated that a *substantial* delay might cast a spectre of doubt over the genuineness of the claim.

It is worthwhile to note that the Scots Law 'equivalent' of Tort is Delict. *In terms of relevance to the construction industry, the most notable difference between Tort and Delict is the absence, from the latter, of laws relating to* Trespass.

2.2 Negligence

In legal terms, *negligence* is defined as *a breach of a duty of care resulting in injury or damage.*

Thus, in a claim for negligence, the first duty of the court is to establish the existence or otherwise of a *duty of care*, given the circumstances surrounding the case.

Both Case Law and Statute Law contain numerous situations in which a duty of care is deemed to exist. For example: equipment manufacturers are considered to owe a duty of care to the users of their equipment; similarly, skilled practitioners owe a duty of care to those who employ them to practise their particular skills. In the particular context of the construction industry it is interesting to note that, in respect of their inspections to ensure compliance with the building regulations, building inspectors are deemed to owe a duty of care to the eventual owners of a building. In the event that defects in the building appear as a result of non-compliance with the regulations, the building inspector's employers (i.e. the local authority) may be held vicariously liable for his negligence in performing the inspections.

Two specific situations of relevance to the construction industry require mentioning at this point, the first of which concerns the duty of care imposed on occupiers of premises.

The Occupiers Liability Act 1984 imposes a *statutory* duty of care on occupiers of premises to ensure that persons entering the premises are reasonably safe whilst using the premises for the purpose for which they are permitted to be there.

In interpreting the statute, the term 'premises' may be taken to include construction sites and the term 'occupier' may be applied to any persons exercising control over the premises irrespective of whether or not they are the owners of such. In this light, the contractor will usually be regarded as the occupier but the employer (where the term is used in its *contractual* sense) may be regarded as a joint occupier in circumstances where he exercises control by virtue of his presence on site.

It is worth noting that, in addition to his duty of care towards persons *permitted* to be on the premises, the occupier also owes a limited duty of care to *trespassers*. In such cases, the occupier is required to take all reasonable steps to prevent injury if he is aware of any particular dangers on the premises. In this respect, it must be noted that a warning notice drawing attention to the danger is only regarded as a satisfactory discharge of the duty of care providing it is, or is likely to be, effective under *all foreseeable circumstances*.

The second situation of relevance to the construction industry concerns the relationship between an employer and his employees. In addition to the previously discussed *vicarious* liability of an employer for the actions of his employees, and in addition to the *statutory* duty of care imposed on an employer by the Health and Safety at Work Act 1974 (see section 11.3), it must be noted that an employer owes a *common law* duty of care to his employees.

This duty of care has been defined under three heads: the provision of a competent staff of men, adequate material and a proper system with effective supervision.

Whilst the first requirement is largely self explanatory, in that suitably qualified and experienced personnel must be employed for the performance of tasks requiring special skills, it is worth noting that it may extend to cover the employment of personnel who, by virtue of an inherent lack of discipline, are unsuited to the nature of particular operations. Thus, if an injury results from the conduct of an employee who is *known* to indulge in excessive horseplay, the employer may be held liable.

The provision of 'adequate material' means that all materials, tools, plant and equipment provided must be suitable for the work in question and must be safe to use. In the case of items of plant and equipment, it is implicit that they must be in proper working order and maintained in such. It is of importance to note that, in accordance with the Employers Liability (Defective Equipment) Act 1969, an employer is liable for any injury caused by defective equipment *even though the defect is due to the fault of a third party* — for example: the manufacturer. It is even more important to note that the liability is undiminished by the fact that the defect might not have been *reasonably* detectable by the employer!

The term 'system' may be taken to include such things as the general working environment, the sequence in which the work is carried out, the provision of suitable warnings and notices if required, and the issuance of appropriate instructions.

It should be appreciated that these stipulated duties are not absolute in that the requirement is for the employer to take *reasonable* care to ensure that they are satisfactorily discharged. Thus, liability is only incurred by a failure to take *reasonable* care. Additionally, it is of value to note that delegation of the *performance* of such duties does not relieve an employer of his liabilities in connection with such.

> *Party B, an employee at an electrical sub-station, suffered an injury in the course of performing a test because he had removed a safety screen separating the dead and live parts of a switchboard. B had learned to do this from a fellow employee to whom he had pointed out the dangers along with the fact that company regulations, with which all employees engaged on dangerous work were required to familiarise themselves, expressly forbade such a practice. He was told that, as long as no risks were taken, nothing would be done and, additionally, that the common practice was to remove the screen.*
> Held: *B's employers were liable for his injuries since they had not provided a safe system of work merely by instructing him to follow the safety regulations. Furthermore, even if the existence of a safety screen suggested a safe system of work, they had*

allowed the procedures governing its removal to be ignored. B was not found guilty of Contributory Negligence (see under) *since he had followed the example of his superiors.*

In addition to the above duties laid on employers, there are certain *common law* duties of care imposed on employees. Broadly speaking, employees are required to show regard for their own safety and to exercise *reasonable* skill in the performance of their work. Failure to do so may result in liability to the employer.

Once a duty of care is shown to exist, it is worth noting that it extends to cover all persons who might *forseeably* suffer injury or loss as a consequence of the negligent act.

> *Party A was injured in the process of unloading timber which had been negligently stacked by Party B.*
> Held: *B was liable on the ground that he should have foreseen that the timber would eventually require to be unloaded, however badly stacked.*

> *Contractors had excavated a trench along a pavement in a popular part of London. At each end of the trench, a warning notice had been placed, together with a 'barrier' consisting of a sledge hammer with its handle raised and sloping across the pavement. A blind pedestrian, having missed the handle with his stick, fell into the trench and was injured.*
> Held: *the contractors were liable since they should have reasonably foreseen the possibility of blind people using the pavement, and should have taken special precautions.*

Having established the existence of a duty of care, the next matter to be considered by the court is whether or not that duty has been breached.

For purposes of ensuring an objective assessment of such, the Law defines a standard of care and regards as negligent any action which does not conform to such. The standard is that of a 'reasonable man', a legal 'fiction' representing a person who weighs up the cicumstances, takes into account the characteristics of the persons in danger, takes greater care when there is greater danger and never loses his temper. Negligence is thus decided on the basis of how a reasonable man would have acted *under the circumstances*. Since the required standard of care depends on the circumstances, it may be inferred that there is only *one* standard of care appropriate to any given set of circumstances and *any* breach of such, however gross or slight, incurs liability. In the light of this inference, it is worth pointing out that the commonly used phrase 'gross negligence' has no legal significance whatsoever.

It is worth noting that the duty of care, in view of the 'reasonable man' terminology, involves guarding against *probability* and not against *bare possibility*. However, it is generally accepted that, in cases where the risk is greater, such as where children are involved, *reasonable possibilities* must be guarded against.

> *Workmen employed by a demolition contractor lit a bonfire to burn rubbish on a site adjacent to a public park. Despite the fact that the site hoardings had been previously removed, the men failed to keep a strict look-out for children coming on to the site. A five-year old boy, who had been repeatedly chased away from the site, came back to look at the fire and incurred serious burns.*
> Held: *the contractors were liable since they should have taken particular care to guard against the reasonable possibility of children coming on to the site.*

In respect of determining the appropriate standard of care deemed to be acceptable, a distinction is made between the 'man in the street' and the professional who holds himself out as possessing special skills. Such professionals, including practising civil engineers, architects and surveyors, are expected to discharge their duties with a standard of competence consistent with that *reasonably* expected from members of their respective professions. A failure to do so will give rise to the charge of *professional negligence*.

It is normally a defence against such a charge that the defendant followed the practices normally followed by the majority of his profession. However, if those practices are considered in themselves to be negligent, the defendant might well find himself paying for the sins and omissions of his profession!

The required standard of care expected from a professional will be extended in the case of a person who holds himself out as a specialist in a particular branch of his profession. Such a person would be required to display a level of competence consistent with that *reasonably* expected from those who specialise in such a branch. An appreciation of such a legal point should serve as a warning to those who implicitly hold themselves out as specialists and, as a consequence, may be required to take on work beyond their actual capacity.

The same principle applies at the other end of the scale in the case of an unqualified person who purports, on the basis of his experience alone, to be qualified. Since it is not entirely unreasonable to assume that any party employing such a person will be doing so in the (erroneous) belief that he is qualified, a possible charge of professional negligence could not be deflected by a 'lack of qualifications' excuse. Such a person would be deemed, in the eyes of the law, to possess the skills suggested by his purported qualification and would be judged accordingly.

It is of value to appreciate, in the context of professional negligence, that the duty of care covers the entire spectrum of professional conduct. Thus, the professional not only owes a duty of care in respect of his *level of competence*, but also owes a duty of care in respect of *verbal or written statements made by him in his professional capacity*. In the event that he gives unsound advice, misleading information or erroneous opinions which result in damage or injury, he can be liable for a *negligent mis-statement*.

> *A bricklayer working on a site was instructed by a qualified architect to cut a chase in a gable wall. Although the bricklayer should have realised that it was unsafe to do so without strutting and shoring the wall, he went ahead with the task and was injured when the wall collapsed.*
> Held: *the architect owed a duty of care in respect of his instruction to the bricklayer and was therefore liable for the injury sustained as a consequence of his negligent mis-statement.*

This is particularly important in the instance of a so-called 'special relationship' whereby a party makes decisions based *entirely* on the specialist advice and information provided by a professional.

> *An employer entered into a contract with an earthmoving contractor, for bulldozing work at an hourly rate, on the faith of the contractor's estimate of the time required for the work. In the event, the work took much longer than originally suggested, resulting in the employer incurring costs considerably in excess of those envisaged on the basis of the contractor's estimate.*
> Held: *the contractor was held liable for the excess costs incurred as a result of his negligent underestimate of the time required to complete the job.*

As a warning against adopting the policy of: 'if in doubt, say nothing' in an attempt to avoid the possibility of making a negligent mis-statement, it is worthwhile noting that the professional's duty of care extends to cover information or advice which *should* be given but is *not*, irrespective of whether the omission is deliberate or otherwise.

> *A hydro-electric company substantially underestimated the cost of electrically heating a proposed house extension which was to contain a swimming pool.*
> Held: *the company was negligent on the ground that it failed to point out its lack of experience in estimating heat loss from a building containing a swimming pool.*

Finally, it should be noted that, in cases involving alleged professional negligence, the court's assessment of what constitutes an acceptable standard of care in respect of a particular profession is usually based on 'expert testimony' given by members of that profession.

These then are the standards of care appropriate to the 'man on the street' and the person who holds himself out as having special skills or knowledge, and such standards will be applied in determining whether or not the defendant has committed a breach of his duty of care.

Before moving on to consider the final element in a claim for negligence, it is worthwhile to note a particular point in respect of the so-called 'burden of proof' for establishing whether or not such a breach has been committed.

It is a common feature of most civil and criminal justice systems that the burden of proof rests with the accuser. Thus, in an action for negligence, it is incumbent upon the plaintiff to prove the guilt of the defendant. There are, however, certain circumstances in which the situation is reversed in that the defendant is required to prove that he was *not* negligent. Such a situation pertains when the doctrine of *res ipsa loquitur* (literally: 'the facts speak for themselves') is invoked in circumstances where the damage or injury results from something in the control of the defendant and must, therefore, be regarded as directly attributable to his negligence. Thus, if a section of scaffolding collapses and injures a passer-by, the scaffolder is required under the doctrine of *res ipsa loquitur* to demonstrate that he was *not* negligent in his practice.

It is worth noting that the doctrine of *res ipsa loquitur* also applies in circumstances where there is an absence of *direct* evidence. Thus, if the only evidence available in the case previously cited was the injured person lying underneath a heap of collapsed scaffolding, nobody having actually *seen* the collapse, the scaffolder would *still* be required, under *res ipsa loquitur*, to prove his innocence in respect of negligence.

Once negligence has been established, the final element of the claim to be considered is the damage or injury sustained as a result of the breach. The principle here is that the defendant's liability extends to *all* damage resulting from his negligence, *provided such damage was the consequence of an occurrence against which his duty of care was intended to guard.* It is therefore reasonable to infer from this statement that if the occurrence was foreseeable, and in this respect the distinction is again drawn between the professional and the 'man in the street' for obvious reasons, the defendant will be held liable for all resultant consequences. It naturally follows from this, however, that an action for negligence will fail in instances where the occurrence was, to all intents and purposes, completely unforeseeable.

> *An asbestos cement lid was negligently dropped into a cauldron of molten sodium cyanide. After a short while, there was an explosion due to some unforeseeable chemical reaction between the two materials.*
> Held: *there was no liability since the events were not foreseeable.*

The concept of 'foreseeability' only applies to a *limited* extent to the damage or injury sustained as a result of the act of negligence. In the particular case of an injury, the defendant is liable providing the injury is *of the same kind* as that which was reasonably foreseeable irrespective of whether or not that *particular* injury could have been foreseen. In the context of foreseeable injuries, it is worth noting that no account of the existing physical state of the injured person is taken by way of limiting the liability.

> *Party A negligently inflicted a burn on Party B's lip. This resulted in B's death from cancer since the tissues of his lips were in a pre-malignant condition at the time, and the burn acted as the catalyst which promoted the cancer.*
> Held: *Party A was liable for B's death and damages were awarded accordingly.*

It is of value to note that, since only one action can be brought in respect of a given act of negligence, the amount of damages awarded should take into account all foreseeable future losses which might result. Thus if a skilled tradesman suffers an injury which will effectively preclude him from practising his particular trade in the future, the award of damages should take into account the predicted loss of his earning capacity.

Up to now, coverage has been limited to situations in which a charge of negligence may be brought and the defendant held *fully* liable for the resultant damage or injury. There are, however, a number of circumstances under which the charge may be completely and successfully refuted or, failing that, mitigated to the extent that the resultant liability is reduced.

The most obvious circumstances are those in which the damage or injury is incurred as a result of some natural but totally unforseen phenomenon, such a phenomenon being legally classified as an *Act of God*. Thus, if an engineer designs a cofferdam which is overtopped by a completely unprecedented river level, resulting in severe damage or even death, *Act of God* may be raised as a defence against a charge of negligence provided it can be shown that the cofferdam height was sufficient to provide protection against all *foreseeable* river levels.

An additional situation in which a charge of negligence may be voided is that in which a defence of *volenti non fit injuria* is applicable. Broadly speaking this means that, if a person *knows* that a risk of injury exists, but nevertheless *voluntarily* assumes the risk, he cannot complain of any injury he receives as a consequence.

The keyword here is 'voluntarily' which implies that, for the defence to succeed, the willingness to assume the risk must be *demonstrably* free from any form of compulsion or obligation.

Thus, if an employee is aware of a degree of risk attached to the job he is required by his employer to undertake, but nevertheless agrees to run that risk from fear of possibly losing his job by refusing so to do, it will be deemed that his assumption of the risk was motivated by a perceived economic pressure.

A workman was engaged in his work at the foot of a rock cutting. He was aware of a crane swinging heavy stones above his head at intervals but continued with his work. On one occasion, a stone fell from the bucket of the crane and injured him while he was working.
Held: *despite knowledge of the risk of falling stones, it was clear that the man had continued with his work in order to keep his job. Thus, the employer's defence of* volenti non fit injuria

was rejected.

Similarly, if the injured person feels under a *moral* obligation to assume the risk, the element of willingness is negated.

> *An employer had undertaken to clean a well and had installed a pump. His workmen were aware of fumes in the well, had reported this and had been instructed not to re-enter the well. One of them did and was overcome. Wondering what had happened, a second workman entered the well and was also overcome. When a doctor arrived on the scene, he was urged not to enter the well since it was obviously unsafe for him to do so. The doctor, however, said that he felt obliged to do what he could for the men and entered the well. Although overcome by the fumes, he was rescued but died on the way to hospital.*
> Held: *the defence of* volenti non fit injuria *raised against his widow's claim for negligence was rejected on the ground that the assumption of risk was not 'voluntary', in the required sense of the word, since it was motivated by a moral and professional obligation.*

It is worth noting that, in such a case, it is not sufficient for the moral obligation to be percieved by the injured person alone. Unless there is a *clear* and *unavoidable* moral obligation to assume the risk, the defence of *volenti* will stand.

Whilst the above circumstances are those in which a charge of negligence may be *completely* voided, liability may be *reduced* in certain circumstances if it can be proved that an injury sustained was partially due to the negligence of the injured party himself. Thus, if it can be shown that an employee's injury was sustained as a partial result of a breach of his employer's duty of care and partially due to his own negligence, the courts will consider this to be a clear case of *Contributory Negligence*. In such a case, the court will apportion liability to the extent that it considers just and equitable in the circumstances.

> *Whilst fighting a fire in a factory, a fireman was fatally electrocuted as a consequence of the main power switches being defective.*
> Held: *the occupiers of the factory were liable for 10% of the damages since the factory manager was negligent in not knowing that the switches were defective. The previous occupiers of the factory, however, were liable for the remaining 90% of the*

damages since they negligently failed to point out to the current occupiers that the switches were, in fact, obsolete.

A final point must be noted in the context of negligence. As mentioned previously, a *statutory* duty of care is imposed on certain specified persons by enactments such as the Occupiers Liability Act 1984 and the Health and Safety at Work Act 1974. It is vital to be aware of the distinction drawn between a *statutory* duty of care and a *common law* duty of care since a breach of the former, unless otherwise specified in the relevant statute, constitutes a *criminal* offence and is punishable by a fine or, in the event of an exceptionally serious breach, incarceration.

Thus, in the event of a claim being brought for an injury allegedly sustained as a result of a so-called *Breach of Statutory Duty* which renders the defendant liable to *criminal* prosecution, the matter will be dealt with as two separate issues: the *criminal* action for the breach and the *civil* action for damages.

For the criminal action to succeed, the prosecution is required to show three things: firstly, and not altogether surprisingly, it must show that the injury was sustained as a consequence of a breach of a *statutory* (as opposed to a *common law*) duty of care; secondly, it must show that the injury sustained was of the kind that the statutory duty was intended to prevent and, lastly, it must show that the injured person is one of the persons the statute intended should be protected by the duty of care;

Following establishment 'beyond reasonable doubt' of the defendant's guilt in respect of the statutory breach, and following the imposition of the appropriate punishment for such, the question of damages is dealt with in a separate *civil* suit as discussed earlier in this section. In this respect, it should be emphasised that a breach of statutory duty does not automatically imply negligence. Consequently, the defendant's negligence must be established in the normal way before a possible award for damages will be considered.

It is worthwhile to note that a statutory duty is *mandatory* in that the person accorded the duty cannot relieve himself of his obligation by delegating the duty to another party, irrespective of that party's competence. It is not a valid defence, therefore, for the defendant to claim that he had appointed reasonably competent persons in compliance with his *common law* duty of care and that the injury sustained was the consequence of the plaintiff's negligence. Similarly, a defence of *volenti non fit injuria* will not be entertained in cases where a statutory duty of care has allegedly been breached although *contributory negligence* might be raised as a defence to reduce the amount of damages awarded as a result of the subsequent *civil* suit.

> *A contractor, in breach of safety regulations, failed to provide safety belts for men working on a scaffold. A workman fell from the scaffolding and was killed.*
> Held: *the contractor was not liable for civil law damages in that he demonstrated, to the satisfaction of the court, that the man would not have worn a safety belt even if he had been provided with one.*

2.3 Nuisance

A person may be liable under the tort of *nuisance* if he carries out *any activity which directly or indirectly interferes with the use or enjoyment of another person's land.*

As might be imagined from the above definition, the tort covers a very wide range of activities of varying degrees of 'interference potential'. At one extreme of the scale, the interference may be of sufficient severity to result in physical damage to property — for example: vibrations from building operations which result in damage to the foundations of neighbouring property. At the other end of the scale, the level of interference may only be sufficient to cause discomfort or inconvenience to the occupiers of adjacent land or premises thereon — for example: noise or dust from building operations.

In respect of the latter case it should be noted that, for an action in *nuisance* to be successful, the interference must be regarded as being of a sufficient degree to cause *substantial* discomfort or inconvenience to the *average person*. Obviously, the test of acceptability is flexible and depends to a certain extent on the locality. For example, a town dweller cannot expect the same standard of air purity as might be expected by a person living in a country area. There are limits, however, to the extent to which locality is taken into account when assessing the acceptability of interference. For example, the fact that a person lives in a manufacturing area of an essentially industrial conurbation does not preclude his right to bring an action in *nuisance* if a nearby factory operates a steam hammer in the middle of the night thereby making sleep impossible!

In view of the fact that the test of acceptability is based on the *average person*, it may be correctly inferred that actions in nuisance take no account of special sensitivity.

> *A manufacturer of brown paper sued for damage to his stock allegedly caused by heat produced by a neighbouring manufacturer.*
> Held: *the neighbouring manufacturer was not liable for nuisance since the damage resulted not so much from the heat as from the excessive sensitivity of the stored product. An ordinary business would not have been affected.*

For purposes of deciding whether or not a nuisance has been committed, the question of reasonableness may also be taken into account from the standpoint of the defendant. In other words, the question may be asked as to whether the defendant had taken all *reasonable* precautions to minimise the potential inconvenience to the plaintiff. In the context of the construction industry, the question of reasonableness is usually decided by reference to the offending operation in the light of modern construction techniques. However, common sense and thoughtfulness are also factors which are not entirely discounted!

> *In the course of refurbishing an occupied residential building, contractors were extending existing stanchions to support a girder in the first floor of the building.*
> Held: *both the contractors and their employer were liable for dust affecting a second-floor tenant because they had not used dust sheets when it would have been reasonable to do so. Additionally, they were liable for noise because they had made no effort to arrange the work to suit the tenant, a music teacher.*

In cases where the nuisance is of an essentially transient nature, such as that associated with construction, the general interpretation of the word 'reasonable' involves an assessment of the cost (whether in money, time or trouble) of the measures necessary to prevent the nuisance against the duration and severity of the nuisance itself. Thus, the defendant will not be held liable for committing a nuisance if it can be shown that the cost of preventing such was *unreasonably prohibitive* when viewed in the light of the magnitude and duration of the potential discomfort and inconvenience that would result.

Although liability for nuisance is generally associated with actions which constitute *direct* interference with adjacent land or property, such as in the cases previously cited, it is noteworthy that liability may also arise from *indirect* interference. Such will be the case if the defendant's actions are considered to be the cause of a 'state of affairs from which damage is likely to result'. For example, a builder who allows a build-up of rubble to block a culvert and cause water to be diverted on to neighbouring land would be held liable for nuisance.

In addition to direct or indirect interference with land, the tort of nuisance also covers interference with certain *rights* over land. For example, a *riparian owner* (an owner of any stretch of river bank) has the right to object to any obstruction of the river or stream which prevents it reaching him fully or which causes it to overflow on to his land. Similarly, he may object to any pollution which reduces the quality of the water in respect of any purpose for which it could be used in its natural state.

An important right, in the context of the construction industry, is that to *lateral* and *subjacent* support for *unweighted* land. Thus, if subsidence to neighbouring land occurs from the removal of lateral support by nearby excavations, or if it results from the removal of subjacent support through undermining, the owner or occupier of the land may bring a claim for nuisance against the offending party.

The abovementioned rights, together with the right of an occupier to use and enjoy his land without unreasonable interference, are classified as *natural* rights which are *automatically* attached to land and property, and which do not have to be specifically acquired. This is to distinguish them from so-called *easements* which must be specifically *acquired* either by a *grant* from the owner of neighbouring land, or by *prescription* (meaning, in effect, that a legally valid easement is considered to have been created by a 20 year period of uninterrupted enjoyment of the right).

An important example of such concerns the right of lateral and subjacent support to buildings which, unlike the *natural* right of support to *unweighted* land, requires acquisition by grant or prescription.

One of two adjoining buildings had been converted for use as a coachbuilding factory, thus requiring increased support from the other building. This continued for over 20 years. Following demolition of the second building, which effectively removed the necessary support for the factory, part of the factory collapsed. Held: *the claim for the incurred damage was upheld since the right of support had been acquired by prescription under the 20-year rule.*

It may be inferred that if no such easement has been acquired, the owner of a building damaged as a consequence of the withdrawal of support has no right of action in nuisance. However, in the particular case of damage resulting from subsidence, liability will be held for the injury to the land *and* the building if it can be shown that withdrawal of support would have caused subsidence even if the land had not been built on.

In respect of this particular easement, it is worth noting that there is no liability in law for subsidence caused by the abstraction of subterranean water.

> *In the course of road construction, excavations carried out below the water table were kept dry by means of pumping. The plaintiff, a neighbouring landowner, complained that the pumping had reduced the level of the water table in the vicinity and that, as a result, buildings on his land had suffered severe settlement.*
> Held: *the plaintiff had no cause of action in law since there was no liability for the withdrawal of subterranean water.*

Before moving on to consider other relevant easements, it must be stated that, for such a defence to apply, the subsidence must result from the withdrawal of support provided by water *alone* and not that provided by something which may be loosely regarded as akin to water!

> *In carrying out excavation work on their property, the defendants had cut through a stratum of running silt. This had resulted in subsidence of the plaintiff's land.*
> Held: *the silt was in the nature of wet sand rather than muddy water and the 'subterranean water' defence was not applicable — the defendants were therefore liable for the subsidence.*

Another important easement in the context of the construction industry concerns the so-called 'right to light' under which the occupier of a building has a right to a *reasonable* amount of light. In respect of deciding whether or not that right has been infringed upon, it is noteworthy that the decision is based entirely on the amount of light *remaining* and no account is taken of the amount of light *taken away*. With the advent of modern technology, many tests have been suggested to determine the level of light which may be considered *reasonable*. For example, it has been suggested that if 50% of a room is lit to a level exceeding the so-called 'grumble point' (1/500th part of the light received from the sky on a dull or overcast day), there should be no actionable nuisance. However, the courts feel that such rigid tests are out of place in that they would restrict a flexible approach based on circumstance and common sense.

It is of value to note that the 'right to light' should not be confused with a perceived 'right to a view'. Such a right does not exist in law any more than does a so-called 'right to privacy'. Thus the fact that a recently erected building obstructs what was hitherto a perfect view from one's lounge window does not (regrettably!) render the situation actionable unless it results in a diminution of the quantity of light!

The remaining easement of relevance to the construction industry concerns *rights of way*. Although the topic of *customary rights of way* has been discussed in Chapter 1, it should be noted that a *right of way* or *right of entry* may exist by virtue of an easement. It is sufficient in this case to state that, if the easement has been acquired by deed or grant, a developer wishing to purchase a plot of land for building purposes will have notice of the easement in most cases. However, he might not have such notice if the easement has been acquired by prescription.

A final point worth mentioning in respect of easements concerns the manner in which they may be lost. The *extinguishment* of an easement may not only occur through its being *expressly* given up, but may also occur through its *implicit abandonment*. Such is the case if a building with a *right to light* is demolished with a view to rebuilding. If the rebuilding is unduly delayed, this may be construed as evidence of abandonment with the result that the *right to light* is lost. It is noteworthy that even if rebuilding takes place immediately, the *right to light* may be lost if the arrangement of windows in the replacement building differs to a significant extent from the original window arrangement on which the *right to light* was based.

In connection with the coverage of easements, it is useful at this point to mention the converse of such − *restrictive covenants* − which are clauses contained within the title deeds of a property placing restrictions upon the owner of such in respect of his use of the property. For example, if a landowner develops his land as a residential estate, it is possible for him to insert such covenants within the deeds of the individual plots restricting the use of such for residential purposes only. Thus the owners of the plots are bound to each other for their mutual benefit and a breach of such a covenant would give cause for a charge of nuisance. In the event of a breach of such, it should be appreciated that the obligations imposed thereby are *equitable* obligations and are not binding in law, thus limiting the remedies open to a court by way of compensation.

It is generally the case that the only person who can sue for nuisance is the occupier of the land. However, this can be extended under certain circumstances to include persons having a direct interest in the land. For example, a landlord may sue in the event that a permanent injury to his land is threatened by an act of nuisance such as a withdrawal of support which would result in subsidence.

Whilst the person liable is usually the creator of the nuisance, it should be noted that the doctrine of *vicarious liability* applies as much to nuisance as it does to negligence. Thus, whilst the contractor will generally be liable for interference with neighbouring property caused by construction operations, the employer may also be liable under certain circumstances.

> *An independent contractor was employed to carry out refurbishment work on the employer's premises. In the course of the work, the contractor withdrew support from a neighbouring property causing damage to that property.*
> Held: *the employer was liable for the action of the contractor. He was under a duty to see that no mischief resulted from the work and could not abdicate this responsibility by employing someone else.*

As with negligence, the defence of *volenti non fit injuria* may be raised against a claim for nuisance.

> *A landlord allowed a tenant to carry on coal mining activities on the portion of land he had leased. These activities resulted in coal dust being deposited on adjacent land also owned by the landlord.*
> Held: *the defence of* volenti *was rejected on the ground that the permission to mine coal did not carry with it the implicit authorisation to commit a nuisance in the process.*

An additional defence which may be raised is that of *Statutory Authority*. Work that is carried out under statutorily granted powers cannot be subject to civil suit *if the authority for that work is absolute*. Thus, if a Water Authority damages property in the course of repairing a damaged water main, it cannot be charged with nuisance since it has a *statutory duty* to carry out such work.

However, where such authority is *permissive* in that the work is carried out on a *discretionary* basis, the defence of Statutory Authority will only be entertained if the authority was exercised *reasonably*. In other words, if it could be shown that a more reasonable course of action would have avoided the nuisance, liability will be held.

As with negligence, the concept of *foreseeability* will be taken into account when assessing liability for nuisance. Thus, if the defendant's actions result in totally unforeseeable consequences, he will not be held liable for nuisance.

> *The defendant built a retaining wall on the bank of a river. It was badly constructed and became undermined. As a result, it moved and pressed up against a sewer which became damaged.*
> Held: *the defendant was not liable for nuisance since he did not know, nor should have known, of the existence of the sewer.*

In the context of potential defences against a charge of nuisance, it is worthwhile to mention two defences which will not be considered valid.

Firstly, it is no defence to claim that the action causing the nuisance was 'in the public interest' if statutory authority has not been granted.

> *The plaintiff complained of noise emanating from a cement manufacturing works near his house.*
> Held: *The defendants were liable for the nuisance despite their claim that the factory was the principal supplier of cement to the building trade and that an injunction to stop the work would result in a national shortage of cement.*

Secondly, it cannot be claimed as a defence that the plaintiff came to the nuisance.

> *A confectioner had conducted noisy operations on his land for more than 20 years. A doctor moved to the adjacent property and opened consulting rooms. On finding that he was constantly disturbed by the noise, the doctor sued.*
> Held: *the nuisance only arose when the consulting rooms were opened, thus negating the defendant's claim to have a prescriptive right to be noisy since he had been doing so for over 20 years. Since it was not adjudged unreasonable for the plaintiff to carry on a medical practice in the area, the defendant was served with an injunction restraining him from continuing to commit the*

nuisance.

Where an actionable nuisance has resulted in the incurrence of a loss, the plaintiff is entitled to claim full restitution from the defendant, by way of damages, for such loss. The keyword here is *restitution* which implies that the plaintiff is not entitled to make a *profit* from the defendant's act of nuisance. For example where the nuisance has resulted in physical damage to the plaintiff's property, the amount of damages will be limited to the monetary difference in value of the property before and after the damage was incurred – in effect, the cost to the plaintiff of repairing the damage *resulting from the nuisance.*

Such a principle has an important bearing on the award of damages to a business which has suffered a purely financial loss as an alleged consequence of a nuisance. In such a case, the amount of damages will be strictly limited to the proportion of the overall loss directly attributable to the nuisance, such a proportion being decided by the court in the light of the circumstances prevailing at the time.

> *Dust and vibrations from building operations interfered with a neighbouring hotel resulting in an alleged loss of custom.*
> Held: *a proportion of the building work was carried out with reasonable care and, to this extent, no damages for loss of custom were payable. An assessment was made by the court of the loss of custom which was directly attributable to the* actionable nuisance *(the* excess *of noise and dust) and damages were awarded accordingly.*

An alternative to damages, which may be particularly appropriate in the case of an actionable nuisance, is for the court to serve an *injunction* on the defendant. Such an injunction may be *prohibitive*, in that it prevents the defendant from continuing to carry out the actions causing the nuisance, or it may be *mandatory* in that it compels him to carry out certain actions to prevent the continuance of the nuisance. An example of the latter may be an order to demolish a building which has been shown to impinge on a neighbour's *right to light.*

Although it is generally the case that injunctions are served in the event of the defendant's guilt being established *in court*, it is worth noting that the court has the power to issue a so-called *interlocutory injunction* in advance of the trial. If considered appropriate, such an injunction could be issued within days, or even hours, of the complaint arising and would serve to provide the plaintiff with a measure of temporary relief from a particularly severe nuisance until such time as the matter came to court. An interlocutory injunction may also be served in cases where irreparable damage might occasion from continuance of the nuisance.

It should be pointed out that an injunction is an *equitable* remedy which is subject to the court's discretion. Thus, an injunction will not be served if it would be considered *oppressive* – for example, to stop the construction of a building over an unused right of way – or in a case where damages would be considered an adequate remedy. Neither will an injunction be served where a merely *equitable* right has been infringed, such as a breach of a restrictive covenant.

Two specific points should be mentioned in respect to the use of injunctions as an alternative to damages, the first of which applies to circumstances in which a nuisance has not actually been committed but is merely *threatened*. In such a case, a *quia timet* injunction could be served restraining the defendant from continuing an act which might result in physical damage to the plaintiff or the plaintiff's property. It should be noted, however, that such an injunction will only be served where there is *'extreme probability of irreparable injury'*.

Secondly it should be appreciated that, although an injunction is a *discretionary* measure, its use will not be withheld merely because the potential loss is small or that compliance with such would be inconvenient or expensive – providing the use of such would not be oppressive (see above).

A final point worth noting about remedies for nuisance is that injunctions and damages are not *mutually exclusive* alternatives. Thus, damages may be awarded to compensate the plaintiff for any losses incurred as a result of a nuisance, and an injunction may be issued preventing the defendant from continuing the nuisance.

In view of the fact that Scots Law does not recognise the distinction between Common Law and Equity (see section 1.2.3), an injunction per se is not a remedy available in Scotland. Notwithstanding this, however, a Scottish court has the discretionary power to serve an interdict which, although obtained on different principles from an injunction, has much the same effect. The Scots Law equivalent of an interlocutory injunction is an interim interdict.

It is of particular value to note that contravention of the terms of an injunction or interdict amounts to *contempt of court*, one of only two offences (the other being *perjory*) under British Law for which a person involved in a civil suit can find himself sentenced to a term in jail!

The particular instances cited thus far, in which the action for nuisance is brought by an occupier or owner of land or property, are examples of *Private Nuisance* which is an offence covered by *civil* law. Where the interference is caused to a wider group of persons, however, such conduct may constitute a *Public Nuisance* which is a *criminal* offence punishable by the State. Examples of public nuisance relevant to the construction industry are obstruction of the public highway and the creating of dangers on, or near, the public highway.

In respect of these particular examples, the phrase 'public highway' implies that a prosecution for *public nuisance* may only be brought if the highway in question has been dedicated or taken over by the appropriate highway authority. It follows, therefore, that an offence committed on or near a highway *under construction* can give rise *only* to an action for *private nuisance*. An additional point worth noting is that, although public nuisance is primarily a matter for prosecution by the State, an individual has a right to bring a *civil* action in respect of the offending act providing he can demonstrate that he has suffered damage *over and above that suffered by the public at large.*

It should not be inferred from the above distinction that any given instance of interference will constitute *either* a public nuisance *or* a private nuisance. The two categories are not mutually exclusive and there are a number of instances where a given nuisance falls into both categories. For example, a pile of builder's rubble outside the driveway to a house would constitute a public nuisance if it obstructed the highway. If at the same time it obstructed the driveway to a neighbour's house, the neighbour would have the right to bring an action for private nuisance. It should be noted, in this respect, that a successful prosecution for public nuisance does not automatically imply success in the private action − the private nuisance requires establishment as a separate issue.

In addition to *public* and *private* nuisance, attention should be drawn to so-called *Statutory Nuisance* as defined by the Public Health Act 1961. Such nuisances include:

(a) Any premises in such a state as to be prejudicial to health or a nuisance....

(b) Any accumulation or deposit which is prejudicial to health or a nuisance....

(c) Any dust or effluvia caused by any trade, business, manufacture or process and being prejudicial to the health of or a nuisance to the inhabitants of the neighbourhood....

As might be imagined, the last category is of particular relevance to the construction industry. In the event of a *statutory nuisance* being committed, the local authority has the statutory power to serve an abatement notice and may acquire powers to carry out any necessary remedial work in the event of non-compliance with such.

In the context of the construction industry, it is also relevant to note the powers invested in local authorities by the Control of Pollution Act 1974. The Act covers waste disposal, water pollution, atmospheric pollution and, by virtue of it repealing the Noise Abatement Act 1960, noise. Under the terms of Part III of the Act, a local authority (in the form of an Environmental Health Officer) has the power to protect people in the locality of a construction or demolition site from excessive noise, and may serve a notice which will:

(a) specify the plant or machinery which may, or may not, be used;

(b) specify the hours during which work may be carried out;

(c) specify the maximum noise levels which may be emitted from any particular point or during the specified hours;

(d) provide for any change in circumstances;

The Act also permits for anyone intending to commence construction or demolition to obtain the prior consent of the local authority, such consent constituting an acceptable defence in the event of a notice being served as mentioned above.

Although the crime of *Breach of Statutory Duty* is usually connected with acts of negligence, it is useful to note that it is also relevant to the tort of nuisance. Section 5 of the Health and Safety at Work Act 1974 (see section 11.3.1) imposes a statutory duty on controllers or owners of premises to use the best practical means of preventing the emission into the atmosphere of noxious or offensive substances, and of rendering harmless or inoffensive any that do escape. Along the same lines as previously discussed in connection with negligence, it must be noted that a proven breach of this statutory duty does not automatically imply the success of a subsequent civil action in nuisance.

2.4 Trespass

It is commonly perceived that *trespass* is committed solely by *unauthorised entry into land or property in the ownership of another person*. Whilst this is certainly true, it must be emphasised that trespass is also committed by *unauthorised intrusion into the airspace above land owned by another person* (for example: by a protruding crane jib) or by *unauthorised intrusion beneath such land* (for example: by tunnelling or excavation). Interestingly, the tort also extends to *causing anything to come into contact with land belonging to someone else*. Thus, a builder who stacks building material against a boundary wall belonging to a neighbouring property can be sued for trespass.

Whilst the most usual remedy is an injunction to prevent further trespass, the plaintiff may claim full restitution for any loss incurred as a result of the trespass. In this repect, however, it is useful to note that the incurrence of a loss is not a mandatory requirement in an action for trespass. For the plaintiff to sue for trespass, it is sufficient for him to have a *legal* estate and *exclusive* posession of the land or property in question.

He does, however, have to show that the intrusion results from a *direct act in the control of the defendant*. Thus, an encroachment by tree roots might well constitute a nuisance but would not constitute an act of trespass. In this context it is worth noting that, providing the act or the means of preventing such is within the control of the defendant, a claim that the act was a mistake will not be entertained as an acceptable defence.

> *Rubbish placed on the defendant's land rolled on to the plaintiff's land as a result of natural causes.*
> Held: *the defendant was liable since the trespass resulted from a direct act (the placing of rubbish) within his control.*

A final point must be mentioned in the context of whether or not the intrusion constituting the alleged trespass was authorised. The fact that a person has had previous permission to be on the land or premises in question does not preclude the right of the owner of that land to sue for trespass at a later date.

A builder had permission to leave rubbish while demolishing a

building. During the demolition process, the plaintiff became the tenant of the land on which the building was sited. After completion of the work, the builder failed to remove the rubbish.
Held: *the builder was liable for trespass and the plaintiff was awarded damages to compensate for the costs incurred in removing the rubbish.*

Although laws relating to trespass are notably absent from Scots Law, it should not be inferred that unauthorised entry on to private property in Scotland carries no liability whatsoever. Whilst a person can accrue no civil liability merely from his unauthorised presence on property, he can *accrue liability in the event that he causes damage to the property whilst there.*

2.5 Strict Liability

The doctrine of *Strict Liability* arises from the fact that, when a party keeps a particularly serious source of potential danger on his land, or carries on a particularly dangerous operation, the ordinary laws of negligence and nuisance offer only limited protection. To avoid extending the common law duty of care beyond the bounds of reasonable credibility, the law has defined certain activities and certain circumstances which give rise to *strict* liability paying no regard to lack of care.

In recognition of the celebrated case in which this type of liability was formulated, 'strict liability' is often referred to by the *cognoscenti* as *Rylands v. Fletcher Liability* or *Liability under Rylands v. Fletcher.*

F employed reputable independent contractors to build a reservoir on his land. All the necessary authority and permission had been gained. Due to the negligence of the contractors, a disused mine shaft connecting with R's mine beneath the reservoir was not blocked up. As a consequence, R's mine was flooded when the reservoir was filled.
Held: *although F was not personally negligent, nor vicariously liable for the negligence of the independent contractors, he was liable for the full extent of the plaintiff's loss.*

In presenting the *ratio decidendi* at the conclusion of the trial, the presiding judge summed up the essence of *strict liability* thus: *'We think that the true rule of law is that the person who, for his own purposes, brings on his lands and collects and keeps there anything likely to do mischief if it escapes, must keep it at his peril and, if he does not do so, is* prima facie *answerable for all the damage which is the natural consequence of its escape'.*

To incur *strict* liability, the source of potential danger must be *non-natural* in the sense of it being an *accumulation* of something *which, in the land's natural condition, was not in it or on it.* It may be inferred from this requirement that, whilst there is *strict* liability for a reservoir, the owner of a natural lake would only incur liability under the torts of nuisance and negligence.

Additionally, there must be an *'escape'* of such from the defendant's land. Thus, a quarry workman injured at work by flying blast debris would, notwithstanding the cause of his injuries being *prima facie* the accumulation of something essentially non-natural (the explosive), find an action for *strict* liability against the quarry owner defeated by virtue of the fact that there had been no 'escape' from the defendant's land. It follows, however, that the quarry owner's liability for injury or damage sustained outwith the boundaries of the quarry would be *strict*, since there would then have been the required 'escape', irrespective of whether or not negligence or nuisance could be established.

It is useful to note that the above pre-requisites have been construed to accord *strict* liability in circumstances where damage or injury has resulted from *'the escape of something brought upon the land ... as a consequence of some non-natural use of that land'.*

> *Vibrations from pile-driving operations caused damage to an old house in the vicinity of the site.*
> Held: *given the unusually frail state of the house, nuisance was not established. Nevertheless,* strict *liability was incurred by the contractor since the vibrations resulted from a use of the land which was adjudged to be* non-natural.

The defences which may be raised against a claim for *strict liability* are much the same as previously discussed in connection with 'ordinary' tortious liability. *Statutory Authority* is a suitable defence as is a variation of *volenti non fit injuria* in circumstances where it can be shown that the plaintiff consented to the accumulation of that which escaped and caused him damage or injury. *Act of God* may also be claimed, as may the contributory negligence of the plaintiff in assisting the escape. In the latter case, liabilty will still be held, since *strict liability* is absolute, but the amount of damages awarded against the defendant will be reduced to the extent deemed appropriate in the light of the plaintiff's negligence. It is also worth noting that a claim for *strict liability* may be deflected to a certain extent by establishing that the escape was due to some malicious act of an unknown third party.

CHAPTER 3

CONTRACTS AND CONTRACT LAW

In whichever capacity a practising civil engineer is working within the construction industry, irrespective of the type of organisation by which he is employed, it is likely that he will be operating in a contractual environment.

For example, a party commissioning a construction project (the Employer or Promoter) will retain the services of a Consultant Engineer who will provide, under contract and in return for an agreed fee, the necessary technical expertise in respect of the design and supervision of construction of the project. In addition to these duties, the Consultant Engineer will be required to draw up and administer a contract between the Employer and a Contractor who will undertake to carry out the actual construction work in return for an agreed payment. In turn, the Contractor will enter into separate contracts with a number of organisations who, in return for agreed amounts of money, will provide him with the resources he requires in the course of the construction process.

At the heart of each of the above commercial relationships is a legally enforceable agreement – a contract – whereby each party undertakes to discharge certain obligations in return for the accrual of certain benefits. A failure to discharge any of the obligations so undertaken usually involves some sort of sanction either expressed or implied in the terms of the contract or imposed by the law of the land. In this light, therefore, it is essential that all parties involved in the execution of a contract are fully aware of the responsibilities and obligations imposed upon them by the contract, together with the implications of failing to fulfil them satisfactorily.

3.1 The Form of Contract

As stated above, a contract is *a legally binding agreement between two parties, or an exchange of promises whereby one party undertakes to provide something in return for something else from another party.*

With this definition in mind, it can be seen that almost every transaction made in the course of a person's day-to-day life involves a contract of one form or another. For example, a person travelling on public transport enters into a contract whereby the transport company, in return for the appropriate fare, agrees to transport him *safely* from A to B. Similarly, the purchase of goods at a shop involves a contract in which the vendor agrees to sell goods *in a satisfactory condition* in return for a specified payment from the purchaser. It can therefore be inferred that, contrary to a widespread and popular belief, legally enforceable contracts do not always necessitate complex documentation couched in barely intelligible 'legalese'. In the above examples, there is unlikely to be anything in writing other than a ticket or a receipt for cash paid. Nevertheless, the parties involved in the transactions are as legally bound by contractual obligations to one another as are the Employer and Contractor in a formal contract for the construction of, say, a nuclear power station.

Additionally, and again contrary to popular belief, it is worth noting that contracts do not always have to be in writing to be legally enforceable. A contract may be considered to exist by virtue of a *verbal* agreement between the parties involved or it may be be inferred from a *gesture* such as a handshake. Whilst such contracts are as legally binding as written contracts, there exists the problem, in the event of a dispute between the parties, of establishing *exactly* what was agreed and what the parties intended at the time the contract was formed.

It is worth noting that Scots Law requires proof of the existence of a verbal agreement before it can be enforced. For example, the existence of consideration, *although not a legal requirement under Scots Law (see section 3.2.2), would constitute acceptable proof as would the evidence, under oath, of a witness to the agreement.*

Exceptionally to the foregoing, it is noteworthy that some contracts cannot be legally enforced if made verbally. These may be classified into two categories, the first of which consists of contracts which must be completely in writing. Examples of such are promissory notes, bills of exchange, share transfers, marine insurance and, in some instances, hire purchase agreements. The second category consists of those contracts which need not be *completely* in writing but must be *'evidenced in writing'* in that all the material terms of the agreement, including the signatures of the parties thereto, must be documented on at least one note or memorandum. In the context of the construction industry, notable examples of such are contracts relating to land purchase or disposition, or contracts of guarantee whereby a party contracts to guarantee the debt or default of another party. In the case of a civil engineering contract being executed under the I.C.E. Conditions of Contract, the so-called Form of Bond (see section 6.8) constitutes a contract of guarantee in that a specified party guarantees to reimburse the Employer in the event of the Contractor's default. It is worth noting that, whilst such a contract which does not comply with the above requirement may not be enforced under Contract Law, it may be enforceable in Equity (see section 1.2.3) if the circumstances so dictate.

Since construction contracts do not fall into any of the above categories, it is sufficient, from a purely legal standpoint, for them to be entirely verbal. However, and for obvious practical reasons, this would not be advisable and, as a consequence, all such contracts are formalised in writing. It should be noted that 'in writing' does not necessarily imply a formal document in that an exchange of letters is sufficient providing the requirements demanded by contract law (see section 3.2) are met.

For legal purposes, contracts are classified as belonging to either of two categories: *simple contracts* or *contracts under seal*.

A *simple contract* may be in any form as suggested above, requires *consideration* (see section 3.2.2) and has a *limitation period* (see section 3.6.7) of *six years*. If the contract is in writing and contains the signatures of the two parties, it is classified as 'under hand'. On the other hand, a *contract under seal* (or *deed contract*) must be a formal written document to which the seals of the parties are affixed. It has a limitation period of *twelve years* and does not require consideration. A noteworthy feature of such contracts is that statements of fact contained therein cannot subsequently be called into question.

A legal distinction is also drawn between *unilateral* and *multilateral* contracts.

Multilateral contracts are the most common type of contract and are formed when a number of *specified* parties agree to be legally bound to each other on *specific* terms. For example, a *bilateral* contract consists of an agreement between *two* parties whereby one party agrees to provide something in return for something else from the other party. A *unilateral* contract, on the other hand, obtains when a *specified* party offers to provide an *unspecified* party with something of benefit in return for that party's performance of a *specified* action. An example of such would concern the placement, by a *named* party, of an advertisement promising a reward for the return of a lost item of property. In such a case, the person finding and returning the lost item would be entitled to claim the reward even though he was not named in the original offer.

3.2 Formation of a Contract

In order that a contract may be regarded as legally binding on the parties concerned, the Law of Contract requires that certain conditions be complied with.

Firstly, there must be agreement between the parties regarding the purpose of the contract, and the rights and obligations which the contract will create. Agreement is considered to have been reached when an *offer* made by one party is *unconditionally accepted* by the party to whom it was made. Secondly, except in the case of a *contract under seal*, each party must contribute something to the bargain to demonstrate good faith, such contribution being known as *consideration*. Lastly, there must be an *intention* by the parties *to be legally bound* by the terms of their agreement.

3.2.1 Offer and Acceptance

As stated above, a fundamental requirement for the formation of a legally enforceable contract is for one party (the *offeror*)to make an offer which is *unconditionally* accepted by the party to whom the offer was made (the *offeree*).

The *offer*, which may be made to a specific person, or to a group of persons, or to the public at large, constitutes a statement that the party making the offer is willing to enter a legal relationship and be legally bound by specific terms. This definition, however, raises certain questions as to whether all statements of this nature amount to offers and whether, under certain circumstances, such a statement may have any contractually binding effect.

A notable example of such a situation, in the context of the construction industry, concerns notices which appear from time to time in the trade press whereby contracting firms are invited to submit tenders for forthcoming local authority projects. Whilst the local authority in such a case is, in effect, offering contractors the opportunity to compete for the work, such a notice should not be construed as constituting an *offer*, in the contractual sense of the word, which is open to acceptance by all firms submitting tenders. Obviously the consequences of such would be too horrendous to contemplate since a multitude of contractors could all claim to have entered into valid contracts with the local authority! To avoid this possibility arising, such a notice is classified as a so-called '*invitation to treat*' whereby contractors are invited to submit individual tenders (each of which constitutes an offer to carry out the required construction for a specified price) which may be accepted or rejected by the local authority.

In view of the aforesaid, since the party issuing the *invitation to treat* is under no obligation to accept any of the tenders, the cost of preparing a tender must be borne by the party submitting it. An exception to this rule, however, will apply in circumstances where the tenderer, at the request of the employer, undertakes work outwith the normal scope of tendering. Such a request carries an implied undertaking that the tenderer will be paid a reasonable sum of money for the extra work.

> *A contractor prepared estimates for the reconstruction of a war-damaged building in the belief, shared by the owners of the building, that he would be awarded the contract for reconstruction. The estimate, and various other calculations made by the contractor, were used by the owners to support a claim for compensation from the War Damage Commission. Following the award of compensation, no contract was placed with the contractor.*
>
> Held: *since the work carried out by the contractor exceeded that normally required in the course of preparing a tender, it carried an implied promise of payment. The contractor was, therefore, entitled to reasonable remuneration.*

A further example concerns the submission of a so-called 'estimate' in response to an enquiry from a potential customer. Broadly speaking, in such cases, the courts will take into account the intentions of the parties involved when assessing the legality of the situation. Thus, if an 'estimate' has been submitted in the form of an *offer capable of acceptance*, and the courts clearly percieve it as such, its acceptance will be deemed to have given rise to a legally binding contract.

> *An architect wrote to a contractor asking if he would be prepared to submit a tender 'in competition for the work'. Following receipt of the necessary information concerning the tender, the contractor submitted his 'estimate' which was accepted. He subsequently withdrew the 'estimate' claiming that he had made no offer capable of acceptance.*
> Held: *the contractor's 'estimate' had been an offer capable of acceptance and a binding contract had been formed by its acceptance. He was therefore liable for the difference between his estimated cost and that incurred in having the work carried out by another contractor.*

The converse is, of course, true in that an 'estimate' accompanied by a *clear* and *unequivocal* statement that it should not be regarded as a definite *offer capable of acceptance* will not result in the formation of a legally binding contract. However, since an 'offer' thus qualified would undoubtedly be rejected out of hand by anybody in their right mind, the situation is probably of academic interest only!

In view of the fact that a binding contract requires both offer *and* acceptance to have taken place, it is generally the case that an offer may be withdrawn, *without obligation, at any time prior to its acceptance*. A noteworthy exception to this rule arises in circumstances where the offeror has agreed, in return for payment from the offeree, to keep the offer open for (usually) a specified period of time. If such is the case, withdrawal of the offer during the specified period would constitute a breach of the terms of the agreement (which, in view of the financial transaction, would probably have been formalised) and would give rise to the possibility of legal action on the part of the offeree.

It is worth mentioning that the law imposes no limitation on the period during which an offer may remain open to acceptance. This being the case, offers very often contain conditions limiting their period of validity. Apart from the fact that it would be unwise to leave an offer open *ad-infinitum* (since it may well be accepted at a time when the offeror is no longer in a position to fulfil his tendered obligation, perhaps having forgotten to withdraw the offer), such conditions are often included as a commercial ploy to pressurise the offeree into speedy (and, perhaps, ill-considered) acceptance. If a time limit is placed on the validity of an offer, it should be noted that the offer automatically lapses upon the expiry of such and is not capable of acceptance thereafter.

In the case of such a time limit being placed on the validity of a tender for a construction project, a possible revival of the tender after the expiry of the time limit would fairly entitle the tenderer to demand a revision of the terms of his tender.

Finally, it should be noted that the offer, together with its possible withdrawal, only takes effect *from the time it comes to the notice of the offeree.*

It bears repeating that the formation of a *legally binding* contract requires the *acceptance* to be *unqualified* and *unconditional.* Any acceptance which contains phrases like: 'we accept your offer *if* ....' or 'your offer is accepted *provided that* ....' constitutes a *counter-offer* which must be *unconditionally* accepted by the original offeror before a binding contract is formed. It may, therefore, be correctly inferred that the commonly used phrase: 'acceptance subject to contract' is a *qualified* acceptance which effectively prevents the agreement from becoming legally binding until the contract is formalised in writing – house purchasers/sellers beware! It is of interest to note that such a qualified acceptance amounts to a *counter-offer* which may destroy the original offer and preclude its possible acceptance at a later date.

It is not unknown for an offer to include a statement to the effect that, should the offeror hear nothing to the contrary within a given period, the offeree's acceptance of such will be assumed. It should be noted that, notwithstanding the inclusion of any such statements, the acceptance *must be communicated to the offeror* before a binding contract is formed and that *silence does not constitute acceptance.*

That said, however, there are certain exceptions to the rule. For example, if the offeree expressly or implicitly states that his silence may be taken as confirmation of his acceptance of the offer, then such will be the case. Additionally, if the offer is made in clear terms, defining specific work, the offeree's allowing the work to proceed may be regarded as implicit acceptance of the offer, thus allowing the offeror to claim payment for the work done.

Whilst it may reasonably be suggested that such an approach would be somewhat inadvisable in the case of a construction contract, it is worthwhile to note that delayed acceptance of an offer may have a retrospective effect if that is clearly the intention of the parties.

> *A sub-contractor was asked, by the main contractor, to start work*
> *before the terms of the sub-contract were finally agreed.*
> Held: *since the parties fully intended the work to be governed by*
> *the sub-contract as eventually made, the work carried out prior to*
> *the sub-contract becoming legally binding was retrospectively*
> *subject to the terms of that sub-contract.*

In cases where a *formal contract* has a retrospective effect, there is
generally no confusion as to the terms and conditions under which the
contract is executed. However, in cases where the acceptance is
implied, there may be some difficulty in establishing what was, or was
not, agreed or, indeed, whether there is any legal basis whatsoever to
the arrangement.

Consider the case where a sub-contractor submits a tender, subject
to *his* standard terms, to the main contractor whose acceptance thereof
is subject to *his* standard terms. If, as is very often the case, the
sub-contractor's standard terms are inconsistent with the main
contractor's standard terms, a period of correspondence ensues during
which some terms are agreed and others are not. If, at some stage in
the negotiations, the sub-contractor starts work at the request of the
main contractor, the legal principle which will be applied is that the
last letter received before commencement of the work, from whichever
side it emanates, is deemed to be accepted.

> *The seller of an item of machinery included his 'conditions of*
> *sale' on the back of a quotation, such conditions opening with a*
> *clause to the effect that they were to prevail over any terms and*
> *conditions contained in the buyer's order. The buyer replied with*
> *an order containing his terms and conditions, the order including*
> *a tear-off slip by which the seller could acknowlege receipt of*
> *the order. The acknowlegement slip also contained a statement to*
> *the effect that the buyer's terms and conditions were accepted.*
> *The slip was completed and returned by the seller.*
> Held: *the signed acknowlegement slip, being the last*
> *communication received, resulted in the contract of sale being*
> *subject to the* buyer's *terms and conditions and* not *those of the*
> seller. *Consequently, a price-fluctuation clause in the seller's*
> *conditions did not apply.*

Notwithstanding the aforesaid, however, no legal contract is formed
if the parties demonstrate, by their continuing negotiation, that they do
not regard themselves as legally bound. Similarly, if disagreement
persists on some important term fundamental to the issue, no legal

contract is formed. In such a case, the party carrying out the work is entitled to payment of a reasonable sum or *quantum meruit* (see section 3.6.6).

As has been stated, an offer may be withdrawn at any time *prior to its acceptance*, and withdrawal takes effect *from the time it comes to the notice of the offeree*. Similarly, the acceptance *must be properly communicated to the offeror*. In view of the importance of timing, particularly in respect of when the acceptance was deemed to have been made (since it effectively seals the contract), the vagaries of the postal system are likely to give rise to certain difficulties in connection with communications through the post. With this in mind, *English* law lays down the following rules with respect to contractual communications through the post:

(a)   An *offer* takes effect when it is *received* by the offeree;

(b)   The *withdrawal of an offer* takes place when it is *received* by the offeree;

(c)   An *acceptance* is complete when it is *posted* by the offeree;

It would thus appear advisable to make use of *registered mail* for all postal communications in connection with the offer and acceptance since this would ensure the availability of the appropriate dates of postage and receipt in the event of a possible dispute.

It goes without saying, of course, that the application of such rules is bound to give rise to the occasional anomalous situation in which a degree of hardship and injustice is incurred.

> *Party A posted an offer in Cardiff on 1st October. He subsequently posted a letter of withdrawal. The offer reached party B in New York on 11th October. B accepted by telegram and confirmed by letter posted on 15th October. The letter of withdrawal reached B on 20th October.*
> Held: *a valid contract had been formed since the withdrawal was received* after *the acceptance had been made.*

Despite the occurrence of such situations, however, it should be recognised that considerably more confusion and apparent injustice would arise in the absence of any rules whatsoever, and that a different set of rules would merely lead to different anomalies!

One final point should be made in connection with both the offer and the acceptance, namely that they should be sufficiently *certain* to have practical meaning.

> *A contractor submitted two tenders for the construction of a freight terminal. One tender was for a fixed price and the other contained a price fluctuation clause. The employers wrote to the contractor that they had accepted "your tender". The contractor started work before a formal contract had been signed. Shortly after commencement of the works, there was a steep rise in costs and the contractor wrote to the employers requesting sympathetic consideration of a revision of prices in the fixed-price tender. This was refused on the grounds that the tender had been accepted.*
>
> Held: *since the employers had not made it clear which of the two tenders had been accepted, no contract had been concluded and the contractor was entitled to payment on a* quantum meruit *basis (see section 3.6.6) for the work completed.*

3.2.2 Consideration

As a result of its having been originally developed to meet the needs of the commercial world, the *English* Law of Contract stipulates that only those agreements which contain an element of bargain will be enforced.

Broadly speaking, this means that the law will not enforce gratuitous promises. Thus, for an agreement to be legally enforceable, each party to the agreement is required to give 'something' to the other party in return for the benefit he is likely to receive from the agreement. That 'something' is, in legal terms, *consideration*.

On the part of the employer in a construction contract, the consideration is his promise to pay for the works as eventually constructed – known as *executory consideration*. On the part of the contractor who is employed to construct the works, the consideration consists of his undertaking to do so (again, *executory consideration*) or his actual construction of the works – known as *executed consideration*.

In order that it should be adjudged as sufficient basis for a legally enforceable contract, the consideration must conform to certain rules:

(a)   The consideration must have *some* economic value, although the law is not concerned with the *adequacy* of such. Thus, if an

Employer promises to pay a Contractor £1,000 in return for the Contractor's undertaking to construct a ten-span bridge, the contract will be enforced by the courts – despite the fact that the Contractor would plainly require his head to be examined!

(b) The consideration must involve some kind of sacrifice. Thus, a party's promise to perform an act which he was already obliged (either legally or contractually) to do would not constitute acceptable consideration.

(c) An act claimed to constitute the consideration must not have been performed already. Thus, if the performance of a service by one party *precedes* the other party's promise of payment for such, the promise of payment is not enforceable since the party performing the service would not have provided consideration *in return for the promise of payment*.

(d) The consideration must be given by the party seeking to enforce the contract. It therefore follows that if *both* parties desire a legally enforceable contract, consideration must pass *both ways*.

It is worth noting that, if rigidly and strictly applied, the doctrine of consideration may occasionally give rise to a certain amount of injustice.

In the context of the construction industry, consider the case of an Employer who has led the Contractor to believe that he will not insist on the tendered completion date (which is an integral part of the tender – see section 6.2) being adhered to. In view of this 'waiver', the Contractor has not made any attempt to submit claims for extensions of time. Since the Contractor has given no consideration in return for the Employer's 'waiver' in respect of the original completion date, the Employer is (strictly speaking) entitled to change his mind and claim liquidated damages (to the extent specified in the contract) in the event of delayed completion.

Apart from the fact that such a change of mind would not do a great deal for the Employer's reputation in respect of his treatment of contractors, the possibility of his being allowed to do so is patently unjust.

In this light, and in the event of such a situation occurring, the rules of Equity (see section 1.2.3) would come into play and would take precedence over the law. The equitable rule in such a situation is that, where a binding agreement between A and B is waived by A's subsequent promise not to insist on his contractual rights, A cannot sue on the contract if B had acted in reliance of A's promise.

It should be mentioned that the above equitable rule not only applies in the event of one party deliberately misleading the other party, but also applies where the behaviour of one party gives the other party good reason to believe that he has waived his contractual rights. Additionally, it should be noted that the application of the equitable rule does not necessarily extinguish those rights. In certain cases, the aggrieved party may only be given the opportunity to put himself back in the position in which he would have been had he not relied on the representation. In the case of the unjustly treated Contractor cited above, this would result in him being allowed to submit the appropriate claims for an extension of the tendered completion date.

Finally, it should be noted that, for historical reasons beyond the scope of this text, contracts *under seal* are *not* required to be supported by consideration.

There is no doctrine of consideration in Scotland and, in that jurisdiction and country, a 'bare promise' is enforceable. The doctrine of consideration has been extensively criticised by legal commentators throughout the United Kingdom, with many suggesting that a 'bare promise' should also be enforceable in England. At the time of writing, however, this state of affairs has yet to be realised.

3.2.3 Intention to be Legally Bound

Although a legal contract necessitates agreement between the parties as to the rights and obligations it will create, a mere agreement *per se* will not be enforceable at law unless the parties thereto *fully intend to be legally bound by its terms and conditions*.

In the majority of commercial agreements, there is a clear and obvious intention by the parties to create a legal relationship. Indeed, the courts will generally presume this to be so unless the agreement contains a clear statement to the contrary. If such is the case, the agreement will not be enforced. Similarly, the agreement will not be enforced if the parties demonstrate by their continued negotiation that they do not regard themselves as legally bound.

It is noteworthy that, despite the existence of offer, acceptance and consideration, one party may allege that the agreement is not legally binding since there was no intention to create a legal relationship. This is often the case with family arrangements where the courts will presume that domestic agreements are not made with the intention of being enforceable at law. In such cases, the person seeking to enforce the agreement is required to rebut the presumption. If the consequences of a failure to legally enforce the agreement will give rise to a degree of injustice, however, the rules of Equity may affect the situation and, if applied, will prevail over the rule of law.

> *A mother-in-law lived with the family and paid for some building work to the house.*
> Held: *although there was no enforceable loan under contract law, Equity required that the value of the work done be held in trust for the mother-in-law.*

3.3 The Terms of Contract

As stated in section 3.2, a fundamental requirement in the formation of a legal contract is the intention of the parties to be legally bound by its terms.

Whilst the parties are free to decide those terms, providing they are within the law, great care should be taken over their drafting since a possible dispute will be resolved by a court's implementation of the contract in *strict* accordance with those terms. If the terms are written down – *express* terms – the sole task of the court will be to decide what the terms, as written, *mean*. It will *not* be the task of the court to speculate on what the parties may, or may not, have intended by their inclusion.

Broadly speaking, this means that the court is not permitted to consider *extrinsic* evidence which may be introduced by way of explaining the *intent* of the parties, but must reach its decision on the basis of the contract documentation *only*.

It should be stressed, however, that this is a *general* rule which may be relaxed on occasions. For example, the court may be permitted to make use of extrinsic evidence *as an aid to translation* in the following instances:

(a) to identify persons named, or objects mentioned, in the contract documentation;

(b)   where   scientific   or   technical   terms   require   explanation   or
      clarification;

(c)   to demonstrate the meaning of trade usages or terms;

(d)   where the contract documentation is in a foreign language;

Similar relaxation will be applied, permitting the court to admit
extrinsic evidence *in support of a case*, where such evidence is
required:

(a)   to demonstrate that the contract was fraudulently or illegally
      entered into (see sections 3.4.2 to 3.4.5);

(b)   to prove that a *collateral agreement* was made at the same time
      as the contract in question;

(c)   to show that the contract has been *rescinded* (see section 3.6.4);

(d)   to show a subsequent variation of the original contract;

(e)   to show that the contract was only to have become binding after
      performance of a condition which was, in the event, not
      performed;

(f)   to prove trade customs or usage;

It is quite possible that, even after it has applied all the usual rules
of construction, the court will find that the words in dispute are *still*
ambiguous. In such a case, it will apply the rule of *contra preferentem*
and construe the words *against* the person who seeks to rely on them,
unless the effect of such would be to injure third parties.

Despite the fact that the majority of construction contracts make
use of standard forms of contract (see section 6.5), certain
characteristics of a particular contract may give rise to the necessity for
additional conditions to be appended to the standard conditions, or for
a number of those standard conditions to be modified. Such
modifications or additional conditions will usually be drafted by the
Engineer (see section 4.2) either at the behest of the Employer (see
section 4.1) or as a result of negotiations between the
Employer/Engineer and the Contractor. In drafting these conditions or
modifications it should be appreciated that, whilst the Engineer's
implementation of such may be based on their *intended* meaning, a

court's decision in the event of a dispute over such will be based on their *literal* meaning – which may be significantly different from that which was intended!

This being the case, it is not beyond the bounds of reasonable possibility for a decision of the Engineer, on the basis of which the Works have been completed, to be subsequently overturned in court to the acute embarrassment (financial and otherwise) of all concerned. To avoid the occurrence of such a possibility, it would appear advisable to retain the (moderately extortionate) services of a lawyer at the drafting stage rather than the (unbelievably extortionate) services of a battery of lawyers in the event of subsequent litigation.

One final point is worth noting in respect of additions or amendments to standard conditions of contract. If the standard document is *printed* and the additions or alterations are *handwritten*, the courts will accord preference to the *handwritten* words in the event of a dispute, notwithstanding possible inconsistency with other *printed* words in the same document.

It is a popular misconception amongst those involved in the administration of construction contracts governed by standard forms and conditions of contract that everything they require to know about the contract is contained within the standard documents. This is, in fact, not so in that, dependent upon the circumstances, courts may sometimes insist upon the observance of terms which are not contained within the agreement at all. Such terms are known as *implied* terms and are as legally binding upon the parties as *express* terms.

The principles of how such terms can be recognised was established in a shipping case, in 1889, known as the 'Moorcock' case.

> The 'Moorcock' was a barge which tied up at a jetty in the tidal reaches of the Thames. Her owners had a contract with the jetty owners whereby the barge's cargo could be unloaded in return for the payment of an agreed landing fee. Both parties to the contract were aware that the barge would ground at low tide. On one occasion, the barge grounded and suffered damage resulting from the presence of a hard ridge just under the surface of the mud. The barge owners sued the jetty owners for the cost of repairs.
> Held: *despite the absence of a guarantee of safety at the mooring, the suit was successful. In the light of both parties' awareness of the possibility of the barge grounding at low tide,*

the court implied *an undertaking by the jetty owners that the river bottom was, so far as reasonable care could provide, in such a condition as not to damage the grounded vessel.*

In a subsequent case involving an implied term, the presiding judge defined such a term as one which '*....if at the time the contract was being negotiated, someone had said to the parties: "What will happen in such a case?", they would both have replied: "Of course so–and–so will happen; we did not trouble to say that; it is too clear."*'

It is reasonable to infer from the above statement that implied terms and conditions must be *genuinely self–evident* to merit inclusion within a contract. Consequently, it would be unwise to assume that every missing provision can be read into a contract as an implied term, or that one party to a contract can be considered bound by conditions which do not appear within the contract on account of their being perceived as 'blindingly obvious' by the party drafting the contract document.

A 600mm water main ran through the site of a large industrial development project. The nature of the proposed development required that the existing main be diverted to follow another route through the site. The tender documents contained full details of both the existing main and the proposed diversion, together with an instruction requiring the Contractor to provide facilities for water–board personnel to carry out the connection operations in respect of the diverted section. The project designer considered that the obvious undesirability of shutting–off such a large water main for more than a very short time dictated the need for the new section to be complete before *the water supply could be shut off and the old section abandoned. He felt that such a sequence of operations was so obvious that no instructions to this effect were included in the contract documents. However, it was not so obvious to the Contractor who asserted that his programming and estimated costs were based on the assumption that the water supply could be shut–off for a considerable period, allowing the old main to be abandoned* before *building the new section. He had considered that any other sequence of operations would be excessively disruptive and therefore more costly.*
Held: *even if it had been the general practice of the water–board to permit only the briefest interruption to large mains, the failure to include* express *reference in the contract documents to such an important restriction rendered the Employer liable for the*

reasonable additional costs incurred by the Contractor in re-phasing his work.

Implied contract terms may arise in three ways. Firstly, the parties to the contract are presumed to have formulated their agreement with reference to prevailing *customs* or *trade practices*. Thus, in accordance with prevailing trade practice, a tender for a construction project is prepared and submitted on the *implied* understanding that the costs of preparation will be borne by the tenderer whether or not the tender is accepted – subject, of course, to the constraints referred to in section 3.2.1.

Secondly, there are contracts into which terms are implied *by statute*. For example, the Sale of Goods Act 1979 requires that goods for sale be fit for the purpose for which they are being sold and that they must be of merchantable quality. The principles of this Act have been extended, by the Supply of Goods and Services Act 1982, to apply to contracts for the supply of services and will thus apply to construction contracts. Broadly speaking, the Act requires that such contracts be subject to implied terms which impose upon the supplier of the service the obligation to perform the service with *reasonable skill and diligence* and, in the absence of express terms covering completion time and costs, within a *reasonable time* and for a *reasonable charge*.

Thirdly, a court may imply a term to give a contract *business efficacy* in circumstances where it feels that there is no other way in which the contract is workable.

The design of a steel lifting-bridge, carried out as part of a lump sum 'design and construct' contract, was subject to the requirement that the bridge deck could be supported by only one of its two hydraulic rams in the event of failure. In the course of construction, tests revealed that the deck distorted to a damaging degree when unevenly supported by a single ram. The Contractor claimed reimbursement for the additional costs incurred in carrying out the subsequent extensive modifications required to alleviate the problem.
Held: the requirement that the ram system be powerful enough to support the deck in the event of partial failure was largely redundant without the existence of a second implied requirement that the deck be sufficiently rigid to resist damage in this situation. The claim for additional payment was rejected.

In respect of the insertion of implied terms to give a contract business efficacy, it is worth noting that such terms will *only* be inserted if adjudged *obviously essential* to make the contract meaningful. Thus, if a contract is workable and has business efficacy *without* the implied terms, no implication will be made. Similarly, a court has no brief to insert an implied term merely to render a contract *more* workable. In other words, it cannot *improve* the contract by inserting implied terms. It is also noteworthy that a court will refuse to imply a term in the event that such a term would be inconsistent with express terms elsewhere in the agreement.

In view of the fact that contractual disputes are matters of civil law and, therefore, subject to the doctrine of *judicial precedent* (see section 1.2.2), it is quite possible for terms which have been implied in decided cases to act as precedents and give rise to so-called *common law contractual duties* beyond the express terms of a contract. In the context of the construction industry, notable examples of such include the Employer's obligation to give the Contractor possession of the site within a reasonable time following finalisation of the contractual formalities. Similarly, the Contractor has a common law contractual duty to carry out the works with proper skill and diligence. Additionally, the materials supplied for incorporation into the Works must be of acceptable quality and fit for the purpose for which they are intended. Such requirements, however, are usually covered by express terms in a construction contract thus precluding the necessity for further implication.

Finally, it should be noted that there are some terms which *cannot* normally be implied into construction contracts. For example, the Employer gives no implied guarantee as to the nature and suitability of the site, or as to the 'buildability' of the design.

> *The foundations of a river bridge were designed to be put in with caissons. The Contractor found this to be impossible and eventually had to abandon the attempt. The work was completed in accordance with a revised design which avoided the use of caissons.*
>
> Held: *the Contractor was entitled to the specified contract payments for the work done on the basis of the original design and, under the particular form of contract, to additional payment for work completed on the basis of the more costly revised design. He was not, however, entitled to reimbursement for the additional costs and delays incurred in the course of his attempts to use the original method.*

On the other hand, if the Contractor completes his work strictly in accordance with a detailed specification, he gives no implied guarantee that the finished product is fit for the purpose for which it was intended.

A house was built in strict accordance with a detailed specification which called for external walls to be constructed of 9 inch solid brickwork with no rendering. Following completion, one room became uninhabitable as a consequence of rain entering the house through one un-rendered wall.

Held: *the purchaser's claim against the builder was rejected on the ground that the builder had done exactly that which he had contracted to do – build a house according to the given specification – and that he had given no implied guarantee of fitness of the finished product. It was pointed out that to have made the walls waterproof would not have been in compliance with the specification and would, therefore, have amounted to a breach of contract.*

It is worth noting in respect of the above case that, had the purchaser relied on the expertise of the builder in drawing up the specification, a failure on the part of the builder to have drawn attention to such an obvious shortcoming would have constituted a negligent mis-statement (see section 2.2) with the possibility of an action in tort.

One final point remains to be discussed in respect of contract terms and that concerns the insertion, by one party, of terms limiting or excluding his liability to the other party – the so-called *exclusion* or *limitation* clauses popularly referred to as the 'small print' clauses.

The rule here is that a party is only bound by such conditions *if he knew of their existence, and agreed to them, before, or at the time, the contract was made.* The effect of this rule is that if the exclusion or limitation clauses appear *within the body of the contract*, the parties signing the contract are deemed to have read, *and* agreed to, the conditions and are legally bound thereby. In the event of a suit on the contract, therefore, a plaintiff's claim that he had not read the 'small print' will not meet with any sympathy from the court – unless he can prove that his signature had been obtained fraudulently or by misrepresentation (see section 3.4.2).

It is noteworthy that *written* exclusion or limitation clauses can be over-ridden by *oral* statements. Thus, if party A is told *verbally* by party B that the exclusion or limitation clauses in the contract between them do not apply at all, or only apply to a limited extent, party B can only rely on the relevant clauses to the extent stated verbally. This state of affairs can, of course, give rise to a number of problems. For example, there is the problem of misinterpretation where A is under the impression that B's verbal statement means something completely different from that which B intended it to mean. Even assuming that there are no problems of interpretation, the burden of proving what was actually said rests with the plaintiff A and that might prove to be an insurmountable barrier in the absence of corroboration.

To avoid these sorts of problems occurring, standard terms usually contain a proviso to the effect that verbal statements will not affect the written conditions unless confirmed in writing.

It is very often the case that exclusion and limitation clauses are not contained within the body of a contract document but are documented elsewhere. In such cases, the party seeking to rely on the clauses must demonstrate that the whereabouts of the relevant documentation had been *clearly* communicated to the other party *and* that the other party had agreed to those terms before, or at the time, the contract was made.

> *M shipped his car on D's ship which sank with the car. M had used this facility on many previous occasions and had always signed a 'risk note' containing a list of standard conditions governing the carriage of goods on D's ships. One of the conditions excluded D from liability for any loss 'wheresoever or whensoever occurring'. On the occasion in question, M had not been asked to sign the 'risk note'.*
> Held: *despite D's claim that M was familiar with all the conditions of carriage of goods, since he had signed so many 'risk notes' in the past, D was adjudged liable for the loss of the car. The court considered that the lack of M's signature on the 'risk note' on this occasion, irrespective of whether or not M was familiar with the conditions contained therein, made the agreement a straightforward contract of carriage without any exclusion clause.*

As might be imagined, the use of exclusion and limitation clauses is open to a certain amount of abuse by commercial and public bodies in a position to dictate their terms, however unfair, with a view to taking advantage of consumers who very often fail to read the 'small print' until it is too late! Fortunately, however, the opportunities for such are strictly limited by the possibility of statutory intervention in the shape of the Misrepresentation Act 1967 (see section 3.4.2) and the Unfair Contract Terms Act 1977.

In the particular context of exclusion and limitation clauses, the latter Act invalidates any attempt by a contract term to exclude or limit liability for *death or personal injury resulting from negligence.* Additionally, terms excluding or limiting liability in respect of *any other loss resulting from negligence* are enforceable only insofar as they are adjudged *fair and reasonable.* For example, a clause excluding or limiting the defendant's liability in respect of a risk which the plaintiff could normally have been insured against would be considered reasonable by a court.

A third provision of the Act states that, where a party is in breach of a contract made on the basis of *his written standard terms of business*, he cannot rely on terms therein excluding or limiting his liability in the event of said breach. In this respect, it is not necessary for a party seeking to invalidate such a term to establish the conditions as being *unique* to the offending party, but merely to establish the other party's *frequent use of such.* In the context of the construction industry, this provision of the Act may therefore be reasonably inferred to apply to Employers who make habitual use of, *inter alia*, the I.C.E. Standard Conditions of Contract (see Chapter 9). On the other hand, an Employer who is using a particular standard document for the first time would not appear to be subject to the Act.

In the event that 'standard usage' cannot be established, it is noteworthy that breach-related exclusion and limitation clauses are, in general, only enforceable to an extent adjudged *fair and reasonable* in the light of prevailing circumstances. In this context, the burden of establishing fairness and reasonableness lies with the party seeking to enforce the terms.

3.4 Avoidance of the Contract

There are certain circumstances in which a contract has been formed in accordance with the legal requirements detailed in section 3.2, but the agreement is not legally enforceable.

The most common instance of such is where there is a lack of *genuine consent* between the parties. This will be the case, for example, when one or other (or, indeed, both) of the parties is labouring under a factual mistake concerning an issue fundamental to the agreement, or when one party has been induced to enter into the contract as a result of a misrepresentation by the other party. A lack of genuine consent is also evident in the case of one party entering into the contract under duress or as the result of undue influence.

Less common, but nevertheless noteworthy, instances concern contracts which involve a certain amount of illegality, or contracts where one party is under some legal incapacity to contract.

These points are dealt with in turn, together with the related doctrine of Privity of Contract whereby only parties to a contract may sue upon that contract.

3.4.1 Mistake

Broadly speaking, from the point of view of available remedies, mistakes may be classified as: those that are considered to be fundamental to the agreement, and those that are not. In the event of a *fundamental* mistake being discovered subsequently to the contract being made, the contract may be voided whereas *non−fundamental* mistakes, such as errors of judgement, must be borne by the contracting parties. In the case of a mistake voiding a contract, it is noteworthy that the absence (by definition) of malice precludes the award of financial damages by way of compensation.

Where the mistake relates to the *existence* of the object(s) of the agreement, it is considered (for obvious reasons) to be central to the agreement and the contract can be voided. An example of such would concern the sale of an object which, unbeknown to either party, had been destroyed by fire.

Similarly, a mistake over the *identity* of the subject matter would be regarded as sufficient to void a contract.

> *An Employer negotiated with a Contractor for the construction of a building having some alternative design points. Following discussions between the Employer and the Contractor, the Employer decided to adopt the more costly alternative, but did not make this absolutely clear in the final specification.*
> Held: *since the Contractor was able to demonstrate good reason for believing that he had contracted to construct the lower priced alternative, the contract was voided.*

It is worth noting, however, that mistakes (whether mutual or otherwise) concerning the *quality* of the subject matter are not considered sufficiently fundamental to void the contract. In such cases, the courts will apply the principle of *caveat emptor* (literally: 'let the buyer beware') and require the parties to stand by their mistakes.

> *An art collector purchased a painting of Salisbury Cathedral from a gallery. Both the collector and the gallery believed the painting to be by John Constable, the artist of renown. The collector subsequently discovered that the painting was, in fact, by a lesser artist. Understandably 'miffed', he sued the gallery.*
>
> Held: *there was no mistake as to the identity of the subject matter since both parties had agreed it to be a specific painting entitled 'Salisbury Cathedral' and had agreed their terms on this basis. A valid contract of purchase had, therefore, been made and the mistake over the* quality *of the painting was not sufficient ground for avoidance.*

Mistakes also arise out of the parties having different intentions at the time of making the contract. For example, it may be the case that one party signed the contract in the erroneous belief that he was signing something of a different nature altogether. Thus, a party signing a contract of purchase in the belief that it was only an application to examine the goods on an 'approval or return' basis would be entitled to plead *non est factum* (literally: 'that is not the case') and have the contract set aside.

Such a plea, however, would only be accepted if the mistake was considered to be such that the party did not appreciate the *fundamental character* of the document he was signing. It therefore follows that a mistaken belief as to the *terms* contained within the document would not be sufficient grounds for avoidance in a case where the signed document was of the *same nature* as that which the signatory perceived it to be.

> *Party A signed documents which he believed to be deeds of transfer of land. One of the documents was a mortgage deed on which a claim was subsequently made against A.*
>
> Held: *since the document was of the same class and character as a deed of transfer, there had been no mistake as to its nature. A's claim of* non est factum *was, therefore, rejected.*

In the context of the parties' intentions, it is worth noting that drafting mistakes may result in the written contract not accurately reflecting the intentions of the parties. In such cases, and subject to certain rules, the equitable remedy of *Rectification* (see section 3.6.5) may be applied. It is also worth noting, in this context, that a party cannot accept an offer to contract in the knowledge that the terms of that offer are clearly not intended by the other party. Failure to comply with this ruling will result in the subsequent contract being set aside.

> *Party A offered to sell property to Party B, naming a price of £1,250. B accepted the offer but knew perfectly well that the quoted price was a slip for £2,250.*
> Held: *the court refused to order Party A to sell at £1,250.*

A final type of mistake worth noting is one concerning the identity of a party to the contract. For example, an Employer might enter into a contract with a firm of contractors whom he considered reliable. Unbeknown to the Employer, however, the reliable contracting firm might have been taken over by another company who, for a variety of reasons, was unacceptable to the Employer. In such a case, the Employer would have sufficient grounds to require the contract to be set aside.

3.4.2 Misrepresentation

A *representation* is a *statement of existing or past fact, made before or at the time a contract is made, which relates to, but is not a term of, that contract.* If that statement is untrue, for whatever reason, it constitutes a *misrepresentation*. If the misrepresentation is relied upon by the party to whom it was addressed, and *has the effect of inducing that party to enter into the contract*, the party making it will incur liability to an extent dependent upon whether the misrepresentation was made innocently, or with fraudulent intent.

In this context it must firstly be noted that, for a statement to constitute misrepresentation, it must be as to *fact* rather than *opinion*. Secondly, it should be noted that silence may be construed as misrepresentation. Examples of such would be where a previously made representation becomes false before the contract is concluded, without that material fact being drawn to the attention of the party relying on the representation, or when a material fact is not disclosed prior to the contract being concluded.

Misrepresentation is considered *fraudulent* when the party making the representation does so in the *full knowledge* that it was false. A defence that the party making the representation honestly believed it to

be true at the time it was made will be rejected if the courts consider
that no *reasonable person* would have done so in the circumstances. If
the representation is in the form of an *opinion* (as opposed to a *fact*)
which a court adjudges that no *reasonable person* would hold, that
opinion will be treated as a *fact* and will give rise to liability for
misrepresentation

> *A vendor described a house owned by him as 'let to a most*
> *desirable tenant' when, in fact, the tenant had been long in*
> *arrears with his rent.*
> Held: *no reasonable person would have considered the tenant as*
> *being 'desirable'. The statement, therefore, amounted to a*
> *fraudulent misrepresentation of fact.*

It is worth mentioning, at this point, that a statement which is
literally true, but has the effect of seriously misleading the party to
whom it was made, may constitute fraudulent misrepresentation.

> *A prospective purchaser of a tract of land enquired of the*
> *vendor's solicitor whether the land was subject to any restrictive*
> *covenants (see section 2.3). Without troubling to inspect the*
> *relevant documents, the solicitor replied that he was not aware of*
> *any.*
> Held: *although the statement was* literally *true, it was misleading*
> *to an extent which constituted fraudulent misrepresentation.*

In the event of fraudulent misrepresentation being proved, the
injured party may elect to have the contract affirmed, whereby it will
continue unaffected by the misrepresentation, or rescinded (see section
3.6.4). Additionally, he will also have the right in tort to claim
damages for *deceit*.

It is noteworthy that a disclaimer of responsibility for
misrepresentation will not exclude liability for fraud.

> *The contract drawings, supplied by an Engineer to a Contractor,*
> *showed a wall in a position which the Engineer knew to be*
> *incorrect. As a result of this error, the Contractor incurred much*
> *needless expense. The contract contained a clause to the effect*
> *that the Contractor was required to satisfy himself as to the*
> *dimensions, levels and nature of existing works and that the*
> *Engineer did not hold himself responsible for the accuracy of*
> *information given in this respect.*
> Held: *the disclaimer within the contract was no defence against*

the Engineer's liability for fraudulent misrepresentation.

A misrepresentation is adjudged *innocent* if the party making it can prove that he had reasonable grounds to believe, and honestly *did* believe up to the time the contract was made, that the facts represented were true. It is noteworthy, in such cases, that the burden of proof lies with the party claiming the misrepresentation to be innocent and *not* fraudulent.

In accordance with the Misrepresentation Act 1967, an *innocent* misrepresentation may be described as *negligent* or *non−negligent*. In either case, the injured party may elect to rescind or affirm the contract. However, in the case of misrepresentation adjudged to have been made with *a negligent indifference as to its truth or otherwise*, or with *a negligent disregard of the other party's interests*, the injured party is entitled to claim additional compensatory damages.

Another feature of the Act precludes the use, in contracts, of clauses excluding or limiting the rights of the injured party to avail himself of any remedy open to him in the event of misrepresentation, and invalidates any such clauses appearing in a contract. In this respect, however, it is worth noting that such clauses may *exceptionally* be allowed if adjudged by the courts to be *fair and reasonable*.

3.4.3 Duress and Undue Influence

If a party is *forced* to enter into a contract under *duress*, or as the result of *undue influence*, the contract is voidable.

If the party's acquiescence has been obtained by means of actual or threatened violence to his person (but, interestingly, *not* his goods), he will be deemed to have acted under *duress. Undue influence*, on the other hand, is similar to *duress* except that the improper threats must be for the benefit of the party making them.

In both cases, the burden of proof lies with the aggrieved party. In the latter case, however, the aggrieved party has the additional burden of proving that the influence was applied by, or on behalf of, the party most likely to benefit from his forced acquiescence. It is the court's duty to decide whether the influence claimed actually affected the forming of the contract, and whether or not it was 'undue'.

3.4.4 Illegality

The general principle regarding illegality is that the courts will not enforce contracts which contravene the law of the land.

That broad statement having been made, however, it must be noted that the extent to which enforcement is precluded is largely dependent upon the nature and extent of the illegality and whether or not it is fundamental to the contract.

For example, a contract in which the object of the agreement is illegal is automatically void and no action may be brought by either party. An instance of such, in the context of the construction industry, would be a contract for the construction of a building which contravened a statute, bye-law or statutory regulation.

> *A Contractor completed the contruction of a wooden shop resting on wooden foundations in direct contravention of statutory provisions that required the use of incombustible material throughout.*
> Held: *since the Contractor had acted illegally, the contract was unenforceable and payment could not be recovered.*

If the illegality pertains at the time the contract is drawn up, the contract is void from the outset. If, on the other hand, a change in law renders the contract illegal after performance has begun, the contract will come to an end and both parties will be discharged from their obligations thereafter.

Presumably based on the maxim that ignorance of the law is no excuse, it is of interest to note that such rulings apply whether or not the parties to the contract are aware of the illegality. It is even more interesting to note that the rulings apply even in situations where a party does not wish to rely on the illegality as a defence against a contractual claim by the other party!

It is noteworthy, however, that a plea of ignorance *may* be accepted *under certain circumstances*. By way of illustration, consider the case of a building constructed by a Contractor under contract. If the building *itself* contravened a law, a plea of ignorance on the part of the Contractor would be rejected and he would be unable to recover payment for the work done. If, however, the proposed *use* of that building contravened a law, the Contractor would be able to enforce the contract if he could show that he was in ignorance of the potential infringement of the law.

A similar circumstance pertains in the case of a building constructed without the necessary planning permission required by the various Planning Acts. If the Contractor was aware of this omission, but carried on regardless, he would be required to suffer the consequences of the contract being voided. If, however, the Employer had contractually undertaken to obtain the necessary planning consent but, unbeknown to the Contractor, had failed to do so, the Contractor would be fairly entitled to the normal remedies for the Employer's breach of contract.

Apart from situations such as those illustrated, in which the *subject matter* of the contract is tainted by illegality of one form or another, there are circumstances in which the subject matter is perfectly legal but the *performance* of the contract involves a degree of illegality.

If the illegal aspects of the performance are *fundamental* to the contract, the guilty party will not be able to enforce the contract whereas the innocent party will retain all the normal contractual rights. If, however, the illegality is *not* essential to the performance of the contract, *both* parties will retain their contractual rights. Thus, a contractor contravening the statutory provisions of, say, the Health and Safety at Work Act (see section 11.3) in respect of certain construction operations will possibly be taken to court for doing so but the contravention will not affect his contractual right to claim payment for the work in question.

A specific, and noteworthy, type of contract–related illegality arises in connection with so–called 'restraint of trade' agreements. Common examples of such are:

(a) An ex–employee agrees not to set up in competition with his ex–employer within a certain area or within a certain period;

(b) A seller of a business agrees not to set up in competition with the buyer of his business;

(c) A manufacturer places certain restrictions on the subsequent sale of goods supplied by him;

Such agreements are only legally enforceable to the extent that they are considered reasonable in terms of the parties' interests and that of the public in general, and are not too far–reaching in nature or effect. As with all rulings including the word 'reasonable', the final arbiter in such matters is a court of law.

In this context, it is worthwhile to note the effects of the Treaty of Rome which, amongst other things, expressly prohibits all agreements which restrict, or attempt to restrict, free competition.

3.4.5 Legal Incapacity

Certain parties are either debarred from entering into contractual agreements or are restricted as to their capacity to contract. Notable examples of such parties include persons under 18 years of age (legally classified as 'infants' or 'minors'), drunks, bankrupts, certified lunatics, aliens and convicts under sentence.

In the context of the construction industry, the most relevant of these examples is the infant (although cynics might suggest that *all* the cited examples are relevant to the industry to a greater or lesser extent!) who occupies a privileged position under contract law. Generally speaking, an infant has no capacity to enter into, or be sued upon, contracts other than those of apprenticeship or those for the purchase of goods or services adjudged necessary for his 'condition of life'. Although an infant is not bound by a contract for the purchase of 'non-necessary' goods or services, it is worth noting that he has the capacity to sue the supplier in the event that such goods or services are defective. It should also be noted that an adult's guarantee of an infant's unenforceable contract is of no effect.

Particular mention must be made, at this point, of the contractual powers of *corporations* and *limited companies*.

The extent to which a company or corporation, in its capacity as a 'legal person' (see section 1.5.2), can enter into contracts is strictly limited. In the case of a corporation created by Act of Parliament or Royal Charter, the contractual powers of such will be defined by the Act or Charter, whereas the contractual powers of a limited company formed in accordance with the Companies Act 1948 will be governed by its Memorandum of Association. Any contracts entered into outwith these limitations will be regarded as *ultra vires* (literally: 'outside the power') and will be voided.

However, since it is not incumbent upon a party entering into a contract with such an organisation to check upon that organisation's capacity to contract, it is worth noting that the contractual rights of the other party will not be diminished by the fact that the organisation may have contracted *ultra vires*.

It is also worth noting that a person signing a contract on behalf of a company or corporation must be specifically authorised to commit the organisation. This being the case, it is common practice in contracts involving organisations for the documents to state the names and positions of the signatories, and to include statements to the effect that the signatories are authorised to sign on behalf of the organisations. In the event that a contract is signed by a person *not* authorised to do so, the organisation in question will not be bound by the terms of that contract. However, the other party to the contract may still be able to enforce the contract if he can demonstrate that he *had good reason to believe* that the person signing on the organisation's behalf was authorised so to do.

3.4.6 Privity of Contract

The common−law doctrine of *Privity of Contract* states that *only the parties to a contract may benefit from, or be sued upon, that contract*. Even if the terms of a contract expressly include for the accrual of benefits by a third party, *privity* excludes that party's right to enforce the contract.

In the context of the construction industry, an example of the effects of *privity* requires prior clarification of the contractual relationships which exist in a typical construction contract.

Generally, the principal contract is between the Employer and the Main Contractor who, in turn, has a separate contract with each of his sub−contractors. The Engineer (using the term in the context of the I.C.E. Conditions of Contract) has a contract of *Agency* (see section 3.7.2) with the Employer, whereby he acts as the Employer's agent, but has no contract with the Main Contractor or any of his sub−contractors. Likewise, the Employer has no contract with any of the Main Contractor's sub−contractors. This being the case, the Employer may make use of a clause in the main contract entitling him to make payments directly to the Main Contractor's sub−contractors, but *privity* excludes the right of the sub−contractors to enforce such − since the sub−contractors are not party to the main contract.

In relation to a contractual set−up of the type cited, the situation may result in a certain amount of confusion when, for example, *nominated* sub−contractors are used or when the Engineer is a member of the Employer's staff.

In the first case, the Engineer, in his capacity as the Employer's agent, instructs the Main Contractor to enter into contracts with *specified* sub—contractors for the supply of certain *specified* materials or for the performance of *specified* services. The prices of such materials or services will have been determined by negotiation between the Engineer and the sub—contractors in question and will be specified in the main contract between the Employer and the Main Contractor. Despite the initial arrangements having been made between the Engineer (on the Employer's behalf) and the sub—contractors, the latter will be contractually bound to the Main Contractor and *not* to the Engineer or the Employer.

In the second case, the Employer and the Engineer are effectively one. However, it should be understood that, in the context of administering a contract between the Employer and the Main Contractor, the Engineer is *contractually independent* from the Employer *and* the Main Contractor in spite of the fact that *his* employer is one of the parties to the contract.

Whereas third parties cannot acquire contractual rights in England, the Scots Law doctrine of Jus Quaesitum Tertio *permits the parties to a contract to confer the benefits of that contract on a third party. In the event of default by either party to the contract, the third party may sue upon the contract even though not a signatory to such.*

3.5 Discharge of Contract

Discharge of Contract is a general term whereby the parties to a contract are released from further obligation under the contract. Once a contract has been discharged, neither party can rely on the terms of that contract but can only enforce whatever rights arise out of the discharge.

In view of the fact that a contract does not *automatically* come to an end, unless *statute barred* by virtue of the expiry of its *limitation period* (see section 3.6.7), a situation whereby a contract is considered discharged can only come about through some act of the parties thereto. Firstly, if both parties have fulfilled all their obligations under the contract, the contract is considered to have been discharged *by performance*. Secondly, if circumstances render the contract impossible to perform either at the outset or after performance has commenced, the contract is considered *frustrated* and both parties are released from further obligation. Thirdly, a contract may be discharged in the event of a *breach of contract* by one party.

3.5.1 Discharge by Performance

The legal ruling in respect of this situation is that the party claiming release from further obligation on the grounds of performance must demonstrate that his performance is *complete* and *exact*.

This ruling may be relaxed in certain circumstances, however, the first of which concerns performance which, though not *exact*, is considered to be *substantially exact*. Such will be the case, for example, where the overall performance is affected by a few minor deficiencies. Providing the deficiencies are sufficiently small as to be adequately compensated for by an award of financial damages, a court will consider the performance sufficient to merit the contract being discharged.

> *A Contractor was employed to decorate a one-roomed flat and to provide furniture. The contract price was £1,100 with £750 payable during the performance of the work and the balance of £350 payable on completion. The Employer took possession but, as a result of defects in the work, refused to pay the outstanding balance of £350.*
> Held: *the Contractor was (strictly speaking) not entitled to claim any further payment under the contract, but was entitled to claim on a* quantum meruit *basis for the outstanding balance* less *the cost of rectifying the defects.*

It should be noted, however, that the above *de minimis* ruling will not apply in cases where the deficiencies in performance, albeit minor, go to the root of the contract.

> *A Contractor installed a central heating system which, as a result of defective workmanship, gave off fumes and provided up to 36% less warmth than it should have done.*
> Held: *since the defects went to the root of the contract, in that the central heating did not function properly, the Contractor was not entitled to claim any payment for the work – despite the fact that the cost of remedying the defects was assessed by the court at less than half the contract sum!*

The second situation in which the 'exact performance' ruling may be relaxed concerns contracts which are *divisible* in the sense that payment is due from time to time as the work progresses. Such would be the case in a construction contract which provided for the overall work to be completed, and paid for, in stages. In such a case, the obligation to complete each stage of the work would be severable from the contract involving the completion of the entire works and *substantial performance* could be claimed in respect of each stage completed. It should be noted, however, that any retention (see section 9.1.10) deducted from the contract payments would only be refunded on satisfactory performance of the *entire* contract.

A final point should be noted in respect of deficiencies in performance which come to light after the contract has been discharged. In such a case, the aggrieved party is entitled to sue for such on the proviso that the action is brought within the appropriate *limitation period* (see section 3.6.7).

3.5.2 Discharge by Frustration

In general, the courts are strict in respect of their enforcement of the terms of a contract, notwithstanding the fact that performance of those terms might prove to be very much more difficult or expensive than anticipated at the outset. Quite simply, the law of contract requires the parties to a contract to stand by the bargain they have struck.

In spite of this, however, the law recognises there to be certain circumstances under which the performance, or *continued* performance, of the contract becomes impossible. In such circumstances, a contract is regarded as *frustrated* and the parties thereto are automatically discharged from further obligation.

> *Party A had contracted to build machinery on Party B's premises and to maintain it for two years after completion. The premises were destroyed by fire before the machinery could be completely built.*
> Held: *the contract had proved impossible to perform in its entirety and both parties were released from further obligation. Party A was not, therefore, liable for damages in respect of non-completion.*

The law also recognises the possibility of an event changing the circumstances in which a contract is to be carried out to the extent that enforcement of the contract would result in the parties being held to something essentially different to that which they had originally agreed.

A Contractor agreed to build a reservoir for a fixed price. The

contract, which specified a completion time of six years, included a clause giving the Engineer wide powers to extend the completion time in the event of delays being incurred. Eighteen months after the commencement of work, following the outbreak of war, the government prohibited any further work on the reservoir and ordered the Contractor's plant and machinery to be sold.
Held: *the 'extension of time' provision was intended to apply to the type of delay normally encountered in work of this nature. To have applied it to the delay in question, thereby requiring a resumption of work after the war, would have resulted in the contract being completed in radically different circumstances to those envisaged at the time the contract was made. The contract was, therefore, frustrated from the date of the government order and both parties were released from further obligation thereafter.*

Apart from the requirement of a *radical* change of circumstance, the doctrine of frustration requires the change in circumstance to be completely unforseeable by either party. A consequence of this requirement is that an event will not be deemed to have frustrated a contract if the risk of its occurrence was intended to be carried by one of the parties.

A Contractor agreed to build a number of houses under a 'lump-sum' contract to be completed in 8 months. Due to a shortage of labour and materials, the work took 22 months to complete and resulted in heavy additional expense being incurred by the Contractor. The Contractor claimed that the original contract had been frustrated by the delay and that payment for the work done after the alleged frustration should be re-negotiated under an implied contract.
Held: *the delay had not frustrated the original contract because the whole object of a lump-sum contract was to place the risk of incurring additional expense as a consequence of delays of this nature on the Contractor.*

Similarly, a party cannot claim a contract to be frustrated by a change in circumstances of which he is the cause.

In 1916, a Contractor agreed to build a house for £1,900. Since the work was unprofitable, he deliberately delayed performance until the work was stopped by the Government under wartime regulations. The earliest time at which the Employer could thereafter build was 1919, by which time the cost had escalated to £4,153.

Held: *the Employer could recover £2,583, that being the difference between the two sums plus £825 already paid to the Contractor less £495 for the value of work done prior to the cessation of work.*

As might be reasonably inferred from the foregoing, the legal effect of frustration is that the contract is discharged, both parties are released from contractual obligation thereafter and contractual payments made prior to frustration become recoverable. In the context of a frustrated construction contract, this means that all contract payments to the Contractor made prior to frustration are recoverable by the Employer. Under the terms of the Law Reform (Frustrated Contracts) Act 1943, however, a court *may* adjust the recoverable sums of money to allow for reasonable expenses incurred by the Contractor in the performance of the contract prior to frustration, and also to take into account any valuable benefit obtained under the contract by the Employer.

It should be noted, however, that the provisions of the Act *only* apply insofar as they do not conflict with express terms in the contract covering payment and recovery in the event of frustration. In any event, the financial ramifications of frustration are usually mitigated by insurances taken out by both parties to cover themselves against the more common risks.

3.5.3 Discharge by Breach

A *breach of contract* (or, more precisely, a breach of the *terms* of contract) occurs when one party fails, either wholly or in part, to discharge an obligation imposed upon him by the terms of contract.

From the standpoint of available remedies in the event of a breach, the law classifies the terms of a contract as belonging to either of two categories: *conditions* and *warranties*.

A *condition* is a term considered to go to the root of the contract in that a failure by one party to discharge such would effectively force the other party into accepting something *radically* different from that which he contracted for. Consequently, the law gives the injured party the right, *if he so wishes*, to treat the contract as at an end and to sue the offending party for all damages incurred as a result of the breach. If the injured party does *not* wish to avail himself of this facility, he may *affirm* the contract (whereby it will continue unaffected by the breach) but still claim damages in recompence for the reduction in benefits he can expect to receive as a consequence of the breach.

On the other hand, a *warranty* is a term which is not considered to be fundamental to the essence of the contract in that it is subsidiary to the main purpose for which the parties entered into the contract. This being the case, a breach of such does not give the injured party the right to terminate the contract – it merely gives him the right to claim damages by way of recompence for any disadvantages imposed on him by the breach.

As an example of the difference between *conditions* and *warranties*, consider the inclusion, in a typical construction contract, of a clause requiring completion by a given date. Such a clause will be construed by a court as a clear *condition*, particularly if the importance of timely completion is emphasised by the Employer expressly specifying it to be *'of the essence of the contract'*. In such a case, therefore, delayed completion will amount to a clear breach of contract. Even if such an *express* qualifier is absent, however, it is worth noting that the circumstances of the contract might well give a court grounds to *imply* timely completion to be 'the essence of the contract' if it considers that to have been the clear intention of the parties at the time the contract was signed.

It is commonly the case in construction contracts that delayed completion is associated with the payment of liquidated damages by way of compensation to the Employer. By implication, therefore, timely completion in such a case is subsidiary to the main purpose of the contract in that a delay in completion may be compensated for by monetary payment alone. This being the case, the status of the 'timely completion' clause is reduced to that of a *warranty*, thereby removing the Employers right to terminate the contract in the event of a delay. Notwithstanding this, however, it should be noted that, if completion is delayed much beyond the extent envisaged at the time the contract was signed, the Employer has the right to make time 'the essence of the contract' by serving notice on the Contractor to complete by a specified date. Providing the specified date is such that, in the opinion of a court, it allows the Contractor a *reasonable* time to complete following receipt of the notice, delay beyond that date will constitute a breach of contract.

An additional situation in which a breach of contract occurs is when one party refuses to proceed with the contract, or puts himself in such a position that he is unable to proceed. Such a course of action would constitute a *repudiation* of the party's obligations under the contract and would give the other party the right, *if he so desires*, to terminate the contract and to sue for damages. It may be inferred from the statement 'if he so desires' that one party's repudiation of his contractual obligations, leading to termination of the contract, is only effective if accepted by the other party – in other words, the decision whether or not to terminate the contract lies *solely* with the innocent party.

In the context of a construction contract, an example of such would be a Contractor's refusal to remedy defective work. Aside from any remedies under the contract, such as withholding payment, the Employer has the option of accepting the Contractor's repudiation of his contractual obligations, in which event the contract will be terminated and the Contractor will be expelled from the site. The Employer will then be entitled to sue the offending Contractor for damages in recompense for losses arising from the original breach (the defective work), for losses arising from the termination and for any additional costs incurred in employing a replacement Contractor to complete the contract.

3.6 Remedies for Breach of Contract

In the event of a breach of contract, the ordinary remedy open to the injured party is to sue for damages in recompense for any losses arising out of the breach. Whilst damages provide adequate compensation in the majority of cases, there are certain instances when this will not be so. In such instances, Equity (see section 1.2.3) provides alternative remedies, namely: *Specific Performance, Injunction, Rescission, Rectification* and the principle of payment on a *Quantum Meruit* basis, which may be applied according to the circumstances surrounding the case. Following a discussion of the rules and principles relating to the payment of damages, these Equitable remedies will be covered in turn together with the Limitation rules relating to the period of time within which an action in contract law may be brought.

3.6.1 Damages

The general principle behind the payment of breach-related damages is *compensation* in that, according to the doctrine of *restitutio in integrum* (literally: 'restitution in full'), the innocent party must be restored to the financial position in which he would have been had the offending party discharged his contractual obligations in full. Thus, a claim can *theoretically* be made for *all* losses arising out of a breach of contract.

The problem here, however, is that the consequences of a breach may be endless. For example, a breach of contract on the part of a Contractor who fails to complete the works on time may result in the Employer incurring losses of such magnitude as to drive him over the edge into bankruptcy! In this light, common sense dictates that a line has to be drawn somewhere and that some losses might be too *remote* to be recoverable in law.

The rule regarding remoteness of loss is that compensatory damages are awarded to cover:

either   (a)   those losses which arise *in the normal course of events* from the breach;

or   (b)   those losses arising from the breach which may be *reasonably* supposed to have been *in the contemplation of the parties at the time the contract was signed*;

The two aspects of this rule are known in legal circles as (respectively) the first and second limb of Hadley v. Baxendale (1854) in recognition of the classic case from which they emanated:

> *H was a mill owner who contracted with B to transport a broken crankshaft to the makers for repair. B delayed delivery of the crankshaft with the result that H's mill became idle. H claimed for loss of profit during the idle period.*
> Held: *the stoppage was not the* natural *result of B's delay in delivery, neither was its possibility forseen by B. H's claim was rejected.*

Whilst it may be correctly inferred that losses falling outside the aforementioned contraints will be adjudged too remote to be recoverable, it would be *incorrect* to infer that *all* losses complying therewith may *automatically* be claimed.

In order to reduce the possibility of the defendant being effectively penalised for something outwith his control, the plaintiff is required to take all *reasonable* steps to mitigate his losses. For example, where an Employer is entitled to employ a replacement Contractor to complete work left unfinished by a defaulting Contractor, and to claim from the defaulter any additional costs incurred thereby, the Employer is required to act *reasonably* thriftily. That is not to say, however, that the Employer is legally obliged to employ the cheapest replacement in order to save the defaulter money! It should be appreciated that the Employer is the innocent party who has been placed in a difficult position through no fault of his own and, in this light, the courts will not scrutinise his conduct too closely.

As stated previously, the plaintiff is required to be reinstated to the *financial* position in which he would have been had the breach not occurred. This being the case, the measure of damages awarded is normally the *actual* monetary loss incurred by the plaintiff. Thus, where a Contractor is in breach of contract in carrying out defective work, the normal measure of damages awarded to the Employer is the actual, or estimated, cost of carrying out the necessary remedial work. Similarly, where the breach consists of work left unfinished, the amount of damages awarded against the defaulting Contractor will be based on the *extra* cost of completion by a replacement Contractor.

It is worthwhile to note an apparent exception to this general rule of reinstatement. In the event of a building surveyor producing an erroneous report on the condition of a property, the damages claimable will *not* be based on the cost to the plaintiff of rectifying any unreported defects but will be based on the difference between the value of the property *as reported* and its *actual* value.

In cases where the level of damages is based on the costs incurred by the plaintiff in rectifying defective work carried out by the defendant, or in completing work left unfinished by the defendant, inflationary cost increases obviously relate the cost of the necessary work to the date at which it is carried out. To remove the possibility of arguments arising as to the date at which the cost of the work is assessed for the purpose of setting damages, the following rulings apply:

(a) If the remedial work has been carried out prior to the case coming to court, the damages will be based on the *actual* costs incurred by the plaintiff at the time;

(b) If the remedial work has not been effected by the time the case comes to court, providing there is good reason for not doing so, the damages will be based on the costs prevailing at the date of the hearing;

(c) If there is good reason for the remedial work to be delayed until a date subsequent to the court hearing, the damages will be based on the cost levels estimated to prevail at that future date;

For purposes of assessing liability for damages, it is worth noting that the law recognises the concept of *joint liability* whereby several defendants may be adjudged liable for the plaintiff's loss. In such cases, a court has the power to apportion the overall sum of damages amongst the defendants according to the adjudged extent of individual liability. It is noteworthy, however, that the establishment of joint liability does not affect the plaintiff's right to recover *full* damages for his loss from any one of the defendants.

A further point must be noted, in respect of the assessment of liability, concerning instances where the plaintiff is adjudged partially culpable for the loss incurred. Whereas damages relating to an action in Tort may be reduced to the extent of culpability of the plaintiff (see section 2.2), the same principle does not apply to Contract Law. As a consequence, the plaintiff's right to recover full damages in recompense for his loss remains undiminished by his partial culpability.

In all the previously cited cases, where the amount of damages payable in the event of a breach has *not* been specified within the contract, the sum of money claimed is termed *unliquidated* damages. If, as is the case in the majority of construction contracts, the sum of money payable in the event of a breach *is* stipulated within the contract, it is termed *liquidated* damages and can be demanded in full by the injured party.

It should be noted, however, that the courts will not always enforce full payment of such. If a court considers that the stipulated sum is extravagant with regard to the greatest possible loss likely to arise out of a breach, or if it considers that the sum has been inserted by way of a threat to induce the promisor (the party liable for payment of the liquidated damages) to discharge his obligations under the contract, it will regard the damages as *punitive*. In such an event, the court will exercise its equitable jurisdiction and award *actual* damages rather than the stipulated sum.

That said, however, the converse is true in that it is possible for a court to award the *full* amount of liquidated damages in cases where the stipulated sum *exceeds* the actual loss but represents a *genuine* attempt by *both* parties to pre-estimate the losses likely to be incurred in the event of a breach.

3.6.2 Specific Performance

Specific Performance is a court order which is *positive* in nature in that it requires the defendant to do *precisely* that which he had contracted to do.

Being an *equitable* remedy, an order for specific performance *cannot* be claimed *as of right*, as would be the case with damages, but is granted at the *discretion of the court* in circumstances where it considers that financial damages would *not* provide adequate compensation for the plaintiff. For example, and for fairly obvious reasons, damages would not be regarded as adequate compensation for a house-purchaser faced with a vendor who refused to sell, despite being under contract so to do. Additionally, specific performance will be granted in circumstances where a court considers that an award of damages will effectively allow the defendant to 'buy' the right to avoid his contractual obligations.

Aside from the case where an order for specific performance will not be granted on account of damages being considered adequate compensation, there are two specific instances in which such an order will not be considered.

Firstly, specific performance will not be granted to enforce a *contract of personal service*. Apart from the somewhat questionable validity of enforced personal service, Equity does not allow itself to be used in this manner.

Secondly, an order for specific performance will not be granted in circumstances where performance by the defendant would require the constant supervision of the court. An example of such would be a *contract to build*. Notwithstanding this general rule, however, there are certain circumstances in which a *contract to build* may be subject to an order for specific performance:

(a) if the work to be performed is clearly defined within the contract;

(b) if the nature of the plaintiff's interest in having the work done is such that financial damages would constitute inadequate recompense;

(c) if, by virtue of the defendant's being in possession of the land, the plaintiff is unable to employ another person to carry out the works;

*A city corporation bought land in connection with a proposed
improvement scheme. They sold a part to the defendant who
covenanted to demolish old properties on the land and replace
them with new houses. Following the approval of demolition plans,
the defendant refused to build the new houses.*
Held: *the defendant's obligation to build was clearly defined on
the demolition plans; the construction work could not be carried
out by another contractor since the land was in the possession of
the defendant; since the vacant site in the city centre would not
yield rates, damages were considered to be an inadequate remedy.
With these facts in mind, the court ordered specific performance
of the terms of the covenant.*

3.6.3 Injunction

Like an order for *specific performance*, an *injunction* is an
equitable remedy and is, therefore, subject to the same conditions of
applicability as mentioned previously. However, an injunction differs
from a specific performance order in two particular aspects:

Firstly, whereas specific performance is *positive* in nature, an
injunction is *negative* in that it *prevents* the defendant from undertaking
a particular course of action. In a contractual context, therefore, an
injunction would require the defendant to cease a course of action
which would result in continued breach of contract.

Secondly, an injunction *can* be granted in relation to a contract of
personal service. Thus, in a case where the defendant has failed, or
refuses, to perform a personal service for which he is under contract to
the plaintiff, a specific performance order cannot force performance of
the service but an injunction can restrain the defendant from
performing that particular service for another party.

3.6.4 Rescission

Rescission is the *equitable* right to have a contract cancelled with
the parties being restored to the positions in which they found
themselves prior to entering into the contract. It can be ordered in
cases where a contract becomes voidable on the grounds of innocent or
fraudulent misrepresentation, whether by virtue of a false statement or
withholding of a material fact, or when the contract gives the right of
rescission on the occurrence of a specific event which has happened.

Since rescission of a contract requires that the parties to the contract be returned to their original positions, it follows that such a return must be *possible* for the rescission order to be granted. For example, if a Contractor had completed the construction of a building under a contract which was voidable on any of the aforementioned grounds, restoration of the contract parties to their original positions would, strictly speaking, require destruction of the building. For obvious practical and economic reasons, this would not be desirable and, as a consequence, rescission of the contract would not be considered a viable remedy.

It is worthwhile to note three final points in regard of rescission, the first of which is that rescission and an award of damages are *mutually exclusive* options. Thus, a rescinded contract cannot be accompanied by an award of damages to the plaintiff. Secondly, since the remedy of rescission is based on the equitable precepts of fairness and justice, rescission will not be granted if the effects of such would be detrimental to third parties. Finally, the right to subsequently rescind a contract is lost if the contract is either expressly or implicitly affirmed by the plaintiff.

> *The plaintiff bought a lorry from the defendant. He subsequently discovered that the condition of the lorry had been misrepresented by the defendant prior to the contract of purchase having been made, but nevertheless continued to use the lorry.*
> Held: *by his continued use of the lorry, which effectively affirmed the contract, the plaintiff had waived his right to rescind for misrepresentation.*

3.6.5 Rectification

As has been discussed previously (see section 3.4.1), the parties to a contract are *generally* required to stand by the terms of their written agreement notwithstanding the existence of mistakes contained therein. This ruling, however, is subject to the proviso that the mistakes are not fundamental to the agreement.

One instance in which a *fundamental* mistake obtains is that in which, due to a *common* mistake, the written agreement *does not accurately reflect the intentions of the parties*. In such an instance, providing the parties can demonstrate that they had reached a prior final agreement which continued up to the time the contract was put in writing, a court is permitted to correct the mistake by applying the *equitable* remedy of *rectification* with the parties thereafter being bound by the rectified agreement.

It is generally the case that rectification will not be ordered in cases where only one of the contracting parties has made a mistake. However, rectification of a *unilateral* mistake will be considered if the plaintiff can demonstrate that the defendant was aware of the mistake but, in order to take unfair advantage of such, failed to disclose its existence.

> *A Contractor made errors in extension* (multiplying the unit rate of an item by the item quantity to obtain the total price for the item) *in the course of calculating his overall tender price for a lump-sum contract. Whilst some errors were in favour of the Employer, and some against, the net result was a lump-sum price in favour of the Employer.*
>
> Held: *the errors were obvious to an extent which suggested a deliberate intent, on the part of the Employer, to take advantage of such by failing to disclose their existence to the Contractor. The errors were rectified by inserting the correct extensions.*

It is noteworthy that, in cases where a *unilateral* mistake is discovered, the courts will sometimes offer the plaintiff a choice between *rectification* and *rescission* but such instances are rare.

3.6.6 Quantum Meruit

As has been seen previously, a party to a contract can sometimes claim *reasonable* payment for work *partially* completed under that contract. Such payments are made on the *equitable* basis of *quantum meruit* (literally: 'as much as has been earned') and can be claimed in cases where there is no *express* agreement regarding the plaintiff's payment for work done under the contract, or in cases where the plaintiff's remuneration is not fully provided for.

> *A wide variation clause in a contract was qualified by a statement specifying the approximate value of the variations expected to be covered by the clause. Due to a delay in ordering the variations, the Contractor and Employer entered into a Deed of Variation which required the Contractor to carry out the necessary work on a 'cost plus profit' basis subject to a fixed profit limitation based on the estimated value of variations as specified in the original contract. In the event, the value of the variations far exceeeded the originally estimated value.*
>
> Held: *it was unjust to increase the extent of additional work without allowing the Contractor to increase his profits accordingly. The profit limitation clause was cancelled and the*

Contractor was entitled to claim, on a quantum meruit *basis, for a* reasonable *profit for the* total *additional work ordered.*

Where a contract contains *express* stipulation of the terms of payment of the plaintiff, a *quantum meruit* claim will only be considered in cases where the defendant has prevented the plaintiff from completing the work specified in the contract, or in cases where the work has been completed under a void contract.

3.6.7 Limitation

If a party is guilty of a contravention of civil law, it would be somewhat unjust for that party to be for ever after waiting for an action to be brought against him. With this in mind, Equity requires any action to be brought 'within a reasonable time' or be barred for *laches* (delay).

In the particular case of an action in contract, the Limitation Act 1980 lays down precise periods during which such an action must be brought, such periods being: *six* years for an action relating to a *simple contract* and *twelve* years for one relating to a *contract under seal*, with both periods commencing from the date at which the cause of action arises.

In the particular case of a breach of contract, the limitation period is deemed to run from the date of the *last relevant breach*. For example, a Contractor carrying out defective work would, in doing so, be in breach of contract. In the event that he refuses to remedy the defect, he would be guilty of a *further* breach of contract. In such a case, the limitation period relating to a possible action to claim recompense for the defect would commence with the occasion of the *second* breach even if that breach occurred (as would probably be the case) at the end of the maintenance period.

An exception to the normal rules concerning the commencement of the limitation period is provided by section 32 of the Limitation Act 1980 and concerns instances where facts relevant to the plaintiff's right of action have been concealed from him by the defendant. In such cases, the limitation period is deemed to run from the time the concealment is discovered or ought, with reasonable diligence, to have been discovered.

It should be noted, in this respect, that the commission of a breach, in circumstances where the discovery of such is likely to be substantially delayed, amounts to a deliberate concealment of the facts.

> *A builder under contract was required to use a specific type of brick which he knew, at the time, to be unavailable. Unbeknown to the Employer, he used a different type of brick which, eight years later, began to flake badly. Upon investigation, it was discovered that the bricks were of the wrong type.*
> *Held: by using a non-specified material, the builder was in breach of contract. By withholding the material fact that he had used non-specified bricks, the builder was guilty of fraudulent concealment of the Employers right of action. The limitation period was, therefore, deemed to run from the time of discovery of the defect. Since the Employer could not have discovered the defect any earlier, the cost of remedying such was based on the cost levels prevailing at the time of discovery.*

A final point worth noting is that the commencement of the limitation period is postponed if the plaintiff is under some disability which prevents an action being brought at the time. For example, if the plaintiff is a minor, he will be legally precluded from bringing an action until he reaches his majority. In such a case, the appropriate limitation period will commence at the date of his reaching his majority.

3.7 Special Contracts

Up to this point, coverage has been restricted to contracts of a *general* nature such as those between Employers and Contractors. There are, however, some contracts which are *special* in nature in that, although they are subject to the general law of contract, particular legislation or case law has arisen in their regard.

Whilst the practising civil engineer will probably encounter most, if not all, of them in the course of his professional career, two examples of such merit particular mention, namely: *Contracts of Employment*, which cover the rights and responsibilities of employers and employees, and the *Agency* relationship which defines the rights and obligations of the professional engineer acting in his capacity as an Employer's agent.

3.7.1 Contracts of Employment

A contract of employment is an agreement whereby one person places his services at the disposal of another in return for agreed remuneration. Such contracts are subject to statute law and civil law, both of which lay down the rights and obligations of the parties involved beyond those expressly stated within their agreement.

Such agreements are equally effective whether made orally or in writing. In the former case, however, the Employment Protection (Consolidation) Act 1978 requires the employer, within thirteen weeks of the employment commencing, to provide the employee with *written* terms relating to the rate of pay, the hours and the requirements relating to notice, together with details of holiday entitlement, sick pay arrangements and pension provisions.

Special mention must be made of the minimum notice requirements laid down in the above Act, which apply in *all* cases where the period of employment exceeds *four weeks*. In accordance with the Act, an *employee* is required to give minimum notice of *one week* whereas the minimum notice required to be given by an *employer* varies from *one week*, where the period of employment does not exceed *two years*, to *twelve weeks* in cases where the employee has worked for that employer for *over twelve years*.

It should be emphasised that the aforementioned notice periods constitute the *statutory minimum* required and may be extended by mutual agreement if the circumstances so dictate. If *no* period of notice is specified within the terms of the contract, common law requires that any extension of the notice period beyond the statutory minimum does not exceed that which is considered *reasonable under the circumstances*.

Two particular points must be mentioned in connection with notice requirements. Firstly, the rights to notice may be waived by the parties and, in this light, it is common for employees to accept payment in lieu of notice. Secondly, neither notice nor payment in lieu of notice is required in cases where there is a right to *summary dismissal* in instances of, for example, severe misconduct. In such cases, the dismissal is construed as the employer's acceptance of the employee's implied repudiation of the contract.

Aside from any express stipulations in a contract of employment, *common law* imposes certain duties on employers and employees. It should be noted, in this context, that the duties imposed on one party may be construed as the rights of the other party. Thus if a *duty* is placed on the *employer*, for example, the *employee* has a *right* to expect the employer to discharge that duty satisfactorily.

As far as the *employee* is concerned, common law requires him to

exercise *reasonable skill and diligence* in the performance of his duties.
As discussed previously in connection with Negligence (see section 2.2),
the standard of skill and diligence required is that of a *reasonable man*
employed to perform the work in question. If the employee holds
himself out as one possessing special skills, the standard of skill
expected from him is accordingly higher. Additionally, the employee is
under a duty to obey all lawful instructions given to him either by his
employer or by a person so authorised by the employer. Lastly, he
must act in good faith and do nothing which might detrimentally affect
his employers business, such as divulge confidential information or make
use of such information for his own benefit.

> *Party A was employed as managing director of Party B's
> company. While employed as such, A had invented a swimming
> pool which B marketed. Having left B's company, A set up in
> competition with B using knowlege and confidential information
> gained whilst in B's employ.*
> Held: *an injunction was issued to prevent A's company competing
> with B's company in the manufacture of swimming pools.*

If an employee contravenes any one of these conditions, he
becomes liable for dismissal. Whether or not he is liable for *summary*
dismissal will usually depend on the seriousness of his misconduct. For
example, insolence or disobedience will usually *not* result in summary
dismissal, unless it is repeated, whereas dishonesty or theft probably
will.

In terms of common law requirements, the *employer's* primary duty
is to pay the employee the agreed remuneration. Interestingly, and as a
result of nineteenth century employers paying their employees with
tokens which could only be exchanged at the company shop, an
employee is required to be paid in *cash* or an acceptable equivalent
such as a cheque or bank transfer.

In addition to the above duties, both employer and employee must
comply with the common law *duty of care* (see section 2.2) imposed
upon them, together with any *statutory* requirements contained within,
for example, the Employer's Liability (Defective Equipment) Act 1969
(see section 2.2) or the Health and Safety at Work Act 1974 (see
section 11.3).

Mention must be made, at this point, of an employee's right to
bring a claim for *unfair dismissal* against his employer. According to
provisions introduced by the Industrial Relations Act 1971, and now

contained within the Employment Protection (Consolidation) Act 1978 as amended by the Employment Act 1980, an employee whose period of employment is *less than one year* is entitled to be dismissed *without reason*. He must, however, be given proper notice, or payment in lieu of such, unless he is liable for summary dismissal. Should the employee's period of employment *exceed one year*, however, he is entitled to bring a claim against his employer for unfair dismissal if he considers that to be the case. If the number of employees in the company or organisation is less than *twenty*, the demarcation line is drawn at *two years*.

Complaints of unfair dismissal must, in the first instance, be brought before the appropriate *Industrial Tribunal* for consideration. In such an event, the burden of proof generally lies with the employer in that he is required to demonstrate that the dismissal was *not* unfair. He may do this by showing that, for example, the employee was dismissed for reasons connected with his conduct, or with his capability or because he was made redundant. The employer is additionally required to satisfy the Tribunal that he acted *reasonably* in his treatment of the matter.

In the event that the employer does not give an acceptable reason for the dismissal, or fails to satisfy the Tribunal that he acted reasonably, the dismissal will be adjudged *unfair*. The employee is then entitled to compensation as of right, the amount of such being principally based on actual and prospective loss of earnings and benefits. Additionally, the Tribunal has the discretionary power to issue an order for the employee's reinstatement.

Should the employee fail in his claim, he may be given leave to appeal to the *National Industrial Relations Court* (see section 1.4.2) but *only* if the appeal concerns a point of law. Appeal from this court lies to the *Court of Appeal (Civil Division)* but, as before, only on a point of law.

3.7.2 Agency

Agency is a legal term used to describe a relationship between two parties whereby one party, the *Agent*, acts on behalf of the other party, the *Principal*.

In the context of the construction industry, the most obvious example of such a relationship is that which obtains between a Consultant Engineer and the Employer in a civil engineering project whereby the Consultant Engineer draws up the contract between the Employer and the Main Contractor and subsequently represents the Employer's interests during the execution of such. (The relationship between the Employer and the Consultant Engineer is covered in depth in Chapter 4.)

The most common way in which *agency* arises is *under contract* whereby the Principal *expressly* authorises his Agent to perform certain specified duties on his behalf in return for an agreed remuneration. In addition to his *express* authority in such cases, it is common for the Agent to be invested with the *implied* authority to act on the Principal's behalf in respect of matters adjudged by a court to be *reasonably incidental* to those covered by the written agreement. The extent of implied authority, however, is probably only of academic interest in the case of the Engineer/Employer relationship cited previously, since those aspects of the Engineer's authority not expressly stipulated in the agency agreement are generally detailed in the contract between the Employer and the Contractor. This being the case, an Employer would find it very hard to deny the Engineer's authority in the event of being sued by the Contractor.

Whilst an *express* appointment is the usual way of creating an agency relationship, one may also arise by *implication*. For example, the relationship between a partner and his partnership is one of agency in that any given partner has the *implied* authority to act on the partnership's behalf and bind all the other partners by his actions (see section 1.5.2). A similar relationship exists between a company and a director of that company.

An agency relationship may also be implicitly created in circumstances where the Principal acts in such a way as to clothe a person with an *ostensible* authority to act on his behalf. In such a case, the Principal will be bound by acts done within that authority. Additionally, an *agency of necessity* may arise in circumstances where it becomes an urgent necessity for a party to perform an action on behalf of another party whose instructions cannot be obtained. For example, a person in charge of perishable goods in transit may lawfully sell those goods on behalf of the owner if they are in danger of perishing before the owner's instructions can be obtained. In such a case, the owner would be bound by the terms of sale negotiated by his 'agent'.

A final situation in which an agency is created without express authority is that in which a person holds himself out as authorised to act on behalf of an identified Principal. In such an event, it is open to the Principal to ratify the Agent's act, within a *reasonable* time, and be bound by it. In the case of the Agent making a contract on the Principal's behalf, ratification of the Agent's action relates back to the date on which the contract was sealed *providing* the Principal was competent to make the contract at the date of the Agent's act *and* at the time of ratification. For example, a building insurance taken out by a broker acting without the authority of the building's owner could not be ratified by the owner after the building had been destroyed.

In a contractual situation, the usual role of the Agent is to act on behalf of the Principal in making a contract with a third party. The general rule in such a case is that, having made the contract, the Agent drops out and thereafter restricts his activities, if so specified within the terms of the agency, to protecting the Principal's interests in respect of the contract execution.

The rights of the third party, in respect of his bringing a subsequent action under the contract, depend upon whether or not the Agent has disclosed his agency relationship with the Principal. If the agency *has* been disclosed, the third party can *only* sue, or be sued by, the Principal. On the other hand, however, if the agency *has not* been disclosed by the Agent, the third party has the right of action against *either* the Principal *or* his Agent. In this respect, it should be pointed out that the choice, once made, is *binding*. In the event, therefore, of the third party bringing an unsuccessful action against the Principal (assuming he has discovered the identity of such), he would be unable to bring a subsequent action against the Agent in the hope of getting a better result!

In cases where a contract with a third party has been made by an Agent who *purports* to act for the Principal, but does not have *actual* authority so to do, the contract will only bind the Principal and the third party if subsequently *ratified* by the Principal or if the Principal has acted in such a way as to clothe the Agent with *ostensible* authority. In cases where the Agent, in making the contract, has acted *innocently* or *fraudulently* outwith authority either expressly or implicitly invested in him by the Principal, the contract is automatically void. In such a case, therefore, the third party would have no right of action against the Principal but would be entitled to bring an action for *breach of warranty of authority* against the Agent.

An agency agreement, as with a contract of employment, imposes certain obligations on both parties with one party having the right to expect the other party to fulfil his obligations satisfactorily.

The Agent, for example, has a right to expect the Principal to indemnify him against all liabilities incurred in the *proper* execution of his duties. Similarly, he has a *lien* (a right of possession) over goods belonging to the Principal, but in *his* possession, with the facility of selling such goods to offset any financial losses incurred as a result of the Principal's default on the agreement. If the agency agreement makes no specific reference to payment, the Agent is entitled to remuneration for his services to an extent adjudged *reasonable* or *customary*.

In return, the Agent is expected to act honestly and obediently, and to exercise *reasonable* skill and care in the performance of *all* duties required of him under the agreement. Unless the agreement contains express provisions to the contrary, or unless such provisions may be implied by necessity, the Agent has no authority to delegate performance of those duties. In the context of the construction industry, the Engineer is expressly forbidden to assign or delegate any of the duties imposed on him by virtue of his agreement with the Employer. In a case where the Engineer requires the services of a specialist, two courses of action are open to him: he may request the Employer to retain the services of the requisite specialist or he may make use of such in an *advisory* capacity only, whilst retaining overall liability for any actions based on the specialist's advice.

Finally, a contract of agency may be brought to an end by mutual agreement or, failing that, by operation of law. In respect of the former, agency agreements frequently contain provisions for termination by either party upon reasonable notice. In the event of the death of either party, the agreement is *automatically* terminated, as it would be in the event of the Principal becoming bankrupt. Automatic termination also applies if the agreement becomes frustrated, such as by the destruction of the subject matter, or illegal.

PART 2

CONTRACTS AND CONTRACT ADMINISTRATION

CHAPTER 4

THE PARTIES INVOLVED IN A CIVIL ENGINEERING CONTRACT

The realisation of a typical civil engineering project – from the initial 'bright idea' stage, through the design stage, to final completion of the construction – involves contributions from a multitude of individuals and organisations. It is generally recognised, however, that the *main* contributions are made by three particular participants – the *Employer*, the *Engineer* and the *Contractor* – with the remainder being regarded as incidental.

In broad terms, the Employer – alternatively known as the *Promoter* or, by virtue of popular terminology within the construction industry, the *Client* – is the party who commissions the project, pays for it and undertakes to own it when completed. For the purpose of the project construction, the Employer enters into a contract with a Contractor who undertakes to carry out the construction in return for an agreed price. By way of obtaining the necessary technical expertise in respect of the realisation of the project, the Employer retains the services of the Engineer who undertakes the requisite design work and supervises, on the Employer's behalf, the execution of the construction contract.

4.1 The Employer

Before considering the various types of organisations (or, indeed, individuals) which might commission civil engineering projects, it is useful to consider two important points regarding project promotion in general.

Firstly, it should be appreciated that the majority of construction projects require some form of legal authority to allow them to proceed. Depending upon the status of the Employer and the nature of the project, the requisite authority may consist of parliamentary sanction, legislation, statutory regulation or, at the very least, some form of planning permission.

Whichever form of authority or permission is required, it is the duty of the Employer to obtain it, and warrant to the Contractor that it *has* been obtained, prior to the commencement of construction. It should be noted that, irrespective of the nature of the authority sought, applications for such will rarely, if ever, be considered unless accompanied by details of the scope and nature of the proposed project. Additionally, bearing in mind the workings of the bureaucracy, it is a fairly safe bet to suggest that the procedures associated with such applications will be complex to an extent which makes the start-up procedures for a nuclear power station appear positively straightforward, inflexible to a degree which beggars belief and will *always* take a disproportionate amount of time to be completed! With these points in mind, prospective Employers would be well advised to enlist the services of an expert and initiate the appropriate proceedings at the earliest possible opportunity such that the myriad of other factors associated with the project may be given the attention they deserve.

Secondly, and with particular reference to the prospective contractual arrangements between the Employer and the Contractor, and (in the form of an *agency* agreement) between the Employer and the Engineer, it should be remembered that a construction contract validly entered into carries the *implied* undertaking that the requisite funds are, or will become, available. Consequently, 'lack of funds' will not be entertained as a defence against an action for payment under a contract. Bearing this in mind, therefore, prospective Employers should ensure that the cost of a project is well within their available budget before entering into a contract for the construction of such.

For reasons largely associated with their sources of finance, it is necessary to categorise Employers according to the so-called 'sector' of the commercial world to which they belong: the *public sector* or the *private sector*.

4.1.1 Public Sector Employers

The *public sector*, which includes central government departments, local authorities, nationalised industries, public utilities and new-town development corporations, currently promotes between eighty and ninety percent of all civil engineering work in the United Kingdom.

It therefore follows that the amount of civil engineering work available at any given time closely relates to the prevailing level of public spending. This, of course, results in a very healthy state of affairs for the construction industry in times when the national economy is in good shape, particularly since a publicly–funded project carries little, if any, risk of a Contractor being left unpaid in the event of the Employer becoming bankrupt in the course of the construction period.

On the other hand, it should be appreciated that public spending on construction is generally one of the first targets for reduction in times of financial stringency and government cutbacks. In such times, Contractors who rely mostly on publicly funded projects are likely to be caught (metaphorically speaking) with their trousers down!

Promotion of construction projects by *central government* is generally restricted to those projects, such as roads, bridges, defence installations and the like, which are considered to form part of the national infrastructure. Contracts for such are entered into with the appropriate government department (such as the Department of Transport or the Scottish Development Department in the case of a road project in either England or Scotland) with the requisite funds, derived from taxation or other sources of government revenue, being provided by vote of Parliament. Broadly speaking, this means that Parliament decides whether or not to fund a given project, or programme of projects, and the extent to which funds should be made available to the department concerned.

It is of interest to note that expenditure *in excess* of the initial vote, although possibly sanctioned by the department in its contractual capacity as the Employer, is subject to a *supplementary* vote. It is even *more* interesting to note that a Contractor is under no obligation to enquire whether or not a government department is contracting beyond the funds voted by parliament. If a government department *has* contracted in excess of voted funds, perhaps based on the philosophy that supplementary funds can hardly be refused for a contract under way, and Parliament subsequently *refuses* to allocate the additional funds required, that department would remain contractually liable to provide the requisite funds or be sued, in its capacity as a 'legal person', for breach of contract.

In accordance with powers invested in them by Charters and Acts of Parliament, *local authorities* can enter into contracts and raise funds for payments under them.

Contracts entered into *outwith* these powers are considered to be *ultra vires* (see section 3.4.5) and become *automatically* void. Similarly, and notwithstanding the legality of such in respect of a local authority's capacity to contract, contracts entered into on behalf of a local authority are only binding on the authority if signed by an authority official with *express* powers so to do. Aside from these two important constraints, local authorities are subject to the normal legal doctrines regarding their ability to contract and their liability to be sued upon contracts.

In the context of construction, local authorities generally restrict their promotion of projects to those undertaken for the benefit of local residents. However, wider based projects may also be undertaken with the authority acting as an agent of central government. Although projects of the latter variety are funded directly by central government, contractual payments in the former case are met from local authority funds which are derived from government grants, loans and sources of local revenue such as rates or the so-called 'community charge'.

As previously mentioned, a contract validly entered into carries the implied undertaking that the Employer is in possession of, or will obtain, the necessary funds. In this respect, a local authority's liability to meet contractual payments in the course of an on-going project will not be diminished by a reduction in central government grant during the period of the contract, or by the refusal of central government to sanction a loan. In such a case, the local authority would have no option but to obtain the additional funds by means of levying supplementary rates and/or increasing the level of the community charge.

It is worth noting that, at the time of writing, even *this* option has become subject to restrictions imposed by central government in the form of rate- or charge-capping. This being the case, a local authority contracting beyond its means, even if such contracts were entered into in good faith in the expectation of sufficient funds being available when required, would appear to have no choice but to divert funds from elsewhere in its overall spending programme.

The powers of other publicly-owned bodies, such as *nationalised industries, statutory boards* and *public utilities*, to contract are laid down in the particular Acts of Parliament by which the bodies were constituted and are, broadly speaking, similar in extent to those detailed previously in connection with local authorities. Whilst such organisations are (theoretically, at least!) expected to finance their operations *primarily* out of revenue obtained from consumer payments for the goods or services provided, a certain amount of central government funding is usually available for 'bailing out' purposes!

In the case of *new-town development corporations*, which are responsible for the construction and development of new towns, initial funding is provided by central government with subsequent funding being obtained by the corporation out of revenue and the sale of assets.

Finally, it should be noted that the finances of all public sector organisations are subject to strict governmental control. This being the case, and bearing in mind that governments, by and large, have a vested interest in appearing to the voting public as wise, careful and honest custodians of the public purse, it is almost inevitable that the letting of publicly-funded contracts follows some form of *competitive* tendering procedure (see sections 7.1 to 7.3).

4.1.2 Private Sector Employers

Private sector Employers, who range from multinational corporations right down to private individuals, provide the construction industry with around ten to twenty percent of its work. Unlike their public sector equivalents, private sector Employers are generally free to let contracts on any basis they so desire (subject to the constraints imposed by contract law) and need not necessarily invite competitive tenders if they consider that their best interests would not be served by so doing.

The greater proportion of private sector promotion is undertaken by *incorporated companies*, which include multinational corporations, having either public or private limited liability. Such companies may be incorporated by Act of Parliament or by Royal Charter, or they may be created under the Companies Act 1948. In the first two cases, the contractual powers of the company will be specified within the relevant Act or Charter whereas a company formed under the Companies Act may only contract within the terms contained within its Memorandum of Association (in effect, the company 'constitution' drawn up at the time the company was registered). Contracts formed outwith these specified limitations will be considered *ultra vires* and will be voided.

Incorporated companies obtain their funding from a variety of sources. In the case of the larger companies, the greater proportion of funding is obtained from public or private subscriptions to company shares and by means of debentures. The balance, which is used to provide operating capital on a short term basis, is obtained in the form of loans from banks or other financial institutions.

Contracts may be promoted by *groups of individuals* not registered as trading organisations *per se* – for example: a partnership or a sports club committee.

Broadly speaking, liability for payments under such a contract accrues *jointly* to each group member signing the contract. This being the case, a party entering into a contract promoted by such a group would be well advised to ensure, on the basis of financial references obtained from each potential signatory to the contract, that there are sufficient signatories to ensure adequate *collective* financial status. Should such a contract be signed by certain individuals *on behalf of the group*, a potential Contractor should ascertain *a priori* the authority of such individuals to bind the remaining group members in personal liability under the contract.

Contracts promoted by *private individuals* usually involve small construction projects, such as a house extension or a domestic garage, with a total value not exceeding a few thousand pounds. This being the case, it is highly unlikely that the work will be carried out and paid for on the basis of a formal written contract. Whilst a verbal agreement between the Employer and the Contractor is sufficient to bind the parties in the eyes of the law (see section 3.1), it is usually advisable (from the point of view of *both* parties) to have the agreement evidenced by 'something in writing'. For such purposes, a *written* acceptance of a *written* quotation or estimate will generally suffice.

Although the Employer will thereby be bound to fulfil his obligation to pay for the work when completed, with 'lack of funds' being an insufficient defence against an action for payment, a potential Contractor should remember the maxim about 'getting blood out of a stone' and insist on the agreement being subject to the Employer submitting acceptable financial references beforehand!

4.2 The Engineer

As a result of economic considerations, popular demand or, in the case of certain central or local government sponsored projects, the chance to elicit a few more votes from a grateful electorate, the Employer becomes aware of the need for a new project or the possibility of extending existing works. Although the Employer may have a vague idea of possible alternative solutions to his problem, it is unlikely that he will possess sufficient technical or engineering expertise to enable him to assess, in any great detail, the relative merits and practicability of such. This being the case, it is even *more* unlikely that he will possess sufficient expertise to design the project or to supervise its construction. It is therefore necessary for him to retain the services of a party having the requisite expertise – the *Engineer*.

Broadly speaking, and dependent upon the specific terms of his appointment, the duties and responsibilities of the Engineer, in the context of a fairly typical civil engineering project, will be as follows:

(a) to act as the Employer's principal professional and technical adviser in respect of the development of the project;

(b) to design the works;

(c) to act as the Employer's *agent* (see section 3.7.2) in drawing up the contract for the construction of the project;

(d) to supervise and administer the execution of that contract;

Before discussing the Engineer's duties in greater depth, however, it is useful to consider the matter of his selection and appointment.

4.2.1 Selection and Appointment of the Engineer

Although the title of *Engineer*, used here in its contractual context, may be accorded to a *named* person, it is more usual for it to be *impersonal* with the title of some official post being quoted in the contract documents. The latter approach has the advantage of avoiding any contractual confusion which might arise through the departure of a named Engineer. In this respect, it should be noted that the I.C.E. Conditions of Contract provide for the Engineer to be replaced at any time during the contract providing the Contractor is notified of such replacement.

The office of Engineer is generally filled in one of four ways:

(a) Promoting organisations having their own technical and professional experts may appoint the Engineer from within. For example,

British Rail will usually appoint its Chief Civil Engineer to the post of Engineer to administer and oversee its civil engineering contracts.

(b)   Employers with, or anticipating, a long term or continuous programme of civil engineering works may delegate their promotion and contract administration activities to an 'agent', with the necessary technical staff, who will appoint the Engineer from within. For example, a road construction project in England, promoted by a County Council in its capacity as an agent of the Department of Transport, will usually have the County Surveyor as Engineer. In Scotland, such projects will generally be promoted by Regional Councils and will have the Regional Director of Roads as the Engineer.

(c)   Whether or not in one of the above categories, an Employer may engage a firm of consulting engineers to provide the necessary expertise for particular projects or series of projects, with the title of Engineer being accorded to the firm or partnership in question. In such cases, the relationship between the Employer and the consultancy is formalised by a contract of *agency* (see section 3.7.2) which sets out the duties and responsibilities of the Engineer, together with his remuneration.

(d)   For the purposes of overseeing a large multi–disciplinary project, such as the design and construction of a power station involving civil, mechanical and electrical engineering work, the Employer may elect to appoint a project manager. Such a party may be an employee within the Employer's organisation, or he may be an independent consultant with experience in such matters, and will appoint, or arrange for the Employer to appoint, individual consultants for each of the disciplines involved. One of these consultants will be named as the Engineer for the civil engineering contract(s).

In the context of a civil engineering project, it is generally recognised that the Engineer should be a Corporate Member of the Institution of Civil Engineers or, if the title of Engineer is accorded to a firm of consultants, that the senior members of that firm are so qualified.

Whilst such a status should not be taken as an *absolute guarantee* of professional competence, it is usually accepted that it provides a reasonable indication of such. Additionally, an Employer might take a modicum of comfort from the fact that corporate membership of the Institution renders the engineer concerned subject to the bye-laws, regulations and rules of professional conduct of the Institution and, as such, liable to be 'struck off' in the event of his professional incompetence or misconduct!

4.2.2 Duties of the Engineer

Upon his appointment, the first task of the Engineer is to familiarise himself with the Employer's basic requirements, in respect of the project, in so far as they can be defined at this stage.

In apprising the Engineer of his basic requirements, which should be set down in an initial project 'brief', the Employer should appreciate that *full* advantage of the Engineer's expertise and experience, particularly in respect of the project development, can *only* be taken if the latter is in possession of *all* relevant project information *at the outset*. This being the case, the initial brief should contain *inter alia* full and candid information relating to the Employer's short and long term economic objectives in promoting the project, together with details of any possible political, legal or financial constraints to which the project is, or might become, subject. It is also useful for the brief to contain details of any possible alternative schemes which the Employer might have in mind at this stage.

In addition to the aforementioned details, the initial brief will include a 'scope of work' statement setting out the Engineer's duties and responsibilities as envisaged by the Employer. Despite the fact that the Employer is the 'paymaster' in respect of his relationship with the Engineer, and is thereby entitled to have the final word in regard of the latter's duties, it is reasonable to suggest that, by virtue of his expertise and possible previous experience of similar projects, the Engineer will be better qualified to appreciate exactly what will be required of him, in the course of the project, than will be the Employer. This being the case, it would be in the Employer's best interests for this statement, which will form the legal basis of the relationship between the Employer and the Engineer, to be fully discussed with the latter before being finalised.

Two particular points require mentioning in respect of the 'scope of work' statement, the first of which concerns the freedom of the Engineer to obtain specialist advice if the situation so warrants.

It is a common fault of Employers to expect the Engineer, in his capacity as principal adviser to the former, to possess the necessary expertise to deal with each and every problem which might arise in the course of a project. In view of the wide range of work involved in some civil engineering projects, however, this will not always be the case – indeed, it would be unreasonable to expect such. In this light, therefore, it may become necessary for the Engineer to obtain specialist advice, or to make use of specialist services, from time to time and the 'scope of work' statement must allow him the facility to do so.

If, in the Engineer's opinion, such services are likely to be required on a long term basis, it may be advisable for them to be retained by the Employer under separate agreements. On the other hand, specialist services required on a short term basis for particular aspects of the project are best obtained by the Engineer, on an *ad hoc* basis, with the associated fees being reimbursed by the Employer.

In the first case, the agreement between the Employer and the specialist accords the Employer the right of action against the specialist and, as a consequence, relieves the Engineer of any liability in connection with the latter's performance. In the second case, however, it should be appreciated that the terms of the *agency* agreement (see section 3.7.2) between the Employer and the Engineer will probably preclude the latter's right to assign any of his obligations or responsibilities under the agreement. This being the case, the Engineer will have the facility to engage the necessary specialists in *advisory* capacities only, whilst retaining overall liability for any actions based on such advice.

The second point to be made in connection with the 'scope of work' statement concerns the extent to which it permits the Engineer to consider new developments which might arise in the course of the project.

In the event that certain circumstances, such as the existence of adverse ground conditions, indicate the neccessity for a radical adjustment of the scope of the project, a decision regarding such requires prior consideration to be given to a number of possible alternatives. Obviously, the value to be gained from such an exercise depends on the extent to which the Engineer is permitted to consider alternative solutions. Whilst it would be somewhat unrealistic to expect the 'scope of work' statement to provide the Engineer with *carte blanche* in this respect, it should be appreciated that any recommendation made by him on the basis of a study subject to severe constraints is likely to be, at best, of somewhat questionable validity and might result in a final decision which is not in the Employer's best interests.

This being the case, and to avoid any confusion arising over the matter, the extent to which the Engineer is permitted to consider alternative solutions must be clearly defined in the 'scope of work' statement and should be a matter for discussion and agreement between the two parties before finalisation.

Following the finalisation of the project brief, the first major task required of the Engineer will be to undertake a number of project studies by way of *preliminary investigation.*

The nature and extent of the studies involved in the preliminary investigation will vary according to circumstance and will obviously depend, to a large extent, on the character, size and complexity of the project under consideration. However, it is likely that some, if not all, of the following studies will be required:

(a) a *market study* to assess the likely demand for projects of the type in question. Obviously, the need for such a study depends upon whether or not the Employer has already carried one out in the process of reaching his initial decision to fund the project.

(b) an *economic feasibility study* to determine the viability of the project in financial terms. Broadly speaking, such a study involves a direct comparison between the estimated cost of realising the project and the estimated financial benefit which the Employer can expect to accrue from the finished product and is intended to give the Employer an indication, in financial terms, of whether it will all be worth it in the end.

(c) a *technical feasibility study* to identify any physical or technical problems that are likely to beset the project. Factors to be considered at this stage include the availability of resources (plant, labour and materials) likely to be required for the project construction, the ease (or otherwise) of access to the site, the availability of services (sewerage, water, gas, electricity etc.) and the likelihood of problems arising in connection with the disposal of waste material during or, depending upon the nature of the project, after construction. Consideration should also be given to any statutory or local authority restrictions which might affect the construction or subsequent use of the project, together with any aspects of the project which might infringe adjacent landowners' rights. Finally, the study should involve the collection of any physical or climate-related data, such as the incidence of severe storms or flooding, which might have a bearing on the design, construction and subsequent use of the project.

(d)   a *geotechnical study* to investigate ground conditions on the proposed site of the project and to identify any particular problems which might arise from such.

(e)   an *environmental impact study* to assess the effect of the project on the environment both during and after construction. Although the requirement for such a study will largely be dependent upon the nature of the project, it will also depend, albeit to a lesser extent, on the attitude of the Employer towards such matters. Whilst not wishing to imply that all Employers are 'tarred with the same brush', it would not be unreasonable to suggest that, in many cases, the main purpose of instigating an environmental impact study has more to do with averting the inevitable public outcry that would result from the absence of such than it has to do with a genuine desire to minimise the project's effect on the environment.

Although the Engineer will, in the main, be capable of carrying out the geotechnical and technical feasibility studies unaided, it is unlikely that he will possess a sufficient range of expertise to enable him to perform the economic and environmental studies without recourse to specialist advice. In the event that the Engineer (as opposed to the Employer) retains the requisite specialists to perform the necessary studies, it bears repeating that the studies should be carried out under *his* direction since *he* bears the overall burden of responsibility for any recommendations which he proffers the Employer on the basis of such.

On completion of his assessment of the results of the preliminary investigation, the Engineer will generally be required to submit a report of his findings, along with his recommendations based thereon, to the Employer for a decision as to whether or not the project is to proceed any further.

Following a favourable decision on the matter, the Engineer will embark upon a *preliminary design study*.

Prior to commencing any design work *per se*, the Engineer's initial task will involve the collection and collation of such technical information as deemed necessary for the design of the works. Whilst most of the information relating to such aspects as the structural design of the project can be readiliy obtained from Codes of Practice and other similar publications, a thorough *site exploration* is required to elicit the requisite information concerning the topography, ground conditions and, if appropriate, the geology of the site.

In the majority of cases, the requisite topographical information can be obtained by means of a *topographical survey* of the site carried out by the Engineer or, if circumstances indicate the preferability of such, by a specialist in that discipline. There are, however, certain projects, such as those for the construction of major roads, where the extent of work involved in the survey is far in excess of that which can be efficiently performed using standard 'theodolite and level' techniques. In such cases, details of the site topography are best obtained from an aerial survey of the area carried out by an organisation specialising in such.

A sound knowledge of site ground conditions is generally accepted as a fundamental requirement in respect of the adequate and economic design of any construction project – a maxim which would, no doubt, be confirmed by the Employer and Engineer on a certain structural contract in Pisa, carried out many years ago, if they were alive today! In this light, the importance of a *geotechnical survey* of the site cannot be over-emphasised.

Since the extent of such a survey will depend entirely upon the nature of the project concerned and the geotechnical complexity of the site, it is difficult to lay down hard-and-fast rules in this respect. However, it would be reasonable to assume that any given survey will involve the sinking of *sufficient* bore-holes or trial pits, and the performance of *sufficient* laboratory and/or in-situ tests, to provide the Engineer with reasonably accurate answers to the two $64,000 questions which are asked in connection with *every* construction project, namely: '...what am I constructing on?' and '...how will it behave under loading?'. Since it is unlikely, in the main, that the Engineer will possess the necessary facilities to carry out such tests, the survey will probably be carried out by a specialist organisation working under his direction. Although the survey reports will undoubtedly contain 'expert' interpretations of the test results, it should be remembered that the overall responsibility for design decisions based on such lies with the Engineer.

Depending upon the nature of the project, the Engineer might require a certain amount of information regarding the underlying geology of the site. In many such cases, sufficient information can generally be abstracted from existing large-scale geological maps of the area. However, in cases where a *detailed* knowledge of the underlying geology is *fundamental* to the project design, or in cases where the project is located in an area of geological uncertainty, a *geological survey* of the general locale might be indicated. Such a survey, which may include resistivity and seismographic studies, will probably be carried out by a specialist who, depending on the extent and importance of the survey, will be retained either by the Employer or the Engineer. In the latter case, similar conditions to those discussed previously in connection with specialist geotechnical services apply in regard of the Engineer's responsibility for design decisions based on the survey results.

Subsequent to his collation of the requisite technical information, the Engineer will carry out the preliminary design of the project. Broadly speaking, the design will be sufficiently detailed to enable the Employer to visualise the project as a whole and to be provided with a reasonable estimate of project costs and duration. Unless there are sound reasons to the contrary, the exercise will probably entail the comparison of a number of alternative proposals with a view to selecting the one best suited to the Employer's requirements.

Mention must be made, at this point, of an 'optional extra' which might beneficially be included in the preliminary design study prior to a final decision being reached in respect of selecting the most suitable proposal.

It is generally accepted that, with very few exceptions, construction projects will almost certainly cause a degree of inconvenience, at some stage or other, to members of the public. For example, people living in the immediate vicinity of a proposed project are likely to suffer the inconvenience of dust, noise and traffic disruption during the construction period. On the other hand, people living on (say) the flight path of a proposed airport runway might well not suffer any great inconvenience during the construction of the runway but will undoubtedly make up for that in terms of the level of noise-induced discomfort they will suffer after the runway becomes operational! In such cases, the people concerned would not be unreasonable in expecting the opportunity to voice their objections and to have them taken into account in the decision-making process.

Just such an opportunity may be, and frequently is, provided in the form of a 'public participation' exercise undertaken, in the initial stages of the project development, by the Employer or by the Engineer on his behalf.

The problems surrounding exercises of this nature are many and well-publicised, and appear largely to be associated with the possible antipathy of the protagonists. On the one hand, the Employer/Engineer might regard the public participants as little more than a representative sample of the 'anti-everything-all-the-time' brigade whose perception of 'freedom of speech' tends invariably to stop some distance short of allowing anybody else the opportunity to advance a point of view. On the other hand, the public perception of such an exercise might well be that of a public-relations 'red herring' mounted with the sole intent of diverting attention away from what is, in fact, a pre-judged issue.

This aside, however, it is generally accepted that two distinct benefits can accrue to the project through public involvement at the development stage of such. Firstly, a 'public participation' exercise provides an effective means whereby the Employer or Engineer can allay any public fears which are, as will often be the case, engendered by a lack of hard information relating to the project. Secondly, and based on the fact that 'the public' is by no means as ignorant of technical and engineering matters as is sometimes believed (after all, engineers and other professionals are members of the public too!), it is quite within the realms of reasonable possibility for such an exercise to produce a number of constructive ideas from which the project might benefit.

Finally, in connection with public participation, it should be noted that *statutory* provision for a *public enquiry* may exist in the case of certain projects.

Broadly speaking, such an enquiry is held before an independent Inspector (or 'Reporter' in Scotland), appointed for that purpose, and consists of a public hearing of the objections to the project and the points of view of the Employer and/or Engineer. Expert witnesses may be called if the project is of a highly technical nature and the parties involved are entitled to legal representation if they so desire. Following the conclusion of the hearing, the Inspector submits a report of his findings, along with his recommendations based thereon, to the appropriate government minister for a decision on the matter. It is of particular interest to note, in respect of the validity of such an exercise, that the final decision does not necessarily have to be in accordance with the Inspector's recommendation and that appeals against such will *only* be entertained if they concern points of law.

Following his completion of the preliminary design study, the Engineer will be required to submit a formal report to the Employer for a decision on the matter. Whilst the scope and nature of this report will depend largely upon the nature of the project and the Employer's needs in respect of such, it would be reasonable to assume it to include *inter alia* the following:

(a)   full details of any investigations or 'public participation' exercises carried out;

(b)   full details of the design criteria adopted, along with reasons for their adoption;

(c)   a technical and economic comparison of the alternative schemes studied;

(d)   a recommendation, together with cogent justification for such, of the scheme considered most suitable in terms of the Employer's requirements;

(e)   the Engineer's proposals concerning his organisation and management of the detailed design of the recommended scheme;

(f)   the Engineer's recommendations in respect of the contractual arrangements for the prospective construction of the project.

It is worthwhile, at this point, to note some fundamental requirements in respect of the *form* of this report and, indeed, any other reports which the Engineer might be required to submit to the Employer from time to time.

Since the Employer's decisions in respect of the project are likely to be based almost entirely on such reports, it is vital for the information therein to be presented *clearly*, *concisely* and *unambiguously*. Whilst the inclusion of a certain amount of technical information is unavoidable, it should be remembered that the Employer will probably not be an engineer and, as a consequence of such, will undoubtedly have neither the ability nor the inclination to absorb a mass of 'technicalese'. This, therefore, indicates the additional requirements of *simplicity*, *readability* and *minimal 'jargon content'* together with the facility to be *readily understood*.

It is also worthwhile to note a particular point in respect of any recommendations contained within the Engineer's reports.

Whilst the Employer, in the majority of instances, will base his decisions on the Engineer's advice and recommendations, he is under no obligation to do so and may, if he so chooses, ignore such advice completely. In the event that the Employer adopts the latter course of action, however, he should be left in no doubt whatsoever as to the responsibility he incurs in so doing.

If, on the basis of the Engineer's report, the Employer elects to proceed with the project, he must take the necessary steps to obtain the requisite authority for the project construction (see section 4.1). After this has been obtained, the Engineer will be instructed to proceed with the *final design* of the project.

In addition to a detailed design of the chosen scheme, the Engineer will also be required to provide the Employer with an estimated completion time for the project and an estimate of the funds which he will be required to make available at regular intervals throughout the construction period.

In view of the occasional tendency of certain Employers to regard the Engineer's estimate of the project cost as constituting an accurate prediction of the eventual *tender price*, it is worthwhile to note a particular point in respect of such. At *best*, the estimate will be compiled on the basis of prices for *similar* work *recently* carried out under the Engineer's auspices. At *worst*, it will be compiled on the basis of 'standard' prices contained in some form of published database. Either way, it should be appreciated that an estimate compiled on such a basis will inevitably differ, to some degree, from that compiled by a contractor with the prospect of actually carrying out that *particular* work as a *commercial* (i.e. profit—making) undertaking. The extent to which the estimate may differ from the subsequent tender price is impossible to quantify since it will depend largely on such factors as the prevailing economic climate and the nature of, and circumstances surrounding, the project. However, it is reasonable to suggest that the degree of potential disagreement is likely to be greatest in cases where, for example, the contract price is significantly sensitive to tenderers' proposals regarding construction methods, or where the project circumstances are such that tendered prices are likely to differ from 'standard' prices to an extent which will depend largely on the (virtually unpredictable!) perceptions and attitudes of the tenderers.

A task which generally proceeds concurrently with the final design study is that involving the *preparation of the tender documents* for the prospective construction contract. Whilst the nature of the documentation will depend, to a certain extent, on the type of contract (see Chapter 5) proposed, the task will *always* involve the preparation of a Specification for the Works, the compilation of a document detailing the conditions under which the contract will operate and the compilation of the tender drawings. In the case of a standard 'admeasurement' contract, the tender documents will also include a Bill of Quantities (or Schedule of Rates) which requires to be compiled by the Engineer upon substantial completion of the tender drawings. In view of the fact that an admeasurement contract is the most commonly used type of contract for works of civil engineering construction, the documentation associated with such is discussed in detail in Chapter 6, with detailed coverage of the Bill of Quantities and the I.C.E. Conditions of Contract being given in Chapters 8 and 9 respectively.

Upon substantial completion (see below) of the tender documentation, the Engineer, in his capacity as the Employer's *agent*, will require to initiate the procedure for *letting the contract* to a suitable Contractor. Broadly speaking, the procedure (dealt with in depth in Chapter 7) will entail performance of the following duties in the order shown:

(a) the issuance of invitations to prospective Contractors to tender for the works;

(b) an appraisal of the submitted tenders with a view to selecting the most suitable thereof;

(c) the submission, to the Employer, of a formal report containing details of the appraisal and a recommendation in respect of his (the Employer's) eventual choice of Contractor;

(d) the formalising of the contract between the Employer and the Contractor;

The extent to which the tender documentation is complete before tenders are invited depends largely upon the urgency with which the Works are required to commence and, as a consequence, will vary from project to project. In this respect, however, it is reasonable to suggest that the more complete the tender documentation is, the better the tenderers will understand what the construction entails, the more accurate will be the pricing of the works and, as a consequence of these two factors, the smoother will be the contract execution.

Following completion of the requisite contractual formalities between the Employer and the Contractor, the Engineer's final duty is carried out, namely: *supervising and administering the execution of the construction contract.*

Although not a party *to* the contract, the Engineer will generally have extensive powers and responsibilities *under* the contract. For example, approximately 50% of the clauses and sub–clauses contained in the I.C.E. Conditions of Contract include an *express* reference to the Engineer's powers, duties and responsibilities. In the case of a contract executed under these particular Conditions, the Engineer's functions may be summarised as follows:

(a) to ensure, by means of applying the test of 'satisfaction' to the Contractor's discharge of his contractual obligations, that the works are completed (in terms of quantity, quality and expenditure of time) as specified in the contract documents;

(b) to provide the Contractor with any additional or supplementary information (over and above that contained within the contract documents) which he might require to complete the works;

(c) to instruct the Contractor, as and when the need arises, but *only* within the extent of authority invested in him (the Engineer) by the Employer, to vary the Works or to carry out additional work;

(d) to assess any extensions to the contract completion time, and/or additional costs, to which the Contractor might fairly be entitled as a consequence of the aforementioned additions or variations, or as a result of circumstances or conditions which he could not *reasonably* have been expected to forsee;

(e) to carry out regular valuations of the work completed with a view to certifying interim payments to the Contractor;

(f) to certify the Contractor's final account;

(g) to issue Completion and Maintenance certificates upon the completion of the Works and the expiry of the Maintenance Period;

(h) to adjudicate in any dispute between the Employer and the Contractor;

Finally, it should be noted that, in the case of a contract operating under the I.C.E. Conditions of Contract, the Engineer has the authority to delegate certain of his responsibilities, in respect of the day-to-day supervision and administration of the contract, to an appointed representative, namely: the *Engineer's Representative* (or *Resident Engineer* as he is more commonly known). It should be particularly noted, however, that the decisions and acts of the Engineer's Representative are subject to the ratification of the Engineer and, as a consequence, may be overruled or reversed by the Engineer at any time. Additionally, it should be noted that the Engineer has no authority to delegate his responsibilities in respect of claims adjudication, the issuance of completion and maintenance certificates, the certification of the Contractor's final account and the settlement of disputes between the contracting parties.

4.2.3 The Engineer's Independence

The Engineer's *principal* function during the contruction of the project is that of 'contract supervisor' with the responsibility of ensuring that the contract is executed in accordance with its terms and conditions and that *both* parties to the contract fulfil their respective contractual obligations.

In discharging this function, the Engineer must appreciate that he is *not* a party *to* the contract and thus has no authority to amend or alter the terms and conditions agreed upon by the contracting parties. Consequently, he must implement the contract in *strict* accordance with such terms and conditions, irrespective of whether or not he considers them to be fair or reasonable, and must *only* operate within the extent of authority accorded him under the contract. In the case of *discretionary* powers accorded the Engineer under the contract, which allow him to exercise personal judgement in adjudicating upon contractual matters, it is encumbent upon him to act fairly, and with total impartiality, as to either party.

It is worth noting that, in the case of the Engineer's appointment being governed by the Model Service Agreement drawn up by the Association of Consulting Engineers, the requirement for fairness and impartiality is expressly contained within the Agreement.

In order to be able to carry out his contractual duties in the manner prescribed, the Engineer must adopt a position of total independence from *either* party to the contract and must resist any interference or pressure, *from whatever source*, which might compromise that independence.

The position is slightly confused by the fact that the Engineer is also the Employer's *agent* – indeed, a certain proportion of his authority under the contract derives *directly* from this relationship. In the case of a contract executed under the I.C.E. Conditions of Contract, a noteworthy example of such is the Engineer's authority to grant permission, on the Employer's behalf, for parts of the Works to be sub-let.

Despite the existence of this relationship, however, and contrary to popular belief in certain quarters, the Engineer is *not* the Employer's representative on site. Whilst it is certainly true to suggest that it is the duty of the Engineer to protect the Employer's rights in respect of the construction of the project, it would be incorrect to assume that such a duty derives directly from the professional relationship between the two parties. It derives, in fact, from the Engineer's role as an independent and impartial 'quasi-arbiter' which requires him, in reaching a decision on any given contractual matter, to pay *equal* regard to the rights of *both* the Employer *and* the Contractor.

It is sometimes difficult for an Employer, in his capacity as the Engineer's *Principal* (or employer!), to appreciate the importance of allowing the latter to maintain a position of independence, whereby he is able to base his contractual decisions on a fair and impartial assessment of the facts in question, particularly if such decisions are likely to cost him (the Employer) money!

In this respect, two possible consequences of perceived bias on the part of the Engineer might be mentioned by way of emphasising the necessity for his independence and impartiality.

First and foremost, it should not be forgotten that, in the event that he considers his contractual rights to be under threat by a biased Engineer, the Contractor has the right of appeal to an independent Arbitrator (or *Arbiter* in Scotland) or, if such a course of action is deemed inappropriate by virtue of an Arbitrator's limited powers in respect of the award of damages, a court of law. Whichever course of action is adopted by the aggrieved Contractor, it is likely that the Engineer will, at some stage in the proceedings, be called upon to justify his decisions on the matter in dispute. If it is considered by the Arbitrator (or Judge) that such decisions have been affected by bias, they may be overruled or reversed. It is not unreasonable to suggest that the financial ramifications of such would probably be more severe than would be the case if the original decisions of the Engineer 'went against' the Employer but were made in the spirit of fairness and impartiality.

At this point, it is worth mentioning that, in addition to the 'duty of care' owed by the Engineer to the Employer by virtue of the 'special relationship' (see section 2.2) between the two parties, the Engineer is *also* considered to owe a 'duty of care' to the Contractor '... arising out of their proximity ...' in the execution of the contract (see *Arenson v Casson Beckman Rutley and Co. [1975] 3 All E.R. 901, H.L.*). It would thus appear that the Engineer may be sued by the Contractor for negligence if that 'duty of care' is considered to have been breached as a consequence of the former's bias.

The second possible consequence of bias on the part of the Engineer arises out of the remarkable efficiency of the construction industry 'grapevine'. In the event that a given Engineer was *perceived* (*proof* is generally an irrelevance in construction industry gossip!) to be actively advancing the interests of a given Employer at the expense of the Contractor, that 'fact' would become common knowledge in a very short space of time. Consequently, Contractors in general would not be very keen to tender for future projects thus promoted and supervised and those that *did* tender for such work would reflect their suspicions of bias in their prices!

4.3 The Contractor

In a Civil Engineering (or Building, for that matter) contractual context, the term 'Contractor' is applied to any individual or organisation which undertakes to perform, for a given sum of money, the works specified in the contract documents.

Broadly speaking, civil engineering Contractors may be classified according to three distinct categories: *general* Contractors, *specialist* Contractors and *management* Contractors.

General contractors are those who, by virtue of their knowledge, experience and general capabilities, are able to undertake responsibility for the performance of the *entire* Works. For example, major roadworks contracts are usually carried out by Contractors of this type since the overall technical scope of the work is generally small enough to be well within the capability of a single organisation.

Specialist Contractors, on the other hand, are those who confine their activities to carrying out specific classes of work. In the vast majority of cases, such organisations carry out their work on a sub-contract basis under the auspices of a *general* or *management* Contractor who coordinates the activities of the various specialists.

In the context of *civil engineering* works, notable examples of work often carried out by specialist Contractors include bulk earthworks, piling, tunnelling and demolition. By the very nature of the projects undertaken, the use of specialist Contractors is more widespread in the case of *building* works, with notable examples including electrical installation, plumbing works, brickwork, air-conditioning and lift installation and commissioning.

Unlike their *general* and *specialist* counterparts, *management* Contractors do not actually undertake any *physical* construction. Their activities are restricted to the provision of management services to control and coordinate all site activities on a given project, for which they are paid a fee by the Employer, with the construction work being sub-let to 'construction' Contractors on a competitive basis.

CHAPTER 5

TYPES OF CONSTRUCTION CONTRACT

In order to avoid the possibility of the Works being abandoned prior to their completion, the majority of Employers favour the use of *entire* contracts for Works of civil engineering or building contruction.

In essence, a Contractor entering into an *entire* contract with an Employer undertakes to complete certain specified Works in return for an agreed remuneration which will not be forthcoming unless the Works *are* completed. That said, however, there are certain circumstances in which the contract may be terminated with the Contractor being entitled to payment on a *quantum meruit* (see section 3.6.6) basis, the most noteworthy example of which is a situation in which, for a variety of reasons, the contract is terminated at the request of the Employer. A similar example concerns contract termination under circumstances which were foreseeable at the time the contract was drawn up and which were expressly provided for in the contract documents.

Two common features of construction contracts must be mentioned at this stage since, at first glance, they would appear to detract from the strict principles of *entire* contracts.

In order to avoid the obvious possibility of the Contractor abandoning the partially completed Works as a result of his inability to carry the cost of construction until payment becomes due upon final completion, it is common for contruction contracts to provide for *interim* payments (on account) to be made to the Contractor at regular intervals throughout the construction period. It is similarly common for construction contracts to provide for a specified proportion of the contract payments, known as *retention*, to be retained by the Employer by way of insuring him against any defects which may arise in the work. Despite these apparent anomalies, however, the 'entirety' of such contracts is maintained in that *full* and *final* payment for the Works, including a *full* refund of retention, may not be claimed by the Contractor until such times as he is deemed to have discharged his contractual obligations (i.e. completed the Works to the extent and standard specified in the contract documents) *in full*.

Whilst the final decision regarding the type of contract to be adopted for the project construction rests with the Employer, it is generally the case (see section 4.2.2) that such a decision will be based on the advice and recommendation of the Engineer. Since the Employer is entitled to expect such a recommendation to be supported by cogent reasoning, it is encumbent upon the Engineer to undertake the selection procedure in a logical manner, having regard for the nature of the project in question and the advantages and disadvantages of the various types of contract available.

The various types of contract in use by the construction industry are now examined individually, with each type of contract being considered in terms of its distinguishing features, the circumstances under which it might be used and any specific advantages or disadvantages associated with its usage. Although each type of contract will be considered separately, it should be noted that this approach is merely a convenience for discussion purposes and, since it is possible for any given construction contract to involve two or more of the individual type characteristics, should in no way be taken as implying mutual exclusivity.

For discussion purposes, the range of available contract types may be sub—divided into two broad categories of contract: those which are distinguished by the method used to evaluate the *contract price*, and those which are characterised by some *other* distinguishing feature such as the manner in which the contract is arrived at.

The first category reveals three basic types of contract: *fixed—price* contracts, *fluctuating—price* contracts and *cost—reimbursement* contracts.

5.1 Fixed—Price Contracts

The essential characteristic of a fixed—price contract is that, as the title suggests, the contract price is agreed at the time the contract is signed and does *not* subsequently change in the course of the contract *except* as a result of changes in the Specification – in which case the Contractor would be entitled to revise his prices accordingly – or under certain contractually–specified circumstances such as those surrounding claims (see section 9.1.11), variation orders (see section 9.1.9) or the revaluation of provisional sums (see section 8.2.4).

From the standpoint of the Employer, the principal advantage associated with the use of such contracts is that they enable funding arrangements to be initiated in the reasonably certain knowledge that, unless the incidence of variations and claims turns out to be substantially greater than anticipated at the outset, the final contract price will not differ significantly from the tender price.

A particular feature of fixed–price contracts, which the Employer might perceive as being to his advantage, is that, with such a contract, the Contractor is required to take *all* the risks in connection with inflationary cost increases in the course of the contract. This 'advantage', however, should be regarded with a degree of caution since it can, on occasions, constitute something of a 'double–edged sword':

In assuming the burden of risk for cost increases in the course of the contract, the Contractor will undoubtedly include a *contingency allowance* in his prices to cover himself in the event that the risk materialises. If hindsight shows this contingency to have been over–generous, assuming that the Contractor has not priced himself out of the job by adopting such a conservative approach towards risk evaluation, the Employer will have paid out more than he need have done. It goes without saying, of course, that the fortunate Contractor would be 'laughing all the way to the bank'!

At first glance, it might appear reasonable to suggest that the positions of the parties will simply be reversed in the case of an under–estimated contingency allowance – that it will be the Contractor who suffers the consequences and the Employer who benefits. In reality, however, the situation is likely to be somewhat different owing to the fact that Contractors, generally speaking, are not noted for their readiness to accept impending losses without trying to salvage the situation in some way.

Whilst a number of alternative avenues of approach are open to a Contractor faced with such a situation, the obvious danger here is that he will be tempted into 'cutting corners' and skimping on work in an attempt to reduce his costs over the remainder of the contract.

The *direct* consequences, to the Employer, of such a course of action are fairly obvious. Irrespective of the standard of the Engineer's site-supervision, a certain proportion of the defects resulting from 'corner-cutting' on the part of the Contractor will inevitably remain undetected and, as a consequence, the Employer stands to take possession of a finished product of a standard inferior to that which he has paid for.

The *indirect* consequences of 'corner-cutting' are, perhaps, not quite as obvious as the direct consequences but are equally, if not *more*, serious for the Employer. In the event that the defects *are* discovered, as the vast majority of them will be, the Contractor will be required, on pain of being held in breach of contract, to undertake the necessary remedial work *at his own expense*. This will, of course, increase his losses and might tempt him into *further* nefarious practices with the result that the situation quickly degenerates in a downward spiral.

At *best*, the only major 'casualty' in this situation will be the working relationship between the Contractor and the Engineer's site staff who, despite the fact that they will only be doing their job, will be blamed by the Contractor for anything and everything that goes wrong. The usual consequence of this is that the Contractor's desire to 'get one over on the R.E.' tends to take precedence over other, more mundane, considerations such as, for example, constructing the Works to the standard prescribed in the contract!

Apart from the obvious affect on the project, a side effect of such a situation is that the incidence of claims, usually concerning matters with which the Contractor would not (under normal circumstances) be bothered, tends to multiply alarmingly. Whilst the majority of such claims will be little better than exercises in 'grasping at straws', they will *all* require to be investigated by the Engineer and some, inevitably, will be 'approvable'. Consequently, the 'claims component' of the final contract price will, almost certainly, be in excess of that which would normally be the case in less antagonistic circumstances.

At *worst*, the 'downwardly spiralling' situation could, at the end of the day, result in the Contractor incurring losses to such an extent that he is bankrupted and forced to abandon the contract, in which case the Employer would be left with a partially completed project on his hands. Since such projects are of somewhat limited use, the Employer would be left with no realistic option other than to employ a replacement Contractor to complete the outstanding Works.

Notwithstanding any legal or contractual remedies which might be available to the Employer in respect of offsetting the additional costs incurred as a result of the contract forfeiture, it is highly probable that the *net* cost of completing the Works by such means will be *significantly* greater than would have been the case under the *original* contract.

It would appear from the foregoing, therefore, that an *accurate* evaluation of the Contractor's contingency for inflationary cost increases is in the interests of *both* parties to the contract. The problem is, however, that, on account of it being required at the tendering stage of the contract, an *accurate* evaluation of the contingency requires a degree of *accurate* economic foresight on the part of the Contractor and, as countless Contractors (and Chancellors of the Exchequer!) have found to their considerable chagrin, such foresight is generally characterised only by its conspicuous *lack* of accuracy. Consequently, and irrespective of the attention devoted to the matter, the estimation of the contingency allowance is essentially a gamble – and the longer is the contract duration, the bigger the gamble becomes!

This being the case, therefore, it is advisable for the use of fixed-price contracts to be limited to projects of comparatively short duration. It is difficult to cite an exact figure in respect of the contract duration beyond which the use of such contracts becomes ill-advised, since it will obviously depend upon the magnitude and stability of the prevailing inflation rate. As a guide, however, the cut-off point for government-sponsored contracts is (at the time of writing, with inflation running at around 5% p.a. and showing no *obvious* tendency towards instability) *two* years.

Within the *general* category of fixed-price contracts, two *specific* types of contract may be considered: *lump sum* contracts and *measurement* contracts.

5.1.1 Lump Sum Contracts

In essence, a lump sum contract constitutes the simplest type of construction contract in that the Contractor undertakes to carry out the specified work for an agreed lump sum.

In cases where the extent of the specified work is small, this sum will probably be payable in 'single payment' form upon satisfactory completion. This will, of course, require the Contractor to carry the *entire* cost of the project, a fact which he will, no doubt, take into account when evaluating his tender price. Where the extent of the work is somewhat larger, however, it would not be reasonable (or, indeed, cost-effective) to expect the Contractor to finance the project from beginning to end. Consequently, *incremental* payments (on account) may be made at specified stages of the construction or at agreed intervals during the construction period.

As with all other types of construction contract, the nature and extent of the Works will be depicted in the contract drawings, with the workmanship and materials requirements being detailed in a Specification. What sets this type of contract apart from other contract types, however, is that the responsibility for assessing all the quantities involved in meeting the specified requirements, for the purpose of compiling the tender, rests *entirely* with the Contractor.

Insofar as the Employer or Engineer is concerned, this invests such contracts with the attractive quality of a minimal documentation content, with the result that the pre-contract task of document preparation is substantially reduced along with the associated costs. It should be appreciated, however, that any reduction in costs achieved thereby will undoubtedly be offset, to a certain extent, if not totally, by the Contractor's inclusion, in the tender price, of a component to cover the costs incurred by *him* in undertaking the quantification operation.

It should be *further* appreciated that, in undertaking this task, the Contractor assumes the burden of risk in respect of quantification errors leading to an incorrectly evaluated tender price and will undoubtedly include a contingency for such in his price. It goes without saying, of course, that the probability of occurrence of such errors increases with the extent and complexity of the Works and the extent to which they contain components which are particularly susceptible to imprecise quantification *in advance*. An obvious example of the latter is *earthworks*, the *actual* quantities of which invariably differ from those estimated *a-priori*.

Since an inaccurate evaluation of the aforementioned contingency has much the same consequences as those discussed previously in connection with the contingency allowance for fixed-price contracts, lump sum contracts should advisedly be used only for comparatively small projects in which the greater proportion of work is above ground and clearly defined at tender time.

It might *further* be suggested that, on account of the obvious problems which will arise in valuing additions or variations to the

Works, the use of lump sum contracts is best avoided for projects in which the Works are likely to be varied to any significant extent.

This restriction can, to a certain extent, be alleviated by requiring tenders to be accompanied by a schedule of rates to be used for the specific purpose of valuing such variations and additions. It should be appreciated, however, that, since such a requirement carries with it the obvious implication that additional work *will* be required, there is always the chance that the Contractor will see it as a chance of making a 'killing'. Consequently, there is the distinct possibility of such rates being somewhat on the high side, thereby resulting in additions to the Works becoming a disproportionately expensive prospect.

Finally, it should be noted that, although lump sum contracts would appear to be largely unsuitable for *entire* construction projects, they can beneficially be used for *parts* of a project which are *sub-let* by the Main Contractor – provided, of course, that the extent of work is clearly defined and unlikely to be varied in the course of construction.

5.1.2 Measurement Contracts

The fundamental feature of a measurement contract (alternatively known as a *remeasurement*, *admeasurement* or *measure-and-value* contract), which distinguishes it from a lump sum contract, is that the project is broken down into its constituent components with the Contractor being required to tender a *unit rate* (£x per unit quantity) for *each* work item. In the course of construction, the quantity of work completed by the Contractor is measured and he is paid in accordance with his tendered rates.

Since the overall contract price depends on the amount of work carried out by the Contractor, and may thus vary significantly from the tender price, such contracts would appear to deviate from the general principle of fixed-price contracts. Not so, in fact. Although the overall *price* will vary, the individual *rates* are agreed at the outset, by virtue of the tender being accepted, and remain *fixed* for the duration of the contract despite the fact that they might subsequently transpire to be inadequate.

Measurement contracts, in general, come in two varieties: *Bill of Quantities* contracts and *Schedule of Rates* contracts, with the former currently being the type of contract most commonly used for Works of civil engineering construction.

(a) Bill of Quantities Contracts

As suggested by its name, the Bill of Quantities is a (usually voluminous) document, prepared by the Engineer prior to the contract being let, containing a breakdown of the project into its constituent work items with each item being fully described and quantified.

At tender time, the tenderer inserts a *unit rate* and *extended total* (item quantity × unit rate) against each item, with the *tender total* being the sum of the extended totals. During the construction of the Works, the actual quantities of work carried out are measured and valued at the tendered rates and the Contractor is paid accordingly, the eventual *contract price* being the aggregate of all such payments.

In view of the common usage of such contracts, the compilation of the Bill of Quantities is generally carried out in accordance with a recognised Standard. In the case of civil engineering Works, the most commonly used Standards are: the *Civil Engineering Standard Method of Measurement* (published by the Institution of Civil Engineers in conjunction with the Federation of Civil Engineering Contractors) and the *Method of Measurement for Road and Bridge Works* (published by HM Stationary Office under the auspices of the Department of Transport, the Scottish Development Department and the Welsh Office). For building Works, the Bill of Quantities is generally compiled in accordance with: the *Standard Method of Measurement for Building Works* (published jointly by the Royal Institution of Chartered Surveyors and the Building Employers' Federation).

Since most Contractors are familiar with these standards, Bills compiled in accordance with such are readily assimilable by tenderers and offer little, if any, prospect of being erroneously priced as a result of, say, a tenderer's misinterpretation of the nature and extent of work covered by particular Bill items.

In comparison with other types of contract, Bill of Quantities contracts offer the following *specific* advantages:

(1) The quantified breakdown of the project enables a *reasonably* accurate estimate to be made of the project costs, prior to the contract being let, in the knowledge that no significant costs have been omitted;

(2) Since the same advantages accrue in respect of tender compilation, the Employer derives the benefit of a *realistic* tender price;

(3)   The appraisal and comparison of tenders by the Engineer is a reasonably straightforward exercise since all tenders are priced on a common basis;

(4)   The itemised rates provide a basis for evaluating variations or additions to the Works and, in conjunction with the associated quantities, a basis for evaluating interim payments to the Contractor;

In addition to the benefits which accrue to the Employer/Engineer through the use of such contracts, the Bill of Quantities provides a useful starting point for the Contractor to:

(1)   compile the construction programme;

(2)   obtain and evaluate quotations from sub-contractors;

(3)   calculate incentive bonus payments to operatives in the course of construction;

(4)   calculate the materials requirements for the contract;

With respect to the possible use of the Billed quantities as a basis for further calculation, it should be noted, firstly, that they are *net* quantities and secondly, in the case of a contract executed under the I.C.E. Conditions of Contract, that they are only *estimated* (see section 9.1.10).

In considering the prospective use of a Bill of Quantities contract, the aforementioned advantages must be viewed against the considerable amount of time that will be required to prepare a Bill of sufficient detail and accuracy as to enable *realistically* priced tenders to be submitted. It should also be appreciated that the compilation of the Bill cannot (realistically) be expected to commence until such time as the project design is substantially complete. This will obviously introduce a (possibly significant) delay between the completion of the design and the start of construction and will, even more obviously, preclude any overlapping of the two stages.

Consequently, such a contract may be unsuitable in a case where the extent of the Works is such that it would be neither desirable nor cost-effective to delay the start of construction until the *entire* design had been completed and the Bill of Quantities had been compiled.

To a certain extent, this obstacle may be surmounted by *phasing* the Works and letting each phase under a separate Bill of Quantities contract. However, such a course of action, assuming it to be *possible* in the first place, is not entirely problem—free. If, for example, the separate phases were *competitively* let, this would introduce the possibility of each phase being constructed by a different Contractor. Apart from giving rise to the obvious problems associated with a lack of continuity, this would undoubtedly result in the aggregate cost of the phases being greater than the cost of constructing the *entire* Works under a *single* contract. In the event that the separate phases overlapped, as would probably be the case, there would be the additional problem of coordinating the activities of two or more Contractors on the same site.

In such a case, therefore, it might be less problematical to forego the advantages of using Bill of Quantities contracts and to let the *entire* Works under a *single* alternative form of contract.

(b) Schedule of Rates Contracts

In simple terms, a Schedule of Rates is (effectively) a Bill of Quantities without the quantities and is used as an alternative to the latter in situations where the benefits of a measurement contract are required but the use of a Bill of Quantities is precluded.

For example, a Schedule of Rates would be considered more appropriate than a Bill of Quantities for a maintenance or similar contract in which the work would be ordered on an *ad—hoc* basis and would, therefore, be impossible to quantify, with any degree of accuracy, in advance.

Generally speaking, Schedule contracts tend to come in one of two forms:

(1)   The Employer/Engineer prepares a schedule containing unit rates for each item of work, with tenderers being required to indicate the percentage variation (up or down) of such which they require to carry out the work.

(2)   Tenderers are supplied with a schedule of work items and are required to insert their *own* rates against each item. This is the more usual form of Schedule contract.

The principal problems associated with this type of contract, whichever form is used, arise out of the absence of item quantities. Firstly, the absence of a *total* price with which to compare tenders, and the fact that the unit rates may vary extensively from tender to tender, makes tender adjudication virtually impossible. Additionally, the absence of itemised quantities makes it difficult for tenderers to price *realistically* since unit costs, generally speaking, are subject to economies of scale. For example, the unit cost of concrete provision will be significantly less for an overall requirement of 1000m^3 than it would be for a total quantity of only 100m^3.

To alleviate these problems, to the benefit of *both* parties to the subsequent contract, tenderers are often supplied with *approximate* quantities, generally in the form of an upper and/or lower limit, for each item of work.

The major advantage of the use of Schedules is that they can be prepared very much quicker than Bills of Quantities. This being the case, the construction of an urgently required project could be commenced, in advance of an *accurate* Bill of Quantities being prepared, by letting the contract on the basis of a Schedule of Rates containing *approximate* quantities. The award of the contract should, of course, be subject to the proviso that, following completion of the Bill, the work measured and valued in accordance with the Schedule be *revalued* in accordance with the *Billed* quantities. Since the tender rates will be based on the *approximate* quantities contained in the Schedule, it should be appreciated that the Contractor might fairly be entitled to a revision of such in the event that the *Billed* quantities differ *significantly* from the original approximations.

Alternatively, the project could be phased, as suggested previously, with the initial phase being completed under a Schedule contract. During the execution of this contract, Bills of Quantities could be prepared for subsequent phases which could then be let either competitively (see section 5.4.1) or by negotiation (see section 5.4.2) with the Contractor already on site.

5.2 Fluctuating–Price Contracts

As mentioned previously (see section 5.1), the *principal* disadvantage of fixed–price contracts is that their use is (advisedly) limited to comparatively short duration projects. Consequently, such contracts are unsuitable for a significant proportion of the projects undertaken by the construction industry.

This problem is overcome by the use of fluctuating–price contracts (alternatively known as *price–adjustment* contracts) which, in terms of general format, are identical to the fixed–price contracts previously described, but include a facility which permits adjustment of the contract price in line with the possible variation of certain specified costs which the Contractor incurs.

Broadly speaking, this facility has the effect of substantially reducing the extent of the risk carried by the Contractor in respect of inflationary cost increases. Consequently, fluctuating–price contracts can be used for projects of substantially longer duration than would be the case for a fixed–price contract (but see section 5.3.5). Although such contracts do not provide the Employer with the benefit of a fixed price, this 'deficiency' is outweighed by the fact that he will not be faced with paying out excessive contingencies or, more seriously, the prospect of a disgruntled Contractor trying to recoup losses by indulging in 'corner–cutting' and other undesirable practices.

In the case of civil engineering contracts carried out prior to the early 1970's, the traditional approach towards price adjustment was as follows:

(1) At tender time, the Contractor was required to submit a list of the plant, labour and materials to be used on the contract, together with the prices of such on which his tender was based;

(2) During the construction, the Contractor was required to keep records of the quantities of plant, labour and materials used;

(3) At the end of the contract, these records would be consulted and the Contractor would be reimbursed for all 'allowable' amounts in excess of the original;

As might be imagined by anybody with the slightest experience of the construction industry, this was an unbelievably drawn–out and complicated procedure which was fraught with difficulties and sources of potential conflict between the Contractor and the Engineer. Apart from the tedious procedure of ploughing through mountains of paperwork day after day, there were *always* acrimonious arguments about the 'allowable' quantities of resources ('..why did you take five days to do this job, using a JCB and ten men, when it could've been done in thirty five minutes by one man and a shovel!') and the time of the cost–increase relative to the Works requirement ('..if you'd concreted this foundation on time, instead of six weeks late, you could've used cheaper cement!').

In order to avoid this perennial 'outbreak of hostilities', the *Contract Price Fluctuations (CPF)* clause was introduced into the I.C.E. Conditions of Contract and provides for a much more rational approach to the problem by making use of a price adjustment formula which operates in conjunction with published cost indices.

In the case of civil engineering Works, the appropriate indices are the *Baxter Indices,* compiled by the Department of the Environment and published on a monthly basis in the *Monthly Bulletin of Indices (Civil Engineering Works)*, which concern labour, plant and the most commonly-used materials in civil engineering construction. The equivalent indices for building Works are the *Series Two Osborne Indices* which are published on the same basis as the Baxter Indices but are, fairly obviously, more extensive. The indices themselves are similar to the well known FT100 share-price index in that, when viewed in isolation, they are largely meaningless. When viewed against previous months' indices, however, they provide a measure of the fluctuation of prices over a given period.

In accordance with the CPF clause, the procedure for calculating contract price fluctuations may be summarised as follows:

(1)   At the time the contract documents are being prepared, the Engineer is required to estimate the *monetary* proportions of plant, labour and materials included in the Works.

(2)   Prior to tenders being invited, the details are inserted in the contract documents in accordance with paragraph (4) of the CPF clause (see Figure 5.1). Since the proportional breakdown will be based on the price structure of the Engineer's estimate, which will undoubtedly differ to some extent from the operative (i.e. tendered) price structure, minor differences may exist between the *actual* and *documented* proportions. Notwithstanding any differences, however, the fact that the latter proportions form part of the contract implies acceptance of such by both parties to the contract. It should be *particularly* noted that the breakdown represents 90% of the Works, with the remaining 10% being subject to no adjustment. The obvious implication is that the Contractor is required to assume 10% of the total risk of increased costs.

(3)   Shortly after the contract is signed, the *base* indices are ascertained – these being the Baxter Indices applicable *42 days*

prior to the date of signing of the contract.

For the purpose of calculating the Price Fluctuation Factor the proportions referred to in sub–clause (3) of this Clause shall (irrespective of the actual constituents of the work) be as follows and the total of such proportions shall amount to unity:–

(a)   0.16*   in respect of labour and supervision costs subject to adjustment by reference to the Index referred to in sub–clause (1)(a)of this Clause;

(b)   0.16*   in respect of costs of provision and use of all civil engineering plant road vehicles etc. which shall be subject to adjustment by reference to the Index referred to in sub–clause (1)(b) of this Clause;

(c)   the following proportions in respect of the materials named which shall be subject to adjustment by reference to the relevant indices referred to in sub–clause (1)(c) of this Clause:–

    0.09*   in respect of Aggregates;

    0.05*   in respect of Bricks and Clay Products generally;

    0.05*   in respect of Cements;

    0.03*   in respect of Cast Iron products;

    0.17*   in respect of Coated Roadstone for road pavements and bituminous products generally;

    0.06*   in respect of Fuel for plant to which the DERV Fuel Index will be applied;

    0.03*   in respect of Fuel for plant to which the Gas Oil Index will be applied;

    0.03*   in respect of Timber generally;

    0.03*   in respect of Reinforcement;

    0.04*   in respect of other Metal Sections;

    NIL*   in respect of Fabricated Structural Steel;

    NIL*   in respect of Labour and Supervision in fabricating and erecting steelwork;

(d)   0.10   in respect of all other costs which shall not be subject to any adjustment;

Total   1.00

*To be filled in by the employer prior to inviting tenders.

Figure 5.1 Paragraph (4) of the Contract Price Fluctuations Clause

(4)   After each monthly valuation, the *current* indices are ascertained. In order to balance the base index lag, the current indices are those applicable *42 days* prior to the date of the valuation. It is worth noting that the published indices are generally *provisional* in the first instance, and are subject to subsequent confirmation. This being the case, the monthly price fluctuation *may* require re-calculation.

(5)   The 'price fluctuation factor' (PFF) is calculated, for *each* component, from the following formula:

$$PFF = A \times \frac{(C - B)}{B}$$

Where: A   =   Contract Proportion

           B   =   Base Index

           C   =   Current Index

The *total* 'price fluctuation factor' for the valuation in question is the *algebraic* sum of the individual factors, which, dependent on whether prices have risen or fallen, may be positive *or* negative.

(6)   The 'effective value' of the month's work is calculated, that being the *gross* value of the work less any contributions which are based on 'actual cost or current prices', such as *Dayworks* (see section 8.2.3) or payments to *nominated sub-contractors* (see section 8.2.4), and any previous price fluctuations.

(7)   The price fluctuation for the month's work, for which the Contractor will be reimbursed, is arrived at by multiplying the 'effective value' of the work by the *total* 'price fluctuation factor'.

A typical price fluctuation calculation, based on the proportions shown in Figure 5.1, is shown in Figure 5.2.

It should be noted that the process stops on issuance of the *completion certificate*, with the 'price fluctuation factor' of any work carried out thereafter being calculated on the basis of the *final indices* – the indices applicable *42 days* prior to the date of completion. As a result of this, Contractors usually request completion, during a period of rising costs, to be effective from the latter part of the month, thereby gaining an *additional* month's indices increases.

```
┌─────────────────────────────────────────────────────────────────────────┐
│                                                                           │
│  CONTRACT ................................................................ │
│  CONTRACT PRICE FLUCTUATION FOR VALUATION No. ........................... │
│  DATE .................................................................... │
│  DATE OF BASE INDEX ...................................................... │
│  DATE OF APPLICABLE INDICES .............................................. │
│                                                                           │
```

CONSTITUENTS OF WORK	CONTRACT PROPORTION (A)	BASE INDEX (B)	CURRENT INDEX (C)	PRICE FLUCTUATION FACTOR
Labour	0.16	557.7	618.4	0.017414
Plant	0.16	525.1	572.2	0.014352
Aggregates	0.09	843.4	884.8	0.004418
Bricks	0.05	907.1	971.6	0.003555
Cements	0.05	651.3	674.2	0.001758
Cast Iron	0.03	783.3	785.3	0.000077
Coated Roadstone	0.17	1097.7	1058.0	-0.006148
Derv Fuel	0.06	680.9	634.9	-0.004053
Gas Oil Fuel	0.03	1638.9	1460.2	-0.003271
Timber	0.03	600.9	609.2	0.000414
Reinforcement	0.03	217.7	209.8	-0.001089
Metal Sections	0.04	506.7	479.9	-0.002116
Structural Steel	–	510.5	528.4	–
Steelwork Labour	–	750.3	796.7	–

Total Price Fluctuation Factor: 0.025311

EFFECTIVE VALUE   £   £

Gross value of work to date: · 1,523,748.52

Less   (a) Dayworks to date: 1654.28
        (b) Nominated Sub-Contractors to date: 4832.46
        (c) Items based at cost, etc. to date: 527.07
        (d) Previous Cumulative Effective Value: 1,378,423.94 1,385,437.75

                EFFECTIVE VALUE OF THIS CERTIFICATE: 138,310.77

CONTRACT PRICE FLUCTUATION

£ 138,310,77   ×   0.025311   =   £ 3500.78
(Effective Value)   (PFF)   Increase/~~Decrease~~*

                                            *Delete as applicable

Figure 5.2 Calculation of Contract Price Fluctuation

With respect to the validity of this method, it should be appreciated that, whilst it constitutes an extremely rational and straightforward approach to the problem, the price fluctuations calculated thereby may differ, to some extent, from the *actual* price fluctuations dependent upon:

(a) the extent to which the *documented* cost proportions differ from the *actual* proportions;

(b) the extent to which the *actual* price fluctuations differ from the *average* price fluctuations represented by the differences in successive indices;

(c) the extent to which the pattern of interim payments varies from the pattern of incurred costs;

Finally, it should be noted that, in contracts where the fabrication and erection of structural steelwork predominates, with all other civil engineering work being negligible in comparison, the procedure for calculating the contract price fluctuations is governed by the *Contract Price Fluctuations (Fabricated Structural Steelwork)* clause – the 'FSS' clause – contained within the I.C.E. Conditions of Contract.

The procedures detailed therein are identical to those previously outlined, with the exceptions that the FSS clause makes use of only *two* indices – one for the cost of the steelwork and one for the labour used in its fabrication and erection – and that the dates of the applicable 'current indices' are slightly different, namely:

(a) labour used in fabrication – *56 days* prior to the valuation date;

(b) labour used in erection – *14 days* prior to the valuation date;

(c) materials *specifically* purchased for inclusion in the Works – the date of delivery to the fabricator's premises;

(d) materials (if any) *not* specifically purchased for inclusion in the Works – the date of the last delivery as above;

5.3 Cost–Reimbursement Contracts

The basic feature of this type of contract, which distiguishes it from the contracts discussed previously, is that the Contractor is reimbursed for his *actual* costs and is paid a *management fee* to cover his overheads, supervisory costs and profit.

There are four types of cost—reimbursement contract, which differ from each other by virtue of the method used to calculate the Contractor's fee: cost plus *percentage* fee, cost plus *fixed* fee, cost plus *fluctuating* fee and *target—cost* contracts. Each type will be discussed in turn, noting any *specific* advantages or disadvantages associated with their usage. Following that, coverage will be given to the particular circumstances in which the use of cost—reimbursement contracts in *general* might be considered preferable to that of standard fixed— or fluctuating—price contracts.

5.3.1 Cost plus Percentage Fee Contracts

In this type of contract, the Contractor's fee is calculated as a *percentage* of the *actual costs* incurred by him in constructing the Works.

Whilst the fee may, in many cases, be based on a *single* percentage figure applicable to *all* costs, it is more usual for it to be based on *separate* percentages applicable to *particular* cost elements by way of reflecting the different levels of overhead and supervisory support associated with such. Whichever approach is adopted, however, this type of contract is comparatively simple to operate and, by virtue of the relevant percentages being the only matters requiring pre—contract agreement, offers the facility of substantially overlapping the design and construction phases of a project with a view to getting construction underway at the earliest possible opportunity.

These advantages, however, must be viewed against the principal deficiency of this type of contract which is that the direct relationship between costs and fee provides no incentive for the Contractor to control his costs by adopting efficient work practices – in fact, an unscrupulous Contractor is provided with a positive incentive to *maximise* his costs with a view to maximising his fee!

An additional deficiency which deserves a mention, on account of it often being overlooked, is that the Contractor's fee will fluctuate in proportion to any fluctuations in construction costs *even though such fluctuations might bear little, if any, relation to changes in overhead or supervisory costs.*

Since the deficiencies of this type of contract tend to outweigh any advantages to be gained therefrom, their use is generally regarded with some disfavour and is, in the main, only considered in circumstances where the urgency for project completion overrides all other considerations, or in situations where the full nature and extent of the project is uncertain at the time the contract is signed.

5.3.2 Cost plus Fixed Fee Contracts

As the title suggests, the Contractor's fee in this type of contract is a *fixed sum* based on a *agreed estimate* of the *total cost* of the contract. Depending upon the terms of the contract, this sum may either be payable in 'single payment' form on completion of the Works or in instalments at agreed intervals during the construction period. It may also be subject to adjustment in the course of the contract if inflationary cost increases result in the original agreed estimate becoming unrealistic.

The obvious advantage of this type of contract, in comparison to the 'cost plus percentage' contract, is that the 'fixed' fee discourages the Contractor from being inefficient. This, however, is not the same as *encouraging* him to be *efficient* and, in this respect, the deficiency of a 'fixed fee' contract is much the same as that of a 'percentage fee' contract in that the Contractor is provided with no real efficiency incentive other than the prospect of timely payment of the fee and release of his resources for other work.

In order to establish a realistic fee at the outset of the contract, it is necessary to estimate the likely total cost of the project with reasonable accuracy. It therefore follows that the nature and extent of the project must be known with reasonable certainty before the contract is let. Consequently, the use of a 'fixed fee' contract precludes overlapping the design and construction phases of a project to the same degree that would be possible with a 'percentage fee' contract.

5.3.3 Cost plus Fluctuating Fee Contracts

This type of contract provides for the Contractor's fee to be determined on the basis of a *sliding scale* which relates the *amount of the fee*, as a percentage of the incurred costs, to the *magnitude of the costs*. In effect, this means that the lower is the overall cost of the Works, the higher will be the Contractor's fee (as a percentage of his incurred costs) and *vice versa*.

On the face of it, this would appear to counteract the main deficiencies of 'percentage fee' and 'fixed fee' contracts in that the

Contractor is *encouraged* to be *efficient* and, at the same time, *discouraged* from being *inefficient.* However, a closer examination of the details of a typical 'fluctuating fee' contract would show that this is, in fact, not entirely so.

Consider the following typical arrangement:

Incurred Costs	Percentage Fee
1st £10,000	15.0%
Next £25,000	12.5%
Next £50,000	10.0%
Over £85,000	5.0%

For a total incurred cost of £80,000, the Contractor's fee would be £9,125 which represents a profit of 11.4%, assuming Profit = Fee/Cost. The total contract price, payable by the Employer, would be £89,125. If, by increasing the cost efficiency of his performance, the Contractor reduced his incurred costs by 10% to £72,000, his fee would be £8,325 (representing a profit of 11.6%) and the total price payable by the Employer would be £80,325. Thus, for achieving a 9.9% reduction in the overall contract price, the Contractor is 'rewarded' by a meagre 1.4% (in *relative* terms) increase in profit. On the other hand, an inefficient performance resulting in a 10% *increase* in costs (representing a 9.7% increase in the contract price) would be 'penalised' by a profit reduction of only 2.6% (in *relative* terms.

It can be seen, therefore, that whilst a 'fluctuating fee' contract represents a significant improvement upon the previously detailed arrangements, the deficiencies of such are essentially the same, namely that the Contractor is provided with no significant financial incentive to be cost efficient and is not significantly penalised for inefficiency.

Of course, the latter deficiency could be alleviated to a certain extent by requiring the Contractor to specify a 'guaranteed maximum cost' above which no fee would be payable. This, however, would not solve the problem entirely since it would still provide the Contractor with no *positive* incentive to reduce his costs – in other words, it would provide the 'stick' but not the 'carrot', a somewhat unfair and, therefore, undesirable state of affairs!

5.3.4 Target—Cost Contracts

This type of contract is an improved form of cost—reimbursement contract which seeks to retain the advantages of such whilst overcoming its major deficiencies.

As with the 'cost–plus' contracts previously discussed, the Contractor is reimbursed his incurred costs plus a fee. With a Target–Cost Contract, however, the magnitude of that fee, expressed as a percentage of the incurred costs, fluctuates according to the *difference* between the Contractor's *actual* costs and a *target* cost agreed between the contracting parties at the time the contract is signed. In order to provide the Contractor with the necessary financial incentive to execute the contract as cost–efficiently as possible, the fee fluctuation operates as a *bonus* in the event that his actual costs are *less* than the target cost, and as a *penalty* if inefficient work practices and lack of cost control result in actual costs *exceeding* the target cost.

In determining the *target cost*, it should be appreciated that it is, in effect, the best *estimate* of the *probable* actual cost of the project and is, therefore, subject to a degree of uncertainty. This uncertainty is reflected by the use of a *target band* – a margin either side of the target cost – within which the Contractor's fee (the *target fee*) remains constant. If the project is well defined at the outset, and is unlikely to be *significantly* affected by variations or additions, price fluctuations or unforeseen construction difficulties, the target band may be as low as ±1% of the target cost. On the other hand, a band of ±10% may be appropriate for a project which is likely to be subject to *significant* cost variation in the course of construction.

The usual means by which the appropriate financial incentives or penalties can be incorporated into the Contractor's fee involves the use of a *share–formula* which is agreed at the time the contract is signed and which defines the distribution of cost saving/overrrun between the parties. In order to limit the Contractor's liability for cost overrun, and to avoid the use of excessive 'stick', a *minimum fee* is often specified. Similarly, the Employer's liability for excess costs may be limited by requiring the Contractor to specify a *guaranteed maximum cost* above which no fee will be payable.

The characteristics of a typical target–cost contract are illustrated in Figures 5.3 – 5.5 and are based on the following details:

(a)   A Target Band of ±5% of Target Cost;

(b)   50/50 sharing of cost savings;

(c)   60/40 sharing of excess costs (60% by the Contractor);

(d)   A Minimum Fee of 3% of Target Cost up to a Guaranteed Maximum Cost of 140% of Target Cost;

(e)   A Target Fee of 15% of Target Cost;

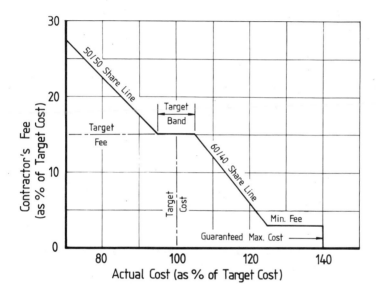

Figure 5.3 Target Cost Contracts: Variation of Contractor's Fee with Incurred Costs

It can be seen from Figures 5.4 and 5.5 that both the Contractor *and* the Employer stand to benefit *substantially* from maximised cost−efficiency on the part of the former. The relative degree to which each party stands to benefit depends, of course, upon the share formula used. For example, the potential increase in the Contractor's profit, as a result of increased efficiency, would be less dramatic than shown if, say, a 10/90 share formula were used for cost savings. The prospective reduction in contract price would, however, increase to the obvious benefit of the Employer. The same 'swings and roundabouts' situation obtains at the other end of the cost scale where a less punitive (from the Contractor's standpoint) share formula would result in a significant increase in the Employer's liability for cost overruns.

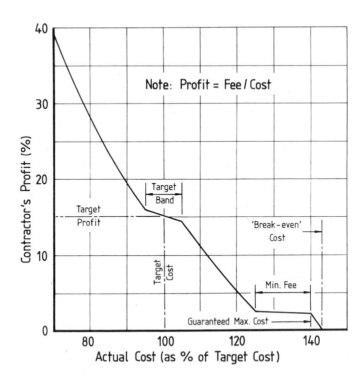

Figure 5.4 Target Cost Contracts: Variation of Contractor's Profit with Incurred Costs

It therefore follows that the determination of the share formulae is largely a matter of compromise. On the one hand, the prospective incentive should be such that the Contractor is *genuinely* encouraged to be efficient *without* being provided with a 'license to print money'. On the other hand, the Contractor should be subject to a prospective cost–overrun penalty of a level which is *sufficient* to concentrate his mind in respect of cost control but which *stops short* of the point at which he might question the wisdom of entering into the contract in the first place!

Of primary importance in respect of the *effectiveness* of this type of contract is the requirement for the target cost to be set, by agreement between the two parties, at a *realistic* level. The temptation to 'beat down' the target in order to maximise the benefits to the Employer should be avoided at all costs since a target which is too *low* will provide the Contractor with little or no incentive to undertake such a contract and, in view of the prospective penalties for cost overrun, might well result in him insisting on a standard 'cost−plus' contract with all its associated disadvantages from the Employer's standpoint.

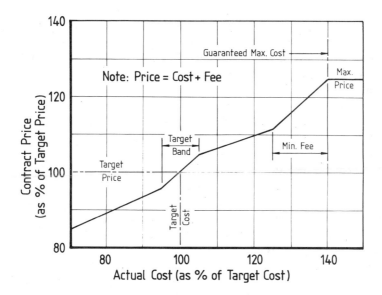

Figure 5.5 Target Cost Contracts: Variation of Contract Price with Incurred Costs

Conversely, a target which is too *high* will be *equally* detrimental to the Employer in that the Contractor will be in a position to make a 'killing' at the former's expense. Consequently, it is vital for sufficient care and attention to be devoted to the matter such that the agreed figure represents (as stated previously) the *best* estimate of the *probable* actual cost of the project.

There is a conflict of opinion regarding the amount of information and detail required to set the target cost. One school of thought suggests the preferability of basing the target on a *minimum* level of detail, for example: a cost/unit floor area for a reasonably uniform building, or an average cost/unit run of piping or cabling, with such costs being determined from experience or from the records of previous projects of a similar nature. On the other hand, there is a considerable volume of opinion which holds that realistic targets can only be set if they are based on a level of detail equivalent to that required for a standard *admeasurement* contract. This, of course, pre−supposes that sufficient detail is *available* at the outset and leads to the obvious suggestion that if it *is*, there is no need to resort to the use of a *target cost*, as opposed to an *admeasurement*, contract in the first place!

In the majority of cases, the target cost will probably be determined on the basis of a level of detail somewhere in between the two extremes, for example: a priced Bill of Quantities containing the 15% or so of items which generally comprise around 85% of the contract value.

Finally, it should be noted that, whilst the target cost may well be realistic at the outset of the contract, changing circumstances might lead to it becoming less so as the contract progresses. Consequently, it is normal practice to include a facility within such contracts for the target cost to be adjusted, as and when the need becomes apparent, to take into account variations or additions to the work content, price fluctuations and delays etc. Bearing in mind that any adjustment of the target cost is subject to agreement between the parties, it is beneficial for the updating procedure to be kept as simple and as straightforward as possible, avoiding any excessive measurement exercises, to enable any adjustments to be agreed 'on the spot' with a minimum of fuss — subject, of course, to the usual 'horse-trading'!

5.3.5 Usage of Cost-Reimbursement Contracts

In the light of the foregoing discussion, it might be reasonable to conclude that, in the majority of cases, the use of a fixed- or fluctuating-price measurement contract would be distinctly preferable to that of a cost-reimbursement contract. Whilst that is certainly true, it is worth noting that there are a number of cicumstances in which it might be advantageous to use the latter contract type:

(a) Inadequate definition of the work at tender time

Since an Employer does not see any return from his investment until such time as his project is 'up and running', it is generally accepted that minimum project completion time is an economic necessity. This being the case, it is becoming increasingly common for the Employer to require the design and construction phases of a project to be overlapped with a view to achieving the earliest possible start to construction. Consequently, there will be occasions where the emphasis placed on early completion is such that the construction contract is required to be let in advance of there being sufficient project definition for a conventional admeasurement contract to be used. In such circumstances, a cost—reimbursement contract, despite its inherent deficiencies, might provide the only viable alternative to such.

Although an exact figure is rarely, if ever, quoted in this regard, it is generally accepted that a conventional contract can only *satisfactorily* accomodate a level of variations and additions up to about 25% of the contract value. Beyond this limit, matters can get a bit complicated owing to the fact that, at this stage, the Contractor is being asked to construct something which is *significantly* different from the subject of his original tender. Consequently, some form of cost—reimbursement contract might be considered as a suitable alternative to a standard admeasurement contract in circumstances where, on account of the uncertain nature and extent of a project, there is a high probability of variations and additions exceeding this level.

(b) Projects of exceptional organisational complexity

Such projects are generally those in which the Works are multi—disciplinary to an extent which requires the construction to be let on a 'multi—contract' basis, with a *separate* contract being used for *each* particular discipline involved.

It may be the case in such situations that, whilst the work content of each contract is definable with sufficient accuracy to permit the use of standard admeasurement contracts, the degree of interaction between the contracts renders it virtually impossible to predict individual work sequences with sufficient certainty to enable the compilation of *realistically* priced tenders of unit rates.

This problem could be overcome, to a certain extent, by the use of an overall 'master plan', drawn up by the Engineer (or Project Manager, depending upon the contractual arrangement) prior to tenders being invited, on which tenderers would be required to base their tenders. This would, however, preclude the possibility of any input, in the form of proposals etc., from the Contractors which may prove to be something of a disadvantage. Additionally, it should be remembered that 'the best laid plans of mice and men gang aft agley' and that, in the context of the construction industry in particular, the more complex is the plan, the greater is the chance that it will 'gang agley'! Consequently, it is almost inevitable that, at some stage in the proceedings, at least one Contractor's operations will be disrupted for reasons totally outwith his control. In view of the interaction between contracts, this will almost certainly have 'follow—on' effects which, in the event of standard contracts being used throughout, will give rise to a subsequent claims 'hassle' too awful to even contemplate!

It goes without saying, of course, that the prospective contractors would be only too aware of the abovementioned potential problems and would 'adjust' their tender prices accordingly. In this light, the use of cost–reimbursement contracts, as opposed to conventional fixed– or fluctuating–price admeasurement contracts, might, despite the inherent deficiencies of such, prove to be the 'lesser of two evils'!

A similar, but often overlooked, circumstance in which organisational complexity might result in a Contractor being subject to delays and disruption for reasons outwith his control is where the construction involves considerable interfacing between the Contractor and Statutory Undertakers. Depending upon the degree of interfacing, the claims potential of a standard contract might be sufficient to suggest the preferability of using some form of cost–reimbursement contract.

(c) Projects involving technical innovation or exceptional technical complexity

The type of projects belonging to this category are those which involve work 'at the forefront of knowledge' or those in which the degree of technical complexity is such that the construction process will require substantial innovation on the part of the Contractor.

These situations carry an *exceptional* element of financial risk since neither the requisite resource levels nor the achievable output are predictable, with any degree of certainty, in advance. Consequently, the use of a cost–reimbursement contract might be preferable to that of a standard contract since it would preclude the possibility of the Contractor overpricing to cover himself against the risk or, conversely, abandoning the contract in the event of his failure to carry the risk.

(d) Construction during periods of high inflation

As discussed previously (see section 5.2), the *Contract Price Fluctuations* clause in the I.C.E. Conditions of Contract provides for reimbursement of inflationary cost increases to be calculated on the basis of 90% of the contract value.

The obvious implication of this is that the Contractor will suffer a 10% shortfall in respect of his recovery of such cost increases and will, therefore, require to allow for such in his tender price. Fairly obviously, the magnitude of the allowance will depend upon the predicted expenditure pattern of the contract, the prevailing rate of inflation and the contract duration.

Assuming that contractual payments will be made in accordance with the 'certificates and payment' clauses (see section 9.1.10) of the I.C.E. Conditions of Contract, the Contractor will *also* include an allowance to offset the 'capital lock–up' costs resulting from the six weeks (on average) delay between the incurrence of costs and their (albeit partial) recovery. As with the 'shortfall' allowance, the magnitude of the 'capital lock–up' allowance depends, for obvious reasons, upon the contract expenditure pattern, the inflation rate and the contract duration.

Since both components of the allowance are sensitive to the rate of inflation, it is clear that a degree of uncertainty surrounds the evaluation of such at tender time.

For example, depending upon the predicted pattern of expenditure during the contract, and the relationship between the rate of inflation and the interest rates governing 'capital lock–up' costs, an allowance of aproximately 1.2% of the estimated total cost will be required to ensure *full* recovery of costs over the course of a 2–year contract subject to 5% inflation. For an inflation rate of 10%, the requisite allowance increases to around 2.6% of the estimated total. Consequently, a Contractor tendering for a 2–year contract, during which inflation might reasonably be expected to vary between 5% and 10%, will be faced with an 'uncertainty element' equivalent to around 1.4% of the estimated total cost of the contract.

Whilst this might be considered an acceptable level of uncertainty, particularly when viewed against an uncertainty of around 6% of the estimated total cost for a 2–year *fixed–price* contract executed under similar conditions, it should be appreciated that the uncertainty increases considerably with increasing contract duration and increasing variability of inflation. For example, a 5–year contract, during which inflation might be expected to vary between 15% and 30% (a not unrealistic potential variation given the time–period in question and the inherent volatility of inflation at such levels), carries an 'uncertainty element', over which the Contractor has no control whatsoever, equivalent to around 16.2% of the estimated total cost of the contract!

From a Contractor's point of view, the impossibility of compiling a *realistically* priced *conventional* tender under such conditions is obvious.

In view of the considerable expenditure of effort (both in terms of time *and* cost) involved in the preparation of such a tender, therefore, it is reasonable to suggest, in such circumstances, the possibility of a somewhat less than enthusiastic response to an Employer's 'invitation to tender'. Since this will effectively decrease the Employer's chance of selecting the most suitable Contractor to carry out the construction, an alternative form of contract might be preferable.

Whilst the obvious choice, in this respect, would be some form of cost–reimbursement contract – since the question of under–collection of costs would not arise – it is worthwhile to mention an alternative approach which would substantially reduce the 'uncertainty element' yet would enable the considerable benefits of a conventional admeasurement contract to be retained.

If the *Contract Price Fluctuations* clause in the I.C.E. Conditions of Contract was modified such that the reimbursement of inflationary cost increases was determined on the basis of 100% of the contract value, as opposed to the current level of 90%, the inflation–related uncertainty would be limited to that associated with the 'capital lock–up' costs, there being no shortfall to consider. In the case of the 5–year contract cited above, this would result in a reduction of the 'uncertainty element' from a totally unacceptable level of around 16.2% of the estimated contract cost to a somewhat less 'off–putting' level of around 5.5% thereof!

(e) Situations involving major unquantifiable risks to the Contractor

Irrespective of the standard and extent of the pre–contract site investigation, *all* work below ground level carries an inherent economic risk due to the possible occurrence of unforeseen ground conditions. This being the case, the use of some form of cost–reimbursement contract might be preferable for such work since it removes the need for the Contractor to price the risk – and also avoids the potentially serious consequences (see section 5.1) of his over– or under–estimating the risk.

There is a conflict of opinion regarding the extent to which work below ground level should be carried out on a cost–reimbursement basis, with one school of thought maintaining that *all* such work should be undertaken thus. On the other hand, a significant body of opinion holds that the prospective use of a cost–reimbursement contract should be *entirely* at the discretion of the Engineer and *only* in circumstances where he considers there to be insufficient information available for the

Contractor to price the risks in a conventional manner. It is reasonable to suggest that whichever attitude is taken depends upon the circumstances obtaining in any particular instance. If, for example, the nature of the project was such that the greater proportion of the Works was above ground, and that which was below ground was not subject to *extraordinary* uncertainty, then it would be reasonable to expect the extent of 'cost−reimbursement' work to be subject to the Engineer's discretion. A tunnelling project, on the other hand, where the *entire* Works were below ground and, therefore subject to *considerable* economic risk, would advisedly be carried out, in its entirety, on a cost−reimbursement basis.

A similar situation involving major unquantifiable risk is that where the performance of the contract is affected by restricted or intermittent access. An obvious example of such a contract would be that for the repair or maintenance of an offshore oil−production platform where the availability of access to the platform support system would be dependent upon the prevailing sea conditions. Since such conditions are, to all intents and purposes, unpredictable, the *realistic* pricing of a conventional tender would present certain difficulties. Additionally, it might be mentioned that, on account of the availability of access being a factor *completely* outwith the Contractor's control, the claims potential of a conventional contract would be a little on the high side, to say the very least!

Finally, it should be noted that, on account of the abnormally high risk of disputes and stoppages, contractors often show a distinct unwillingness to enter into conventional contracts in localities having a bad industrial relations record. In such circumstances, the use of a cost−reimbursement contract would reduce the economic risks − thereby making it a more attractive proposition from a Contractor's point of view − and would provide the Contractor with the necessary flexibility to increase wage levels (subject, of course, to the agreement of the Employer/Engineer) with a view to expediting the work.

It is worth mentioning, however, that this course of action is not without its problems in circumstances where a Contractor is concurrently engaged, in the same locality, in *conventional* contracts which might not be able to support similar wage increases.

(f) Recurrent work

When an Employer repeatedly employs the same Contractor, he obtains the benefit of a consistent approach and standard of workmanship. The Contractor, of course, benefits from the continuous turnover of work. The preservation of this 'special relationship' may, however, require the Contractor to mobilise at short notice on occasions and, under such circumstances, the use of a conventional contract would be impracticable. Consequently, some form of cost–reimbursement contract may be preferable – particularly since the majority of problems usually associated with such (lack of incentive, prospective abuse etc.) will, by virtue of the 'special relationship', probably not materialise. There is the *additional* advantage that such a contract will enable the Contractor to become involved at an early stage in the proceedings – with obvious benefit to the Employer/Engineer.

5.4 Other Types of Contract

The types of contract discussed to date are those which are distinguished from each other by the method of determining the contract price. There are, however, a number of contract types which are relevant to the construction industry but which are characterised by some *other* distinguishing feature.

Before discussing each of these types in turn, it bears repeating that any given contract may exhibit one or more of the type–characteristics as, indeed, it may exhibit one or more of the previously discussed features.

5.4.1 Competitive Contracts

Such contracts are those which are arrived at by a process of formal competitive tendering, by a number of tenderers, against a common Specification.

The obvious advantage of this approach is that it provides the only satisfactory means of arriving at a truly commercial price for a contract. There is always the danger, however, in stringent economic times where a tenderer's desire to be awarded the contract might override all other considerations, or where limitations on the Employer's budget might tempt him into trying to have his project constructed 'on the cheap', of such a contract being awarded on the basis of a tender price which is significantly less than might be considered commercially viable. In view of the potentially serious consequences of such (see section 5.1), for *both* Contractor *and* Employer, considerable care and attention should *always* be devoted to the appraisal of competitive tenders.

In this light, and owing to the fact that the vast majority of construction contracts are let on a competitive basis, the competitive tendering procedure is covered in depth in Chapter 7.

5.4.2 Negotiated Contracts

As implied by the title, such contracts are arrived at by means of negotiation between the Employer/Engineer and a Contractor of their choice, rather than by a process of competitive tendering.

Since such contracts contain no competitive element, it may be inferred that there always exists the possibility of the Employer having to pay out more than might be the case with a competitively-let contract. This aside, however, there are a number of circumstances in which the Employer may derive considerable benefit from the use of a negotiated contract, or where negotiation provides the only feasible means by which the contract can be let:

(a) where specialised plant, materials or services are provided by a single contractor or supplier in the field, or in the geographical locality; This may obviously be taken to include circumstances where the specialist requirements are to be provided under a *nominated sub-contract* (see section 7.5.1).

(b) where, by virtue of previous contracts, a 'special relationship' exists between the Contractor and the Employer/Engineer; This may be taken to include situations where the Contractor is currently engaged with the Employer, albeit for the first time, perhaps, in a contract (competitively or otherwise let) of similar character and extent to the one in question. The most obvious example of such a situation is that where the Contractor is currently constructing the first phase of a multi-phase project and the Employer wishes to ensure consistency of performance and workmanship by employing the same Contractor for the subsequent phases. It goes without saying, of course, that the importance of this requirement must be viewed against the possibility of the negotiated contract price being higher than might be the case with competitive letting.

(c) where the project specification is not sufficiently clear-cut at the outset as to enable formal documentation to be drawn up with a view to letting the construction contract on a competitive basis; In such cases, the use of a negotiated contract will enable the Contractor to become involved at an early stage, thus allowing the

Employer to avail himself of the knowledge and experience of the former in drawing up the final Specification. An additional advantage accruing to the Employer is the (hopefully!) ensured cooperation of the Contractor throughout the entire duration of the project.

Several important points should be noted regarding the prospective use of negotiated contracts, the first of which is that, not unreasonably, the Employer is entirely at liberty to enter into negotiations with other contractors in the event of his failure to negotiate satisfactory terms with his 'first choice'. This being the case, a clear understanding must exist between the Employer and the Contractor, at the outset, as to the extent (if any!) to which the Employer is permitted make *subsequent* use of *confidential* technical or commercial information to which he might become privy in the course of the negotiations.

Secondly, it should be appreciated that, in circumstances where he suspects urgency to be the prime reason for *negotiating* (as opposed to *competitively* letting) the contract, it is not entirely unknown for a Contractor to attempt to prolong negotiations to the point where, by virtue of it then being too late for the Employer to take alternative action, he has the latter 'over a barrel'! This being the case, it is generally advisable for the negotiating procedure to be subject to a strict time-limit agreed by the negotiating parties at the outset.

Thirdly, it is an often-held belief that, on account of there being no 'tender period', the use of a negotiated contract will enable an earlier start to be made on the construction than would be the case if the contract were to be competitively-let. It should be pointed out, however, that experience has shown this not to be so and that there is very little difference between the two types of contract in this respect.

Finally, it is worth noting that it is standard practice, in *competitive* contracts, for minor points of difference between the Contractor and the Employer/Engineer to be resolved by negotiation prior to the contract being finalised. The term 'negotiated contract' should not be applied in such an instance unless the areas of divergence are so wide that the competitive aspect plays only a minor role in establishing agreement between the parties.

5.4.3 Package Contracts

Alternatively referred to as a 'Package-deal' or 'All-in' contract, such an arrangement involves the combination of two or more related jobs, each of which could be let under a *separate* contract, with a view to placement as a *single* contract.

A package contract, which, in the engineering field, generally involves the design and construction of a project, together with the maintenance thereof for a specified period of time, constitutes a departure from 'normal' practice in that the Employer outlines his *broad* requirements, perhaps in the form of a 'sketch' design and an indicative price, and invites prospective Contractors to submit their *complete* proposals, and terms of payment, for fulfilling such.

In the main, civil engineering Works of a *general* nature do not lend themselves to this approach. As a consequence, package contracts are comparatively rare and are limited mainly to specialised multi-disciplinary Works such as power stations, petrochemical Works and the like. In such cases, the appropriate specialist organisations may form a contracting consortium, perhaps retaining the services of a consulting engineer for each discipline involved, which is able to command all the necessary resources to design, construct and equip the project completely – *and* offer a contract for so doing. Alternatively, the project may be undertaken by a single Contractor with a view to retaining the services of the requisite specialist organisations on a sub-contract basis.

From the Employer's standpoint, the principal advantage of such an arrangement is that, having selected a suitable proposal, he is in the somewhat luxurious position of being able to place the *entire* burden of responsibility, for the detailed design and organisation of such, fairly and squarely on the shoulders of the Contractor, with his (the Employer's) subsequent involvement in the contract being limited to the approval (or disapproval) of the finished product – and, of course, paying the bills!

A potential problem is worth mentioning in this regard, however. Since the Contractor is responsible for the *entire* package, the overall price for such will depend on *his* choice of sub-contractors. Furthermore, in the event that he envisages certain *specialised* services being provided on a sub-contract basis, the very nature of his proposal might well hinge around *his* choice of sub-contractors. Not unreasonably, therefore, the Employer (through his Engineer) has no powers of either appointment or sanction regarding the use of such. Consequently, there is always the danger of the Contractor proposing the use of sub-contractors who might not be 'approved' by the Employer/Engineer under other forms of contract. If this is likely to constitute a major problem (i.e. if 'questionable' sub-contractors are proposed for a *significant* proportion of the Works), the matter should advisedly be discussed, with the tenderer, prior to final approval of his proposal – at which point it might be of value to mention, in the event of his reluctance to negotiate, the Employer's powers of *ultimate* sanction with regard to a tendered proposal!

A further problem which should be noted in regard of the use of package contracts is that, since each tenderer will undoubtedly interpret the initial brief in a different way, the appraisal of tenders, with a view to selecting the most suitable, is likely to be a more complicated exercise than the usual straightforward comparison of such (largely) in terms of cost. Whilst an element of commonality could be introduced by the use of a more detailed initial brief, this line of approach would limit the flexibility accorded tenderers regarding the formulation of their proposals and would, accordingly, reduce any benefit which the Employer might derive from the Contractor's experience and practical expertise. Additionally, it is worth mentioning that, by virtue of it reducing the extent to which a tenderer is 'in control of his own destiny', a detailed brief will effectively reduce the probability of any given tender being successful. In view of the high cost of producing a 'design and construct' tender, this would probably result in Contractors being less willing to tender in competition for such work.

A noteworthy benefit, accruing to the Employer through the use of a package contract, is the considerable saving of time *and* cost which is generally recognised to result from the design and construction being carried out by the same organisation. Aside from decreasing the potential incidence of 'designer/constructor' interfacing problems – such as those engendered by delayed or incomplete drawings etc. – such an arrangement ensures that the design is carried out with 'buildability' and construction efficiency very much in mind.

In cases where design or development work is included in the package, the unavoidable lack of a complete Specification suggests the need for a degree of flexibility with respect to the contract price. Consequently, such work is customarily carried out on a cost-reimbursement basis. In view of the ever-present possibility of abuse (see section 5.3) of such an arrangement, whether directly intended or merely resulting from a lack of 'efficiency incentive' on the part of the Contractor, continual control (in a *reactive* rather than *proactive* sense) must be exercised by the Engineer in order to discourage potential time- and cost-overrrun at the Employer's expense.

It might also be mentioned that, irrespective of whether or not the design is carried out on a cost–reimbursement basis, there is always the danger that the Contractor will adopt a design which is favourable to *him*, in terms of increasing the profit–making potential of the construction aspect of the package, but not necessarily in the best interests of the Employer. Consequently, the Engineer must *continually* appraise the design in order to prevent the adoption of unnecessarily costly or elaborate proposals which might result in unwarranted expenditure at the construction stage.

With respect to the documentation for such a contract, it is highly likely that 'standard' Conditions of Contract will be totally inadequate in cases where the work is of a 'multi–disciplinary' nature, Although each discipline involved will have its own particular standard Conditions, direct conflict between such is almost inevitable. Consequently, a 'customised' Conditions of Contract, the content of which will obviously depend upon the nature of the project and the disciplines involved, is probably advisable in the majority of instances.

In compiling the documentation, two particular points require noting with respect to the rights of the parties to the contract:

Firstly, there should be no doubt as to who pays for the necessary re–design or modification of the Works in the event that the selected proposal fails, on completion, to meet the functional requirements of the project. In the particular case of construction projects, the contract documents will generally require the Employer (through his Engineer) to approve and accept the design prior to the start of any construction work and, thereafter, to take full responsibility for any design defects which might arise in the course of construction.

Secondly, it should be appreciated that any project involving design or development work involves a considerable contribution – in the form of either expertise or finance – from *both* parties to the contract. Consequently, the contract documents should *unambiguously* state the conditions (if any) under which the Contractor may re–use his design for other customers and under which the Employer may re–use the design for other projects.

Since these rights and conditions will naturally have a considerable bearing on the contract price, the importance of their *unequivocal* statement *at the outset* cannot be over–emphasised.

In particular regard of the possible re–use of the design, it is of interest to note that, in accordance with the laws relating to Copyright, ownership of a design is vested in the designer (in this case, the Contractor) unless any agreement exists to the contrary. Whilst the designer's employer (*the* Employer, in this case) has the *implied* right (see section 3.3) to make use of this design, this right exists *only* in respect of the project for which the design was produced and does not extend to re-use of the design for *subsequent* projects.

Finally, it should be noted that there may be occasions on which the Contractor is required to finance the *entire* project, from start to finish, and present the Employer with a *fully operational* complex of *proven* performance. In such instances, the contractual arrangement is often referred to (for fairly obvious reasons) as a *Turnkey* contract.

5.4.4 Continuation Contracts

Such a contract is one which is arrived at by a process of negotiation between the Employer and a Contractor currently engaged under contract to him. Although the new contract will be based on the terms and conditions of the *current* contract, the two contracts need not be connected. If there *is* a connection, however, the additional contract is sometimes referred to as an *Extension Contract*.

The principal advantage of such an arrangement is that it provides an element of continuity between the contracting parties. Apart from the benefits accruing to the Employer from dealing with a 'known quantity', such arrangements generally result in less antagonism between the contracting parties than might otherwise be the case which, in turn, leads to a comparatively problem-free contract. Additionally, it might be mentioned that continuation contracts permit pooling or rapid switching of resources between contracts (since an Employer is unlikely to refuse permission for an item of, say, plant to be removed from site if it is to be used on one of *his* projects!) with consequent savings in both cost *and* time.

On the other side of the coin, negotiations regarding the contract price are unlikely to be easy since the Contractor will undoubtedly be aware of the prospective benefits accruable to the Employer from such an arrangement and may well attempt to capitalise upon the strength of his position.

5.4.5 Serial Contracts

In undertaking a serial contract, *both* parties are firmly committed to a series of projects following the initial contract.

Since subsequent projects do not require separate contracts, the principal benefit to be gained from the use of such is that of a saving in cost and time. Additionally, the linking of several projects together in the form of a serial contract is quite likely to result in a lower overall price than would be the case if each project were to be let under a separate contract (in much the same way that tins of baked beans are cheaper if purchased in lots of five hundred than on an individual basis!).

In the context of the construction industry, the use of serial contracts is generally restricted to projects of relatively low individual value.

5.4.6 Running Contracts

Such contracts are those which involve the provision of goods or services, at specified intervals or as required from time to time, over a stated time period. Prices are usually quoted against an estimate of demand but, although a guaranteed minimum value may be specified, no total price is quoted. Contract price–adjustment facilities may be included at the discretion of the parties.

5.4.7 Service Contracts

As might be inferred from the title, such contracts are solely for the provision of *services* and do not include for the provision of *goods*. In the context of the construction industry, the most obvious example of such is the agreement between the Employer and the Engineer.

5.4.8 Management Type Contracts

Over recent years, Employers promoting major construction Works have found that the traditional method of carrying out large projects, in which the design is carried out by an independent Engineer who also supervises the activities of a Contractor chosen by competitive tender, often fails to provide them with the best – or even *acceptable* – value for their money.

Various reasons have been cited for this, not the least of which is that the 'quasi–adversarial' Engineer/Contractor relationship inherent in such an arrangement does not provide the best working atmosphere in which to carry out a project. Additionally, it has been found that designs for large and technologically complex project are often inadequate due to the Engineer (or Architect, in the case of building Works) being stretched beyond the limits of his capability and experience.

Management type contracts (sometimes called *Fast Track* contracts) have evolved from a natural process of attempting to remedy these deficiencies and, in this respect, three distinct systems have emerged: *Construction Management* contracts, *Management* contracts and *Design and Management* contracts.

(a) Construction Management Contracts

The essence of such a contract is that a *Construction Manager* is appointed, in the early stages of the project, to act as the Employer's agent with respect to planning and managing the project construction.

Following his appointment, which is effected by means of an appropriate selection procedure (see section 7.6), the Construction Management Contractor works closely with the design team and provides them with the benefit of his practical experience and construction expertise in matters concerning, for example, the 'buildability' and cost–efficiency of their design. In his *advisory* capacity, the Contractor does not carry out any construction *per se*, leaving that to the *Works* Contractor(s) employed under separate direct contract(s) with the Employer, but restricts his activities to planning, managing and administering the construction phase of the project on the Employer's behalf.

For providing the requisite services, which include managing the tendering phase(s) of the Works Contract(s), the Contractor is paid a fee, which is usually quoted as a percentage of the estimated cost of construction, to cover his on– and off–site costs, overheads and, of course, profit.

In comparison with the other forms of management contract, the principal advantage of a Construction Management contract is that the absence of a contractual relationship between the Construction Manager and the Works Contractor(s) avoids (or, at least, *minimises*) the possibility of the former's discharge of his obligations being affected by his *own* commercial interests. Additionally, it is generally accepted that the direct contractual links between the Employer and the Works Contractors enable the former to exercise stricter quality control than might be possible in management–type contracts where no such links exist.

Although Construction Management contracts are popular in the USA, with a considerable number of organisations offering such a service, they are comparatively rare in the UK. That said, however, a number of UK firms, including engineering and management consultants, have provided Construction Management services overseas. Whilst, in principle, *any* organisation with the required competence may tender for such a contract, actual practice has shown that Employers tend to favour large, experienced national Contractors.

(b) Management Contracts

In terms of its *organisational* aspects, a *Management* contract is identical to a *Construction Management* Contract in that the actual construction work is carried out by *Works* Contractors whose activities are managed and coordinated by a *Management* Contractor appointed in the early stages of the project. The two systems differ, however, in terms of the *contractual relationships* between the participants. Whereas the Works Contractors involved in a Construction Management contract are employed under a series of *direct* contracts with the *Employer*, those involved in a *Management* contract are employed, on a *sub-contract* basis, by the *Management Contractor*.

Whilst such an arrangement may be perceived as advantageous by an Employer wishing to derive the benefits of a management system, but reluctant to assume the risks of *directly* employing the considerable number of Works Contractors required on a large and complex project, it should be appreciated that the absence of contractual links limits the extent to which he is able to exert on-going control of the construction process. This may prove to be distinctly disadvantageous in a case where it is discovered, in the course of construction, that the Management Contractor's control of his sub-contractors is of a lesser standard than might have been hoped for!

The Management Contractor's remuneration generally consists of two elements: a *cost-reimbursable* element and a *fee* element.

The *cost-reimbursable* element, fairly obviously, covers the contractual payments made to the various Works Contractors in the course of their contracts with the Management Contractor. If, as is sometimes the case, the Employer withholds a certain proportion of this element in the form of retention, the Management Contractor will generally withhold a similar amount from the Works Contractors' payments. Additionally, the Management Contractor will be reimbursed any costs incurred in providing services – access roads, tower cranes etc. – for *common* use by the Works Contractors.

The *fee* element, as in the previous case, covers the provision of the planning and management services, the Contractor's overheads and his profit. The fee will also carry an allowance to reflect the risk carried by the Contractor as a result of his contractual links with the Works Contractors and will, as a consequence, be significantly greater than the corresponding fee for a Construction Management contract.

Management Contracts have been extensively used in the UK, since the early 1980's, by the building industry but have, so far, been rarely used for civil engineering Works.

(c) Design and Management Contracts

Such a contract is an extension of a Management Contract (see above) in which the Management Contractor also takes contractual responsibility for the project design.

An initial 'scope design' is carried out, either by the Employer's staff or by an independent designer, and forms the basis on which tenders are invited. Following his appointment, the Management Contractor either carries out the detailed design 'in-house' (if he has the necessary facilities and expertise) or sub-lets it to an appropriate organisation which, in most cases, will provide a facility whereby the Employer is able to monitor the work as it progresses. The arrangements regarding the Contractor's remuneration are similar to those for a Management Contract.

From the Employer's standpoint, the advantages accruing from such a contract are much the same as those associated with a Package Contract.

Although Design and Management contracts have been rarely used in the UK by either the building or civil engineering industries, they have been used on several occasions for building Works overseas.

Apart from the *specific* advantages associated with the *particular* forms of management-type contracts, the following advantages are often cited in connection with the use of management systems *in general*:

(a) a more extensive overlap of the design and construction phases than would be possible with a conventional contract;

(b) better control of design changes and amendments as a result of management continuity throughout the design and construction phases;

(c) improved 'buildability' resulting from the Contractor's involvement at the design stage;

(d) better packaging of the construction work, particularly on large and complex projects, to suit the capabilities of the Works Contractors;

(e) a reduction in the claims potential of the construction process due to the identity of interest between the Employer and the Contractor;

These advantages, however, must be viewed against the commonly–cited *disadvantages*:

(a) a complex and involved tender–appraisal procedure (see section 7.6);

(b) the absence of an *overall* tender price at the outset, that being unavailable until such times as the Works Contracts have been let;

(c) the absence of 'standard' Conditions of Contract; whilst individual Management Contractors generally have their own 'standard' Conditions, the inevitable difference between such tends to *further* complicate the appraisal of tenders;

(d) Management Contractors sometimes lack the *specialist* knowledge required to manage multi–disciplinary projects;

CHAPTER 6

CONTRACT DOCUMENTATION

The Contract Documents, which are prepared by the Engineer, constitute the basis on which tenders will be compiled and, with the notable exception of the *Instructions to Tenderers*, will, collectively, form a legally binding contract, whereby the Contractor agrees to construct the Works under the supervision of, and in accordance with details supplied by, the Engineer, and the Employer agrees to pay for them in the prescribed manner. This being the case, it is essential that the documents detail, in a *clear* and *unambiguous* fashion, *all* the project requirements together with the rights and responsibilities of the parties involved in the execution of the contract.

Although the coverage hereunder specifically relates to an examination of the form and purpose of the documents appropriate to a *Bill of Quantities* type contract executed in accordance with the *I.C.E. Conditions of Contract* (in recognition of the fact that the majority of contracts for Works of civil engineering construction tend to be of this type), it should be noted that much of the documentation is common, in character if not in specific detail, to all types of contract.

6.1 Instructions to Tenderers

The purpose of such instructions is two-fold, namely: to direct and assist tenderers in the preparation of their tenders and to ensure that the tenders are presented in the form required by the Employer and/or his Engineer.

Whilst the instructions will vary from project to project, from

Employer to Employer, and from Engineer to Engineer, it would be reasonable to expect some or all of the following to be included:

(a) a statement of the scope of the contract, giving a brief description of the Works and stating the date by which they are required to be completed;

(b) notes drawing attention to any unusual site conditions (e.g. access restrictions, limitations on working hours, details of any other work being carried out on the site etc.) which might affect the tenderers' costs and, hence, the tenders;

(c) notes drawing attention to any *specific* construction methods or materials required;

(d) notes drawing attention to any *special* Conditions of Contract (see section 6.5) applicable; in the case of a contract executed under the I.C.E. Conditions of Contract, particular reference should be made to the applicability (or otherwise) of the *Contract Price Fluctuations* clause (see section 5.2);

(e) instructions regarding pre-tender site visits for inspection purposes;

(f) an indication of whether or not tenders based on alternative proposals will be considered and, if so, the conditions under which they may be submitted;

(g) instructions on the completion of the tender Bill of Quantities (e.g. that all rates and extensions should be entered *in ink* and that the Bill should be *totalled*);

(h) details of the *Performance Bond* (see section 6.8);

(i) a note of any *specific* insurance requirements;

(j) details of the documents required to be submitted with the tender, for example: a tender programme, a general description of the methods and arrangements proposed by the tenderer to execute the contract and a signed statement certifying that the tender has not been produced *collusively*; in the case of 'open' tendering (see section 7.1), the tenderer will usually be required to submit a statement of his legal/financial status along with details of his previous experience with Works of a similar nature to those in question; in the case of 'selective' tendering (see section 7.2), these details will be required at an earlier stage; a *Tender Bond* (see section 6.2) may also require to be submitted with the tender;

(k) notes on submission of tenders (place, date, time of submission) along with warnings that alterations to the contract documents, improperly completed Bills of Quantities, or 'qualifications' to tenders (see section 3.2.1) could result in rejection thereof;

(l) details of the procedure for dealing with errors in tenders (see section 7.3.2);

(m) a reminder (re-iterated in the *Form of Tender*) that the Employer is not obliged to accept the lowest or, indeed, *any* tender;

It bears repeating that, notwithstanding the importance of such, the *Instructions to Tenderers* do not form part of the Contract Documents *per se* in that, from the standpoints of business efficacy and compliance with the requirements of Contract Law, they are not *fundamental* to the formation of the Contract.

Particular aspects of this document are discussed in greater depth in section 7.3.1.

6.2 The Form of Tender

The *Form of Tender* (see Figure 6.1) is addressed to the Employer and constitutes the tenderer's formal written offer to execute the Contract in accordance with the Contract Documents and within the period stated in the *Appendix to the Form of Tender* (see below).

Although the sum of money for which the Contract will be executed, the *Tender Total*, is central to the issue, in that the tender will be accepted or rejected *largely* on that basis, the Form of Tender makes no mention of a *particular* sum but merely refers to '... such sum as may be ascertained in accordance with the .... Conditions of Contract ...'. This reflects the fact that, with an *admeasurement* contract (see section 5.1.2), the final *Contract Price* (which the Tender Total would effectively become, *were* it to be quoted in the Form of Tender) will inevitably differ from the *Tender Total* and, by definition, will not be available until the Works have been completed.

It should also be noted that the Form of Tender pays due attention to the tenets of Contract Law by including a statement to the effect that *acceptance* of the tender (the *offer*) will constitute a *binding*

All Permanent and Temporary Works in connection with ....

....

Form of Tender

To...

GENTLEMEN

Having examined the Drawings, Conditions of Contract, Specification and Bill of Quantities for the construction of the above-mentioned Works (and the matters set out in the Appendix hereto), we offer to construct and complete the whole of the said Works and maintain the Permanent Works in conformity with the said Drawings, Conditions of Contract, Specification and Bill of Quantities for such sum as may be ascertained in accordance with the said Conditions of Contract.

We undertake to complete and deliver the whole of the Permanent Works comprised in the Contract within the time stated in the Appendix hereto.

If our tender is accepted we will, when required, provide two good and sufficient sureties or obtain the guarantee of a Bank or Insurance Company (to be approved in either case by you) to be jointly and severally bound with us in a sum equal to the percentage of the Tender Total as defined in the said Conditions of Contract for the due performance of the Contract under the terms of a Bond in the form annexed to the Conditions of Contract.

Unless and until a formal Agreement is prepared and executed this Tender, together with your written acceptance thereof, shall constitute a binding Contract between us.

We understand that you are not bound to accept the lowest or any tender you may receive.

We are, Gentlemen,

Yours faithfully,

Signature   ....

Address   ....

                 ....

                 ....

Date......................

Figure 6.1 The Form of Tender

contract between the two parties *notwithstanding* the subsequent signing of a *formal agreement*, and a statement reminding the tenderer that the Employer is not obliged to accept the lowest, or any, tender.

The *Appendix to the Form of Tender*, which forms part of the tender, contains provision for the insertion (by the Engineer, prior to tenders being invited, unless otherwise specified) of the following details:

(a) *Amount of Bond*: in accordance with Clause 10 of the I.C.E. Conditions, this must not exceed 10% of the *Tender Total*; in cases where the Employer is satisfied with the Contractor's previous performance record, and his legal and financial status, a Bond may not be required;

(b) *Minimum Amount of Insurance*: generally calculated at 25% of the *Tender Total* or £500,000, whichever is the greater;

(c) *Time for Completion*: may be inserted by the Engineer or the Contractor; generally the former option is the more satisfactory since it simplifies the comparison of tenders by the Engineer; if, for any reason, *no* completion time is inserted, Equity (see section 1.2.3) requires the Contractor to be given a *reasonable* time to complete the Works; in such a case, any 'liquidated damages' clauses will be voided since there is no *specific* date from which to calculate the damages; provision also exists for inserting the completion time for specific sections of the Works;

(d) *Liquidated Damages for Delay*: the Employer's *genuine* estimate of the pecuniary losses he will incur (per day/week) as a result of the Contractor's failure to complete the Works within the required period; the specified amount of damages will usually cover interest on borrowed capital, additional costs resulting from prolonged site supervision and loss of revenue from the finished product; in accordance with the requirements of Contract Law, not to mention Clause 47 of the I.C.E. Conditions, the specified amount of damages must not be *punitive*;

(e) *Period of Maintenance*: the period, following *substantial* completion of the Works, during which the Contractor is required to remedy any defects which have come to light – generally twelve months for a typical completion time of two years;

(f) *Vesting of Materials not on Site*: a list of goods and materials, for incorporation into the *Permanent* Works, for which payment will be made subsequent to their purchase by the Contractor but prior to their delivery to the site;

(g) *Standard Method of Measurement adopted in the preparation of Bills of Quantities*: details of any method, *other* than the *Civil Engineering Standard Method of Measurement*, which has been used;

(h) *Percentage for adjustment of Prime Cost Sums*: inserted by the tenderer to cover overheads and profit in connection with the performance, by *Nominated Sub–Contractors*, of work items payable on a Prime Cost (see section 8.2.4) basis; this percentage will apply to any Prime Cost Sum for which there is no *individual* percentage adjustment stipulated in the Bill of Quantities;

(i) *Percentage of the Value of Goods and Materials to be included in Interim Certificates*: the Contractor is paid the stated percentage (generally around 80%) of the value of goods or materials, on– or off–site, purchased by him for, but not yet incorporated into, the *Permanent* Works;

(j) *Minimum Amount of Interim Certificates*: generally calculated at around 50% of the *average* monthly value of the contract as estimated by the Engineer prior to tenders being invited; monthly valuations below this figure are carried over into the following month's valuation;

As mentioned in section 6.1, a *Tender Bond* may require to be submitted with the Tender. Broadly speaking, such a bond is the tenderer's undertaking to maintain his tender *unaltered* and *capable of acceptance* during the period of validity stated therein, and to accept any contract which may be awarded him on the basis of such, in default of which a stated sum of damages will be payable to the Employer.

6.3 The Drawings

The drawings are prepared by the Engineer to provide a pictorial representation of the nature and scope of the required work.

For obvious reasons, it is impossible to lay down hard and fast rules as to what drawings are required for any given project (other than to state that a *Site Plan* and/or *General Layout* drawing is essential) and as to what any given drawing should, or should not, contain. As a general rule, however, it is sufficient to state that the drawings should:

(a) be *mutually compatible* (a requirement which is all too often overlooked!);

(b) be suitably *referenced* in accordance with a *logical* and *clearly understood* system;

(c) contain *sufficient* dimensions (in *usable* and *clearly defined* units) and explanatory/descriptive annotation (both *legible* and *free from abbreviation*) to enable *accurate* quantities to be taken–off and to enable the Works to be *correctly* set–out;

(d) *not* contain annotation to such an extent that clarity – a prime requirement for any drawing – is diminished; if necessary, the drawings may be accompanied by appropriately cross–referenced *schedules* such as those commonly used in connection with steel reinforcement (Bar Bending Schedules) or manholes;

(e) be at a *suitable* (and *commonly–used!*) scale which is sufficient to show the appropriate level of detail whilst maintaining the requisite degree of clarity;

(f) be drawn, where appropriate, in accordance with *accepted* conventions (i.e. *not* according to a convention which is invented, and, therefore, only fully understood, by the draughtsman!) such as those detailed in *BS 308 (Engineering Drawing Practice)* or *BS 1192 (Construction Drawing Practice)*;

Particular care should be taken to identify, and distinguish between, *existing* and *proposed* Works, to provide *full* information relating to ground conditions (usually by reference to Borehole Logs and the like), and to indicate (where appropriate) the level and extent of bodies of 'open' water, which may affect the Works, and the locations and levels of existing sewers and services. In including such details on the drawings, it should never be forgotten that the drawings are *legal documents* and, in the eyes of the law, an error (innocent or otherwise) in depicting existing Works and conditions can amount to *misrepresentation* (see section 3.4.2).

Ideally, the drawings should be *complete*, and should contain *full* details of *all* the Works, at tender time. For various reasons, however, this might not be practicable. In such a case, the *Tender Drawings* (which, in any case, should be clearly marked as such and retained for record purposes) should be sufficiently comprehensive as to enable tenderers to understand *exactly* what is required, and sufficiently detailed as to enable the compilation of *accurate* (or, at least, *realistic*) tenders.

In view of the fact that the drawings are (by and large) compiled by human beings, it is unlikely in the extreme that those issued at the outset of the contract will contain *all* the information neccesary for the completion of the Works, or be *exactly* correct in *every* detail. This being the case, the Engineer will inevitably find it necessary, during the course of construction, to issue additional drawings or to revise the originals. This should be expedited without undue delay, as and when the need arises, to permit *uninterrupted* completion of the Works and to avoid the possibility of claims being submitted by the Contractor for delays in the flow of information. In order to avoid possible confusion arising over the issuance of *revised* drawings, which is all too often the case in instances where revision after revision to a particular drawing appears with alarming rapidity, it is good practice to include on such:

(a) the revision *number*;

(b) the *date* of the revision;

(c) *brief details* of the revision;

(d) dates, numbers and brief details of *all previous revisions*;

Finally, it is worth noting that, in the absence of any special agreement with the Employer, the Engineer is *automatically* invested with the copyright of drawings prepared by him, such copyright being infringed in the event of any *unauthorised* copy being made of the drawings or *substantial part* thereof.

6.4 The Specification.

Whereas the Contract Drawings show *what* is to be constructed and *where* it is to be constructed, the Specification *fully details*, for the benefit of the Contractor, the required *methods and manner of construction*, the *standard of workmanship* required of him, the *quality of materials* to be used for the Permanent Works and any *special* responsibilities accorded him in addition to those stated within the Conditions of Contract.

Typically, a Specification for Works of civil engineering construction will consist of three sections:

(a) a *General* section – usually containing a brief description of the nature and scope of the Works, a list of the Contract Drawings, details of access to the site and various clauses relating to such matters as the provision of office and accomodation facilities, the provision of site services (electricity, water, sewerage etc.), traffic control and the like;

(b) a *Materials* section – stating the Employer's requirements in respect of the *basic* materials to be used in the Works, making reference to the relevant British Standards wherever possible;

(c) a *Workmanship* section – covering the contract requirements regarding the performance of activities such as site clearance, earthworks, provision and placement of concrete, formwork manufacture, erection and striking, testing of materials etc., referring to the relevant Codes of Practice where appropriate;

The individual clauses within the Specification will generally fall into one or other of the following categories:

(a) *positive* requirements which specify standards which *must* be met, or practices which *must* be adhered to, for example: 'The methods of transporting and placing concrete shall be subject to the Engineer's approval.';

(b) *negative* requirements which expressly *prohibit* certain practices, for example: 'Structural concrete shall not be placed in any part of the formwork without the Engineer's approval.';

(c) statements which are intended to provide *guidance* by setting down optional courses of action, or ranges of acceptability, for example: 'Upon placement, concrete shall have a temperature of not less than $5\,^{0}$C and not greater than $30\,^{0}$C.';

Irrespective of type or character, however, *all* Specification clauses must *de rigeur* be *clear*, *precise*, *unambiguous*, *complete* and, for obvious reasons of accessibility, suitably *grouped* and *referenced*. Additionally, it is important that clauses are regularly reviewed in the light of technological advances and updated or amended accordingly.

Fortunately, many of the problems of Specification compilation may be alleviated by the use of *standard clauses* which are usually available from databases or libraries of standard clauses. Additionally, it is worth noting that many organisations, such as British Rail, have their own Standard Specification. Possibly the best known, and most frequently used Standard Specification is the *DTp Specification for Highway Works* (known far and wide as the 'Brown Book' – something of a misnomer since it consists of seven volumes of specification clauses accompanied by six volumes of guidance notes – so-called to distinguish it from its predecessors: the 'Orange' and 'Blue' Books), which is so well known that it is generally referred to only by title.

In cases where one of the available standard specifications is used for a particular contract, it is almost inevitable that some clauses therein will not be applicable. Conversely, additional clauses may require to be inserted to cover aspects of the work which are not fully covered by the standard document. At the very least, a number of the standard clauses may require modification in the light of the particular circumstances of the project. Such omissions, additions or modifications should *clearly* be noted in the contract documents (usually in a *preamble* to the Contract Specification) and should be suitably cross-referenced with the standard to avoid confusion or conflict. In compiling any additional clauses, or modifications to existing clauses, it should be remembered that the Contract Specification is a *legal* document (which, as such, will be construed by a court in *strict* accordance with the wording thereof) and that the Contractor's tender is deemed to have been priced in accordance with such. Consequently, considerable care should be taken over the exercise bearing in mind that wrongly- or ambiguously-worded specification clauses could result in inadvertent *over-* or *under-*specification, both of which can have serious financial ramifications:

The result of *over-*specification is fairly obvious in that the Employer will end up with a project constructed to a higher standard than absolutely neccessary. Whilst this may be considered desirable from a 'belt and braces' point of view, it should be remembered that he is also paying for the privilege!

In the case of *under-*specification, it is probable that the deficiency would be discovered by the Engineer at an early stage in the proceedings and rectified by the issuance of a *variation order* (see section 9.1.9) amending the Specification. In view of the fact that the Contractor's tender will have been compiled on the basis of the *original* (inadequate) Specification, however, such an amendment would fairly entitle him to an increase in his rates for the work in question.

Since the Contractor would be well aware of the strength of his position in negotiating the appropriate increase, and would undoubtedly take full advantage of the situation, it is almost inevitable that the eventual cost to the Employer will be in excess of that which would have been the case had the Specification been correct at the time of tender!

There is, of course, a distinct possibility of the Specification deficiency not being spotted until the relevant work is underway. In such a case, the Employer would not only be faced with the abovementioned rate increase but might also require to bear the cost of the Contractor carrying out expensive and time–consuming remedial work on those sections of the work completed in accordance with the *original* Specification.

Finally, it should be noted that there are *two* basic types of Specification:

(a)   the *Method* (or *Recipe*) Specification which provides detailed descriptions of the *methods* to be used in producing various components of the Works, thus placing the entire burden of risk for successful performance of the finished product on the shoulders of the Employer;

(b)   the *Performance* (or *End–result*) Specification in which the standard of the finished product is specified in terms of its expected capabilities (strength, durability etc.); the Contractor is required to ensure that his proposed construction methods, and materials, are of a sufficient standard to achieve the specified end–result and, in doing so, takes all the associated risks;

As might be imagined, there are arguments for and against each approach. By virtue of its detailed requirements removing the need for excessive contingency allowances to be built into the Contractor's cost estimate, the *method* Specification brings a degree of certainty to the contract cost. In doing so, however, it introduces an element of undesirable rigidity to the contract by not allowing the Contractor to devise his *own* solutions to construction problems which, by virtue of his practical experience and expertise, might prove to be more satisfactory, at the end of the day, than those specified. Whilst the neccessary flexibility can be provided by using a *performance* Specification, the Contractor's assumption of the entire burden of risk increases the uncertainty element associated with the cost estimate. With these points in mind, it is rare for a contract Specification to be modelled *exclusively* on one or the other type, it being more common for Specifications to comprise a combination of both.

Whilst it is *generally* the case for individual clauses to be of one type *or* the other, it is *occasionally* the case that a particular clause contains *both* a method *and* a performance approach. A prime example of such a clause is Clause 612 (Compaction of Fills) of the *DTp Specification for Highway Works*, which, after a few specified requirements concerning fill compaction in general, contains two separate sub-sections: 'Method Compaction' and 'End-product Compaction'. If 'Method Compaction' is adopted, the Specification details an extensive selection of techniques, each of which is associated with a specific type of fill material and compaction plant. If the resulting compaction is considered by the Engineer to be inadequate, the cost of remedial work is to be borne by the Employer unless the Engineer can demonstrate the inadequacy to be the direct result of the Contractor's failure to adhere to the specified methods – in which case, the Contractor carries the cost. If, on the other hand, 'End-product Compaction' is the order of the day, the choice of compaction method is left entirely up to the Contractor – subject to the proviso that his proposed method *can* achieve the specified degree of compaction *and* that he can demonstrate that ability to the Engineer.

6.5 The Conditions of Contract

In essence, the Conditions of Contract are the 'rules' governing the execution of the contract. Specifically, they define the *powers*, *responsibilities* and *obligations* of the parties *involved in* the contract. Additionally, they detail the *rights* of the parties *to* the contract in that they specify the various courses of action which are open to either party in the event of the other party's failure to discharge his contractual obligations.

Since the Conditions constitute much of the legal basis on which any court ruling would be made in the event of a dispute, it is imperative that the constituent clauses be written with great care so as to be *clear*, *precise* and *unambiguous*. In particular, the use of teminology which might be open to misinterpretation should be rigorously avoided in that a court, in making its ruling on a dispute, will enforce the contract in *strict* accordance with the written terms regardless of whether it is aware that *both* parties understood what was meant at the time. Additionally, it should be remembered that if a court cannot resolve an ambiguity by applying the so-called 'rules of construction' (see section 3.3), the rule of *contra preferentem* will require it to construe the clause *against the party attempting to invoke it.*

Despite the obvious importance of the Conditions of Contract, it is interesting to note that the absence of such does not invalidate an otherwise legal contract (see Chapter 3). It does, however, increase the chances of a dispute arising between the parties in the course of the contract and severely reduces the probability of that dispute being resolved by mutual negotiation. Admittedly, it is always within the power of the aggrieved party to refer the matter to independent arbitration or a court of law but the procedures for such will *possibly* be slow, *probably* be involved and *undoubtedly* be *very* expensive.

Notwithstanding the considerable degree of freedom afforded under the law for the parties to a contract to decide upon their own contractual terms and conditions, the formulation of customised Conditions for each and every construction contract would be, to say the very least, impracticable. Additionally, it should be appreciated that the complexity of modern construction work, and the plethora of legal problems associated with such, requires complicated contractual provisions to cover all forseeable eventualities. This being the case, it has become commonplace for virtually all construction contracts of any substance to make use of standard Conditions of Contract.

There are two *principal* advantages associated with the use of standard Conditions, the first of which is that the standards are formulated and drawn up by independent groups who, collectively, represent the interests of all sides of the relevant industry. Consequently, the standards are, if nothing else, fair to all parties involved in a contract. There are, of course, the inevitable 'gripes' (invariably from Contractors) concerning alleged bias in favour of the Employer/Engineer but these may be discounted for one very practical reason: if substantial bias *was* evident, any advantage accruing to the Employer therefrom would be negated by a 'compensatory' increase in Contractors' tender prices.

The second, and equally important, advantage is that common and frequent usage of standard Conditions enables the contracting parties to become familiar with, and fully understand, the workings of such, thus reducing the likelihood of disputes arising over misinterpretation.

The most common complaint levied against standard Conditions is that it is difficult (if not impossible) to modify individual clauses to suit particular circumstances. Owing to the considerable degree of interdependence of the clauses in most standards, the modification of one clause neccessitates the modification of several others which, in turn, requires others to be modified etc. etc. In order to avoid opening this 'Pandora's Box', therefore, it is generally considered unwise to modify individual clauses, the practice of which will, with the best will in the world, inevitably lead to uncertainties and disputes. It might also be mentioned that such a practice makes the task of the tenderer more difficult than it would otherwise be in that, in addition to the *normal* problems of compiling a tender within the restricted and relatively short time period available, he is faced with assessing the financial ramifications of unfamiliar contract conditions.

In this light, it is generally recommended that, if the circumstances are such that *special* conditions are *essential*, the use of *additional* clauses is preferable to the modification of individual clauses within the standard document. In drafting these conditions, it is fairly obvious that particular care should be taken to avoid ambiguity (if neccessary, take legal advice!) and (perhaps more importantly) to avoid conflict with the standard Conditions or any other of the contract documents. It is, of course, mandatory for the tendering Contractors' attention to be clearly drawn to any amendments or additions (see section 6.1).

The most common standards used in the construction industry are listed hereunder:

The Institution of Civil Engineers Conditions of Contract and Forms of Tender, Agreement and Bond for use in connection with Works of Civil Engineering Construction (commonly known as the *I.C.E. Conditions of Contract*) is issued jointly by the Institution of Civil Engineers, The Association of Consulting Engineers and The Federation of Civil Engineering Contractors and is the most frequently used standard for civil engineering Works carried out in the UK. Although it is applicable to all civil engineering Works, and may, on occasions, be used for building Works, or electrical or mechanical Works included in a building or civil engineering contract, it is restricted to contracts executed on an *admeasurement* basis using a *Bill of Quantities*. In view of the common usage of such (on around 85% of civil engineering contracts), the I.C.E. Conditions of Contract are covered in depth in Chapter 9.

The General Conditions of Government Contracts for Building and Civil Engineering Contracts (GC/Works/1), published by H.M. Stationary Office, is extensively used by all central government departments for both civil engineering and building contracts. Unlike the majority of other standards, GC/Works/1 is produced solely by the employing agency and not by a joint body representing the interests of the various parties involved in such contracts. A contract executed in accordance with GC/Works/1 generally operates as a *lump sum* contract, with errors in description or quantity being dealt with as variations. Where a *Bill of Quantities* or *Schedule of Rates* is provided, however, the contract will operate as an *admeasurement* contract. A shorter version of this standard, *GC/Works/2*, exists for use in connection with minor Works but, unlike GC/Works/1, has no provision for Bills of Quantities.

The *JCT Standard Form of Building Contract (JCT80)* is probably the most extensively used of all the standards and is, in effect, the building equivalent of the I.C.E. Conditions. It is published by the Joint Contracts Tribunal, representing *inter alia* the Royal Institution of British Architects, the National Federation of Building Trades Employers and the Royal Institution of Charters Surveyors, and comes in separate local authority and private editions, each of which has three versions according to whether a contract makes use of quantities, approximate quantities or does not make use of quantities at all.

The *Conditions of Contract (International) for Works of Civil Engineering Construction*, prepared and issued jointly by the Fédération Internationale des Ingénieurs–Conseils (F.I.D.I.C.) and the Fédération Internationale Européene de la Construction (F.I.E.C.), is the standard document for civil engineering contracts having an international element. The printed document, which contains Forms of Tender and Agreement, is presented in three parts: Part I contains the general Conditions of Contract and substantially follows the I.C.E. Conditions; Part II consists of notes to assist in the preparation of special clauses relating to matters arising out of Part I, such as the ruling language and law governing the contract execution, which are to be specifically agreed between the contracting parties; Part III contains conditions particularly relating to dredging and reclamation work.

Other standard documents which may be encountered, albeit to a lesser extent, are: the *I.Mech.E./I.E.E. Model Forms of Conditions of Contract*, which are designed for use with contracts for the supply and erection of apparatus, machinery or plant, both in the UK and overseas; the *Institution of Chemical Engineers Model Form of Conditions of Contract for Process Plants* for the supply and erection of process plants in the UK; the *British Electrical and Allied Manufacturers Association (B.E.A.M.A.) Conditions of Contract* for use when quoting for the supply of equipment or plant (not necessarily electrical, as the name might suggest) with or without supervision of erection;

Finally, it should be noted that, for use with sub-contracts to a main contract which is executed under the I.C.E. Conditions of Contract, there exists the *Form of Sub-Contract designed for use in conjunction with the I.C.E. General Conditions of Contract* published by the Federation of Civil Engineering Contractors.

6.6 The Bill of Quantities

As detailed previously (see Section 5.1.2), the Bill of Quantities is, in essence, a list of the constituent items of the Works with each item being fully described and quantified. Against each item, the tenderer is required to insert a *unit rate* (the price which he wishes to be paid for completing each unit of the appropriate work) and *extended total* (the product of *quantity* and *unit rate*), the aggregate sum of the latter being the *tender total*.

In view of the importance of this document, its content and structure are covered in depth in Chapter 8 along with details of the method of its compilation.

6.7 The Form of Agreement

Under the Law of Contract, the formation of a valid contract (see section 3.2) requires agreement between the parties regarding the purpose of the contract and the rights and obligations it will create. It is sufficient, under contract law, for this agreement to be evidenced by the *unconditional acceptance* of an *unqualified offer*.

This notwithstanding, however, it is often advantageous or neccessary to confirm the contract by drawing up a formalised written Agreement between the parties. Apart from providing a means of emphasising the formality and importance of the contract, a formalised Agreement may be used to summarise the highlights of a complex contract and to schedule the various items of contract documentation. More importantly, the existence and character of such a document has a considerable bearing on the *limitation period* (see section 3.6.7) of the contract. If the contract is *not* confirmed by a formalised written Agreement, the law regards it as a 'simple' contract which has a limitation period of *six* years. Conversely, a formal Agreement containing the *signatures* and *seals* of the contracting parties confirms an 'under seal' or 'deed' contract which has a *twelve*-year limitation period. It should be noted that, in the case of a written agreement containing only the *signatures* of the contracting parties, the contract is classified as a 'simple' contract executed 'under hand'.

For fairly obvious reasons, all construction contracts of any substance are drawn up 'under seal' and are formalised in a standard Agreement. In the case of a contract executed under the I.C.E. Conditions of Contract, the Form of Agreement, whereby the Contractor agrees to construct the Works in conformity with contract provisions and the Employer agrees to pay for them, is shown in Figure 6.2.

6.8 The Form of Bond

There are four main types of Bond commonly used in connection with engineering contracts: a *Tender* Bond, a *Repayment* Bond, a *Plant Performance* Bond and a *Performance* Bond.

A *Tender Bond* (see section 6.2), if required by the Employer, is enclosed with a tender and constitutes the tenderer's promise to maintain his offer open to acceptance for a specified period of validity and to accept any contract awarded him, on the basis of his tender, during that period. If called for in the form of a *guarantee*, the tenderer must produce a similar promise from his bank, insurance company or other *surety* acceptable to the Employer. In the event of the tenderer's default, he (or his surety, if appropriate) undertakes to provide the Employer with financial compensation to the full extent (typically around 1% of the tender price) specified in the Bond.

In contracts in which the Employer agrees to pay a certain proportion of the contract price in advance of performance by the Contractor, a *Repayment Bond* may be called for whereby the Contractor (or his surety, if the bond is in the form of a 'guarantee') undertakes repayment, up to the maximum amount specified in the bond, of any proportion of the advance payment which, in the event, remains unearned.

In the case of a contract for the supply and installation of an item of Plant, it is common for the Employer to require a guarantee of the Plant's successful performance following installation. In the event that the Plant fails to perform as specified, the terms of the contract will generally require the Contractor to forfeit specified sums of money, dependent upon the degree of failure of the Plant, from the contract price due to him. It may well be the case, however, that the

---

Form of Agreement

THIS AGREEMENT made the .... day of ....
19............ BETWEEN ....
of ....
in the County of .... (hereinafter called "the Employer") of the one part and
.... of ....
in the County of ....
.... (hereinafter called "the Contractor") of the other part
WHEREAS the Employer is desirous that certain Works should be constructed, viz. the
Permanent Works

NOW THIS AGREEMENT WITNESSETH as follows:

1. In this Agreement words and expressions shall have the same meanings as are
respectively assigned to them in the Conditions of Contract hereinafter referred to.

2. The following documents shall be deemed to form and be read and construed as
part of this Agreement, viz:

       (a)   The said Tender.
       (b)   The Drawings.
       (c)   The Conditions of Contract.
       (d)   The Specification.
       (e)   The Priced Bill of Quantities.

3. In consideration of the payments to be made by the Employer to the Contractor as
hereinafter mentioned the Contractor hereby covenants with the Employer to construct and
complete the Works and maintain the Permanent Works in conformity in all respects with
the provisions of the Contract.

4. The Employer hereby covenants to pay to the Contractor in consideration of the
construction and completion of the Works and maintenance of the Permanent Works the
Contract Price at the times and in the manner prescribed by the Contract.

IN WITNESS whereof the parties hereto have caused their respective Common Seals to
be hereunto affixed (or have hereunto set their respective hands and seals) the day and
year first above written

The Common Seal of ....

.... Limited

was hereunto affixed in the presence of:

                or

SIGNED SEALED AND DELIVERED by the

said ....

in the presence of:

....

Figure 6.2 The Form of Agreement

Employer's potential losses, as a result of the failure, are much greater than the proportion of the contract price which remains unpaid. In such a case, the Employer must recover the neccessary damages, from the Contractor or his surety, by means of a *Plant Performance Bond* drawn up at the outset of the contract.

By far the most common type of Bond used in connection with construction contracts is the *Performance Bond*. This constitutes an undertaking by the Contractor that, in the event of his failure to discharge any of his contractual obligations, he will financially recompense the Employer up to the maximum sum stated in the Bond (typically 10% of the contract price). Since the terms of such a Bond will not only cover the *wilful* default of the Contractor, but will also cover his non-performance as a consequence of bankruptcy or financial failure, it is common practice (for fairly obvious reasons) for such a Bond to be called for in the form of a 'guarantee' by an approved surety.

In the case of a contract executed in accordance with the I.C.E. Conditions of Contract, the Employer's confirmation of the contract is conditional upon the signing of such a bond, an extract from the standard Form of which is shown in Figure 6.3.

It is of particular value to note that, although it is usually taken to form part of the contract documentation, the Bond is, in fact, a *separate* contract whose realisation depends upon the successful (or otherwise) performance of the *main* contract and that the *surety*, in the case of the Bond being called for in the form of a *guarantee*, is *not* a party to the main contract.

WHEREAS the Contractor by an Agreement made between the Employer of the one part and the Contractor on the other part has entered into a Contract (hereinafter call "the said Contract") for the construction and completion of the Works and maintenance of the Permanent Works as therein mentioned in conformity with the provisions of the said Contract.

NOW THE CONDITIONS of the above-written bond are such that if:

(a) The Contractor shall subject to condition (c) hereof duly perform and observe all the terms provisions conditions and stipulations of the said Contract on the Contractor's part to be performed and observed according to the true purport intent and meaning thereof or if

(b) on default by the Contractor the Sureties/Surety shall satisfy and discharge the damages sustained by the Employer thereby up to the amount of the above-written bond or if

(c) the Engineer named in Clause 1 of the said Contract shall pursuant to the provisions of Clause 61 thereof issue a Maintenance Certificate then upon the date stated therein (hereinafter called "the Relevant Date")

this obligation shall be null and void but otherwise shall remain in full force and effect but no alteration in the terms of the said Contract made by agreement between the Employer and the Contractor or in the extent or nature of the Works to be constructed completed and maintained therunder and no allowance of time by the Employer or the Engineer under the said Contract nor any forbearance or forgiveness in or in respect of any matter or thing concerning the said Contract on the part of the Employer or the said Engineer shall in any way release the Sureties/Surety from any liability under the above-written Bond.

PROVIDED ALWAYS that if any dispute or difference shall arise between the Employer and the Contractor concerning the Relevant Date or otherwise as to the withholding of the Maintenance Certificate then for the purposes of this Bond only and without prejudice to the resolution or determination pursuant to the provisions of the said Contract of any dispute or difference whatsoever between the Employer and Contractor the Relevant Date shall be such as may be:

(a) agreed in writing between the Employer and the Contractor or

(b) if either the Employer or the Contractor shall be aggrieved at the date stated in the said Maintenance Certificate or otherwise as to the issue or withholding of the said Maintenance Certificate the party so aggrieved shall forthwith by notice in writing to the other refer any such dispute or difference to the arbitration of a person to be agreed upon between the parties or (if the parties fail to appoint an arbitrator within one calendar month of the service of the notice as aforesaid) a person to be appointed on the application of either party by the President for the time being of the Institution of Civil Engineers and such arbitrator shall forthwith and with all due expedition enter upon the reference and make an award thereon which award shall be final and conclusive to determine the Relevant Date for the purposes of this Bond. If the arbitrator declines the appointment or after appontment is removed by order of a competent court or is incapable of acting or dies and the parties do not within one calendar month of the vacancy arising fill the vacancy then the President for the time being of the Institution of Civil Engineers may on the application of either party appoint an arbitrator to fill the vacancy. In any case where the President for the time being of the Institution of Civil Engineers is not able to exercise the aforesaid functions conferred upon him the said functions may be exercised on his behalf by a Vice-President for the time being of the said Institution.

Figure 6.3 Extract from the Form of Bond

CHAPTER 7

SELECTION OF CONTRACTORS

In deciding upon the method of selecting a Contractor to carry out the Works, the Employer is faced with an *initial* choice as to whether the contract should be *competitively* let – whereby a number of contractors compete against one another to submit the most favourable tender – or whether it should be *non-competitively* let.

If the circumstances surrounding the project are such that the latter option is preferable (assuming, of course, that such a course of action is *open* to the Employer), the requisite letting procedure is that by which a *negotiated* contract (see section 5.4.2) is arrived at and is largely a matter for the parties concerned. If, as will be the case in the vast majority of instances, *competitive* letting is the order of the day, the Employer (usually on the advice of his Engineer) must decide whether to use an *open* tendering procedure – in which virtually any number of tenderers may submit competitive bids – or a *selective* tendering procedure where invitations to tender are sent to a restricted number of suitable contractors.

Whichever approach is adopted will largely depend on the type of Employer. For example, a *private sector* Employer (see section 4.1.2), governed only by his organisation's *memorandum of association*, might consider that the interests of his shareholders would best be served by the *latter* approach. In view of the constant need for public accountability, however, his *public sector* equivalent (see section 4.1.1) may consider the difficulties of justifying such an approach (if you're *right*, nobody *remembers* – if you're *wrong*, nobody *forgets*!) to be too formidable to contemplate and might, therefore, prefer the *former* method.

7.1 Open Tendering

With this method, tenders are invited by *public advertisement*. Notices, containing basic information relating to the project (character, size, location, completion time, approximate value and any other information which might be of value to potential tenderers), are placed in suitable publications, such as the national press, trade journals and the like, with an invitation to prospective tenderers to apply for the tender documents with a view to pricing the Works in competition. Following receipt of the tenders, the subsequent contract will, in the majority of cases, be awarded on the basis of the lowest tender.

Whilst open tendering is generally recognised to be a fair way of offering work, and avoids any suggestion of favouritism being levied against the Employer/Engineer, such advantages are outweighed by the many disadvantages associated with the method.

As a consequence of the 'open' nature of the invitation to apply for the contract documents, a considerable number of applications are likely to be submitted, a proportion of which will inevitably be from contractors who do not have the required experience, or who do not have the requisite technical and financial resources to complete the Works, or who *do* have the neccessary resources and experience but are not seriously interested in submitting a priced bid and are really only applying for the documentation out of curiosity. If complete sets of contract documents are sent to *all* these applicants, a great deal of unneccessary expenditure will be incurred by the Employer.

Whilst the latter category may be deterred by the requirement of a deposit to be submitted with all applications, on the understanding that it will be refunded upon receipt of a *bona fide* tender, some form of initial screening is required to eliminate applicants belonging to the former categories. With this in mind, each application is generally required to be accompanied by a statement of the applicant's financial status (a bank reference or similar) and details of his previous experience in the relevant field. With regard to the latter, it is often helpful for applicants to submit a list of the consulting engineers who have supervised contracts in which they have been previously engaged. In addition to verifying the 'previous experience' statements of the applicants, the consulting engineers can also advise against unscrupulous contractors or 'cowboys'. This screening process, of course, takes time and results in a protracted tendering period which not only delays the commencement and, hence, completion of the Works but might also increase the overall cost thereof.

Even after all the *obviously* unsuitable elements have been eliminated by the initial screening process, the remaining applicants, of which there might be a considerable number, will undoubtedly include a proportion of 'unknown quantities' who have not previously been engaged by either the Employer or the Engineer. Whilst it is generally desirable for the work to be carried out by a contractor of whom the Employer or Engineer has had previous experience, lack of personal knowledge is not sufficient reason to disbar a contractor from tendering if the spirit of open tendering is to be adhered to. Consequently, there is always the risk of unsatisfactory performance in the event of the contract being awarded, by virtue of his submission of the lowest tender, to an 'unknown' Contractor of (subsequently discovered) questionable ability. At the same time, it should also be appreciated that, from the point of view of such a Contractor, the Engineer and Employer are *also* 'unknown quantities', a factor which might merit the inclusion of a substantial contingency in his tender price. There is thus the additional risk of the Employer not getting a truly commercial price for the Works.

Whilst some would hold that the severity of competition induced by the sizeable response to an *open* invitation to tender can only be of *benefit* to the Employer, it should be appreciated that it also results in the 'success' rate of individual tenderers being quite low, with much of their work being carried out in vain. Since *all* tenders, successful or otherwise, cost money to produce, this will result in a rise in contractors' overhead costs which will, inevitably, lead to inflated prices for future work. It might also be suggested that, in stringent economic times, excessive severity of competition could lead to 'cut-throat' pricing where contractors are tempted into 'buying' work at unrealistically low prices (the consequences of which, to the Employer, are along the same lines as those detailed previously in section 5.1) merely to keep themselves 'ticking over'.

In view of the inherent disadvantages of this method of tendering, it is, perhaps, not surprising that the majority of Employers, with the possible exception of certain public authorities and international agencies which must be *seen* to give an equal opportunity for *any* firm to win the contract, tend not to regard it with any degree of favour. If, notwithstanding the disadvantages, open tendering is the order of the day, it is recommended practice for tenderers to be informed, at the time the contract documents are issued, of the number to whom the issue has been made.

7.2 Selective Tendering

The essence of *selective* tendering, which tends to be the approach favoured by the majority of Employers and Engineers, is that an invitation is sent to a *select* list of contractors inviting each of them to tender for a particular contract.

In the case of Works of a *general* nature, the listed contractors are usually chosen on the basis of their general reputation and experience. For *specialist* Works, however, the list is restricted to those contractors having the requisite *specialist* experience and, as a consequence, is generally shorter than would be the case for *general* Works.

For fairly obvious reasons, particular care should be taken over the *number* of contractors invited to tender for any given project. If the list is too *long*, the Employer stands to incur much wasted effort and expense in providing an excessive number of sets of the contract documents. Additionally, an excessively long list gives rise to an increased likelihood of abortive pricing by tenderers, the consequences of which have been discussed previously. Conversely, a select list which is too *short* not only tends to provide insufficient competition (too *little* is just as bad as too *much*) for the procedure to result in a truly commercial price for the project but also gives rise to the possibility of *collusive* tendering on the part of unscrupulous contractors.

With this in mind, the National Joint Consultative Committee for Building (NJCCB), in its *Code of Procedure for Single Stage Selective Tendering (1977)* has published guidelines on the compilation of select lists which suggest the following *maximum* number of contractors to be invited to tender for Works of a *general* nature:

Contract value up to £50,000:	5
Contract value between £50,000 and £250,000:	6
Contract value between £250,000 and £1,000,000:	8
Contract value greater than £1,000,000:	6

For Works of a *specialist* nature, which are likely to have higher tendering costs than Works of a *general* nature, the maximum recommended numbers of tenderers for contracts having values between £250,000 and £1,000,000, and those with values in excess of £1,000,000, are 6 and 4 respectively.

In compiling a select list of tenderers, it should be appreciated that some of the contractors thereon may, for a variety of entirely justifiable reasons connected with their prospective workload, or the nature and character of the Works in question, be unwilling or unable to tender. In view of the importance of having a suitable number of tenderers on the list, and also to avoid incurring the needless expense of providing disinterested parties with voluminous and costly sets of contract documents, it is advisable to make preliminary enquiries, of a larger number of contractors than ultimately required, to establish willingness to tender. In doing so, however, and by way of minimising the possibility of contractors indulging in the practice of 'cover pricing' (submitting an unrealistically high tender *either* to ensure its rejection *or* to ensure that, in the extremely unlikely event of its acceptance, the inconvenience will more than be made up for by the potential financial benefits which will accrue), it should be made clear to prospective tenderers that an unwillingness to tender will not jeopardise their prospects of being included on subsequent select lists. Conversely, it should also be pointed out that willingness to tender does not automatically guarantee inclusion on the select list. In the event that this is the case, and that a 'willing' tenderer *is* subsequently excluded from the list, it is generally recommended, by way of preventing his abortive mobilisation of tendering staff in anticipation of imminent work, that he be notified of his exclusion *immediately* following finalisation of the list.

Despite the obvious advantages of *selective* tendering in comparison to *open* tendering, a noteworthy problem associated with the method is the potential restrictiveness of select lists for *general* work if they are not updated on a regular basis. In order to avoid the situation occurring whereby the same contractors get invited to tender time after time, with newcomers finding it virtually impossible to get a 'foot in the door', a system of 'pre-qualification' is recommended in which contractors are invited, by public advertisement, to apply for inclusion on an Employer's select list. The amount of information requested of applicants will obviously depend on the technical content of the Works in question but might reasonably be expected to be along the same lines as detailed previously in connection with *open* tendering.

In the case of Works of a *specialist* nature, the problem of restrictiveness is much more serious and cannot be alleviated by a straightforward pre-qualification exercise. In this case, a newcomer wishing to break into a particular specialist field finds himself in the invidious position of lacking the requisite specialist experience, thereby making his inclusion on such a list unlikely, but only being able to gain such by being included on a select list *and* being awarded a contract. The position of the Employer or Engineer is *equally* invidious: if the Works are let on the basis of a select list of suitably experienced *specialist* contractors, there is the risk of possible exclusion of an experienced *general* contractor who might, despite his being a newcomer to the particular speciality, subsequently prove to be extremely competent in that field; on the other hand, an Employer or Engineer wishing to remedy the situation by letting the Works on the basis of *open* tendering, or by including non–specialists in his select list, bears a significant risk of unsatisfactory performance – which will probably only be discovered when it is too late! Either way, the situation is less than ideal.

Finally, it should be noted that, since the United Kingdom became a member of the European Community, an opportunity must be given for contractors in *all* the Community countries to tender for work throughout the EC. Consequently, proposed *public* Works contracts, to be let by certain specified authorities, *must* be publicised in the *EC Journal* if they are over a specified value – currently 5 million ECUs (European Currency Units) which (at the time of writing) is equivalent to approximately £3.1 million. The authorities subject to the ruling, which applies equally to *open* tenders *and* invitations to apply for inclusion on approved lists for *selective* tendering, include local and regional authorities, new–town corporations, the Commission for the New Towns, the Scottish Special Housing Association and the Northern Ireland Housing Executive.

Having discussed the principles of the two alternative approaches towards tendering, it is neccessary to consider the various procedures to be undertaken by way of inviting tenders from potential Contractors, appraising the tenders upon receipt and awarding the contract to the successful Contractor. It should be stated at the outset that, by definition, the procedures covered relate entirely to *competitively let* (as opposed to *negotiated*) contracts.

There are *two* basic procedures, the choice of which depends on the extent to which the project design and contract documentation is complete at the time of tender.

The most common procedure used to obtain and appraise tenders for Works of civil engineering contruction applies to the case where the design and documentation is *substantially complete* (subject, of course, to the possibility of *minor* amendments or additions becoming neccessary in the course of the contract) at the time of tender. Since it is undertaken in a *single stage* between the design and construction stages of a project, the procedure is known (rather unsurprisingly!) as the *Single-Stage Tendering Procedure.*

In circumstances where the design and documentation is only *partially complete* at the time of tender, an alternative procedure – the *Two-Stage Tendering Procedure* – may be employed.

7.3 The Single-Stage Tendering Procedure

In essence, the procedure may be summarised as follows:

1. The completed contract documentation (see Chapter 6) is sent to contractors along with an invitation to tender for the Works.

2. The tenderers compile priced bids which are submitted to the Engineer, by the specified submission date, for his appraisal.

3. The Engineer carries out a detailed appraisal of the tenders and submits a report, containing full details of the appraisal along with a recommendation concerning the award of the contract, to the Employer.

4. The Employer makes his decision, usually based on the Engineer's recommendation, as to the selection of a suitable Contractor to carry out the Works.

5. The contract is awarded to the chosen Contractor who is then given the instruction to commence the Works.

These operations are now considered in more detail. In view of the fact that the vast majority of UK contracts for Works of civil engineering construction are of the Bill of Quantities type (see section 5.1.2), and are executed in accordance with the I.C.E. Conditions of Contract, coverage is limited to the appraisal of tenders based on such.

7.3.1 Submission of Tenders

Following substantial completion of the design, the contract documentation is completed and sets of such, accompanied by detailed 'Instructions to Tenderers', are sent to prospective tenderers.

Whilst the number of sets of the contract documents supplied to tenderers will depend on a number of variable factors, such as the complexity of the Works and the extent to which specialist sub–contractors are likely to be involved, it is generally recognised that each tenderer should be supplied with *at least two* sets of documents, one copy of the Bill of Quantities requiring to be priced and returned with the tender.

Although the contents of the 'Instructions to Tenderers' have been previously discussed (see section 6.1), various aspects of such require amplification:

As stated previously, the purpose of such instructions is to draw the tenderers' attention to the conditions which apply to the invitation to tender and the subsequent submission of their tenders. In particular, and by way of enabling the Engineer to make a realistic appraisal of the tenders against a common base, it is essential for the instructions to specify the form in which tenders must be submitted and to make it clear that tenders not conforming to the specified format will be rejected.

This notwithstanding, it is generally recognised that there will be circumstances under which individual tenderers might feel the need to submit tenders based on alternative proposals which they consider might be of benefit to the Employer – and, of course, themselves! Such initiative should not be penalised provided the tenderer:

(a) has also submitted a tender based on the *original* proposal;

(b) has not altered any of the *original* documentation in any way;

(c) has provided sufficient supporting information (drawings, calculations etc.) to facilitate a detailed technical appraisal of his proposal;

(d) has fully described his intentions, and the effect of his proposed changes, such that the Engineer may make a comparison between the tender in question, tenders based on the original proposal and further alternatives which may be submitted by other tenderers;

(e) has given sufficient notice, in advance of the closing date for submission of tenders, of his intention to submit an alternative proposal;

If the Engineer or Employer considers the circumstances to be such that alternative proposals *will* be countenanced, a statement to that effect should be included in the 'Instructions to Tenderers', together with the terms and conditions under which such alternatives will be accepted. Details should also be provided of any special design criteria or requirements which could affect possible alternative designs.

In providing tenderers with detailed instructions regarding the date, time and place for tender submission, the 'Instructions to Tenderers' should make it clear that tenders received after the closing date will be rejected and returned unopened *except:*

(a) if a telex, telegram or 'fax', stating the tender sum, has been received on time;

(b) if there is clear evidence (e.g. a postmark) that the completed tender has been dispatched in sufficient time to pre−suppose its timely arrival;

With regard to the latter exception, it is clearly advantageous for tenders to be sent by registered post or recorded delivery.

It is often the case that, lacking confidence in the postal system, or (more likely) having left the finalisation of his tender up to the last minute, a tenderer prefers to deliver the document by hand. In such a case, he should be provided with a timed and dated receipt as evidence of the timely delivery of his tender.

Irrespective of the method of delivery of the tender, it is essential that the document be dealt with *correctly* upon receipt. To ensure that this is the case, it is strongly advised that the package containing the document is clearly marked: 'tender document − not to be opened until....' and, by way of maintaining tendering confidentiality, does not bear any indication of the name of the tenderer.

Since the relevant *tender period* depends on a number of factors, such as the size and complexity of the project, the extent to which it is envisaged that tenderers will have to rely on quotations being received from specialist sub−contractors before finalising tenders and the urgency for early commencement of the Works, it will vary from project to project. Suffice it to say, however, that it must be of sufficient length to enable tenderers to compile *realistically* priced tenders, based on adequate knowlege of the project, such that the risk of subsequent financial 'embarrassment' (to the detriment of *all* concerned) is minimised. With this in mind, a *minimum* tender period of *four weeks* is generally considered to be reasonable. For civil engineering contracts of reasonable substance, the *average* tender period tends to be around *six weeks*.

It should be remembered that, in accordance with contract law, the tenderer is required to submit an *unqualified* tender. This being the case, it is important that any points of doubt, which the tenderer might have with regard to the compilation of his tender, should be referred to the Engineer for clarification *as soon as possible*. If the issues concerned are comparatively minor, such as those which can possibly be resolved by directing the tenderer's attention to a particular Specification clause or drawing detail, the Engineer's clarification need not be repeated to all tenderers. It should, however, be confirmed in writing to the tenderer concerned. If, on the other hand, a tenderer has unearthed an ambiguity or error in the tender documentation, or has discovered the need for supplementary information, the situation is rather different in that the tender documents might require amendment with all tenderers being notified accordingly. In contemplating this course of action, however, the Engineer should realise that any 'generalised' amendment to the tender documents, subsequent to their having been issued, will be extremely disruptive to the tendering process and will impose a substantial additional burden on tenderers. Consequently, this course of action should *only* be considered in the event that the Engineer considers that a failure to undertake such would *seriously* affect the pricing of tenders. If at all possible, amendments should be postponed until after the award of the contract when they can be dealt with by means of variation orders.

In the context of unwarranted and disruptive interference with the tendering process, a particular recommendation (which is, in effect, a *plea* to Engineers on behalf of all the overworked and over-pressurised contractor's estimators of this world!) is appropriate at this point: under no circumstances should the Engineer, or his design staff, be tempted into the all-too-common practice of making 'minor' changes to the project design during the tender period, and notifying tenderers accordingly, no matter how good the idea seems at the time and no matter how 'minor' the changes appear to be! The aforementioned parties should appreciate that, not having 'lived' with the project for, perhaps, the preceding two years, a tenderer's estimator will require to carry out a full appraisal of the effects of the change, the time expended on which he can ill afford, to determine that the change is, in fact, 'minor'. Furthermore, it should be mentioned that apparently 'minor' changes have a nasty habit, upon detailed examination of the tender-related consequences thereof, of turning into major problems!

It should finally be mentioned that, under *English* contract law (see Chapter 3), the tenderer is at liberty to withdraw, or amend, his submitted tender at any time prior to its unconditional acceptance.

Under Scots Law, the situation is slightly different in that the terms of a tender are generally required to include a 'period of validity' during which the tender is deemed to stand firm and open to acceptance. Whilst the length of this period may vary according to circumstance, a maximum *period of two months is generally deemed appropriate.*

7.3.2 Appraisal of Tenders

Upon opening the tenders, the first task encumbent upon the Engineer is that of checking the arithmetical accuracy of each tender.

In the case of a Bill of Quantities type contract, arithmetical errors are, in the main, likely to consist of incorrect extensions (the products of *quantities* and *unit rates*) for certain billed items, incorrect page totals, inconsistency between page totals and tabulated collections or summaries and even, on occasions, incorrect summation in the Grand Summary (see section 8.2.5).

Since such contracts are invariably of the *admeasurement* variety (see section 5.1.2), it might be argued that this exercise is largely pointless since errors of the type described will *automatically* be corrected when the Works are re-measured and valued for the final account. Whilst this is certainly true, it should be pointed out that, since tenders will *primarily* be compared on the basis of their *Tender Totals*, a meaningful comparison requires these totals to be, at the very least, arithmetically correct. Consequently, any arithmetical errors which are encountered should be corrected and the relevant tenderers notified accordingly.

Having completed this exercise, the Engineer will generally make a straightforward comparison of the *corrected* tender sums with a view to rejecting *particularly low* tenders.

Such a course of action is based on the reasonable philosophy that an *unusually* low tender is likely to emanate either from an inexperienced contractor or from an experienced contractor 'buying' work. Either way, the tenderer, if awarded the contract, is likely to incur significant losses in carrying out the work which is undesirable from the point of view of both parties to the contract (see section

5.1).

Following rejection of any such tenders, the lowest three or four of the remaining tenders are selected for detailed appraisal. As soon as possible following this selection, with the principal aim of being fair to any tenderer who might be holding resources available in anticipation of his tender being accepted, the Engineer should notify tenderers individually as to whether or not their tenders have been shortlisted for closer examination. In doing so, it should be made clear to the relevant tenderers that this notification does *not* constitute a 'Letter of Intent' (see section 7.3.5).

The detailed appraisal consists of a check on the selected tenders for *technical errors* which, in the main, are likely to consist of *missing rates* or *incorrect billing*, and *financial anomalies* consisting, by and large, of unusually high or low unit rates for particular Bill items.

Although *technical errors* of the type described tend generally to be the result of *genuine* mistakes on the parts of tenderers, it should be remembered that there is no protection in contract law for such (see section 3.4.1). Strictly speaking, therefore, a tenderer is *legally* obliged to stand by his tender, mistakes included, if it is accepted. There is, however, little mileage to be gained from accepting, and legally obliging a tenderer to stand by, a tender containing genuine mistakes which affect the tender sum to an extent whereby his subsequent financial 'embarrassment' becomes a distinct possibility. Consequently, such errors, if discovered, should be resolved prior to final selection being made. The method by which such errors are to be resolved should be decided upon prior to tenders being invited and full details of such should be included in the 'Instructions to Tenderers'.

In instances of incorrect billing, it is common practice for the tenderer in question to be notified of the error of his ways and to be given the option of confirming his tender, as it stands, or withdrawing it completely. If, as is sometimes the case, the billing error is common to all tenders, and is the result of an ambiguity or error in the contract documentation, the Engineer may deem it neccessary to invite new tenders, with all tenderers being informed of the reasons for so doing, on the basis of the corrected or suitably clarified documentation. In view of the serious consequences (e.g. delayed commencement of the Works, possible breach of tendering confidentiality, considerable 'hassle' and expense etc.) of such a course of action, however, he may alternatively prefer to limit his action to noting the source of the error with a view to its correction by means of a variation order (see section

9.1.9) issued subsequent to the award of the contract.

In the case of missing rates being discovered, the tenderer may be given the additional option of accomodating the omission by an adjustment to his tender sum on the understanding that subsequent appraisal of his tender will be based on the *revised* sum. If the tenderer agrees to this course of action, the priced Bill of Quantities should contain an endorsement, signed by *both* parties to the contract, indicating that all rates and prices inserted *by the tenderer* are to be considered as adjusted by the (stated) percentage that the overall adjustment constitutes of the original total of priced items. Alternatively, in the case of a Bill drawn up in accordance with the *Civil Engineering Standard Method of Measurement*, the error may be accomodated by increasing the 'adjustment' item in the Grand Summary of the Bill (see section 8.2.5). As above, the amendment should be appropriately endorsed by *both* parties to the contract.

Before undertaking this course of action, however, the Engineer should satisfy himself that the omission is *inadvertent* and not *intentional* – as would be the case where, for example, the price for the relevant item has been 'included in' that for a related, and possibly more substantial, item.

The approaches of tenderers towards pricing contracts are many and varied and, more to the point, highly *individual* in that a given tenderer's pricing strategy will depend upon *his* perception of the contract as a commercial venture to be undertaken in *his* particular commercial circumstances. Consequently, it is extremely difficult for the Engineer, in the event that the detailed *financial* appraisal of tenders reveals bill rates which, in *his* opinion, do not realistically reflect the work content of the associated items, to decide whether or not a *genuine* mistake *has* been made.

In a case where, for example, three out of four *identical* 'provision of concrete' items in different sections of the bill are priced at £65.00 per cubic metre, but the fourth item is priced at £6.50 per cubic metre, it would not be unreasonable to suspect that a genuine mistake (probably typographical) *has* been made and that a negotiated correction, along the same lines as indicated above, is in order.

Before undertaking such a course of action, however, the Engineer should exercise careful judgement as to whether he should seek to change a tendered rate in one tender, because of an error he considers to be genuine, whilst there might be equally genuine errors, of which he is unaware, in other tenders. To negotiate in these circumstances would give rise to a breach of tendering parity which would tend to undermine the whole spirit and purpose of competitive tendering. A negotiated rate correction, therefore, should *only* be considered where the Engineer is satisfied that parity of tendering will *not* be compromised.

Whether or not the evidence is as clearcut as in the above example (generally speaking, it is *not!*), the Engineer should not overlook the possibility of *extraordinarily* high or low rates being *intentionally* inserted by the tenderer to 'load' the Bill:

The most commonly encountered type of rate loading is 'front-end' loading whereby the prices of items scheduled for completion in the early stages of the contract are inflated to improve the Contractor's cash flow at a time when it is generally at its worst. In order to keep the tender sum at a competitive level, the prices of later items are compensatorily decreased.

The opposite of 'front-end' loading is 'back-end' loading, a technique whereby the prices of later items are inflated at the expense of earlier items. In the event of an index-linked price fluctuation clause being in operation (see section 5.2), such a technique seves to enhance the Contractor's receipts from inflation-related price fluctuations in the course of the contract. Since it will also worsen the Contractor's cash flow in the early stages of the contract, its use is advisable (from a *Contractor's* point of view) only in times when the 'cost of money' (i.e. the prevailing interest rate) is *less* than the rate of inflation. As a consequence of such instances being comparatively rare − 1975/76 being the only previous occasions on which, to the best of the author's knowledge, this situation obtained − it is fairly unlikely that this type of loading will be encountered.

Where a tenderer suspects that certain work items have been under-quantified in the Bill, he may artificially inflate, or 'spot' load, the rates of such in anticipation of receiving a 'windfall' in the event of re-measurement resulting in a substantial increase of the relevant quantities.

It goes without saying, of course, that any financial advantage accruing to the Contractor from rate loading is gained, by definition, at the expense of the Employer. If the Engineer suspects the use of 'loaded' rates, therefore, he should take the matter into account in his recommendation to the Employer (see section 7.3.3) concerning the latter's selection of a suitable Contractor to carry out the Works.

Whilst he may be tempted into instigating a negotiated rate revision, by way of spreading the Employer's expenditure more evenly over the contract period, the Engineer should bear in mind the very real possibility of the rates in question having been entered for *genuine* commercial reasons, of which he is unaware. That aside, however, it might be suggested that the practice of negotiating changes to tendered rates (other than those which have *clearly* been entered in error) creates what is, in effect, a *two-stage* tendering situation which might, if the practice became widespread, encourage the belief amongst tenderers that their tenders were little better than preliminaries to further negotiation.

7.3.3 The Engineer's Report

Following completion of the appraisal exercise, the Engineer compiles a formal report to the Employer which summarises the results of the detailed tender analysis and contains a recommendation, based on such, as to the latter's eventual selection of a suitable Contractor to whom to award the contract.

Fairly obviously, the length and scope of the report will vary according to circumstance. Whilst it may only extend to a few pages in the case of a small contract, a larger more complex project might neccessitate the compilation of a highly detailed report running into several volumes. Whichever is the case, however, it is *essential* that the report presents a *clear* and *cogently reasoned* case, using minimal 'technicalese', for the Employer's acceptance of the recommended tender.

As a consequence of the variability of circumstance, it is difficult to lay down hard and fast rules as to the format of the report. It is generally recognised, however, that the Engineer's report should contain the following:

(a) a tabular presentation of the salient features of *all* tenders submitted (e.g. tenderer's name, tender sum prior to any *arithmetical* corrections having been made, *corrected* tender sum, proposed completion date, validity etc.) by way of enabling the Employer to gain an overview of the response to his invitation to tender;

(b) details of any tenders which were rejected in the initial screening exercise, together with the reasons for so doing;

(c)   a concise summary of the detailed analysis of the main contenders, including details of any negotiations undertaken with the tenderers and any consequent tender corrections;

(d)   the Engineer's recommendation of (what *he* considers to be) the most acceptable tender, drawing on the conclusions of the detailed analysis;

(e)   a comparison of the recommended tender with the Engineer's pre-tender estimate;

(f)   a recommendation, if appropriate, concerning the possibility of further negotiation (see section 7.3.4);

(g)   a statement of the Employer's funding requirements, based on the recommended Contractor's prices and tender programme, over the contract period;

A particular point should be made in regard of the Engineer's recommendation concerning the award of the contract. Whilst the Employer will naturally be inclined to favour the *lowest* tender, since it is *he* and *not* the Engineer who is paying out the money, the Engineer should not hesitate to recommend a tender *other than the lowest* if he feels it would not be in the Employer's best interest to recommend the *lowest*. For example, a higher-priced tender may be preferred to a lower-priced tender by virtue of a shorter proposed completion time. Similarly, a *prima facie* lower-priced tender may be rejected on account of its loaded rates placing a greater *overall* burden on the Employer's funding requirements than would be the case with a slightly higher-priced tender. In such cases, the Engineer should make it *absolutely* clear as to *why* he is not recommending the lowest tender and, to preclude the Employer's being in any doubt as to the responsibility he bears in the event of his overriding the Engineer's recommendation, should make a full presentation of the consequences of accepting the lowest tender.

7.3.4 Negotiations

In view of the fact that Engineers, by and large, do not have the same level of experience and expertise as that of Contractors in the matter of pricing work, it is generally the case that the Engineer's estimate is based on, at best, a broad range of prices gleaned from previous contracts executed under his auspices or, at worst, some form of published database of (highly idealised) construction prices. Consequently it could be, and very often *is*, the case that the Contractor's tender, priced with the commercial viability of the *particular* contract in question very much in mind, is significantly higher than the Engineer's estimate. (The author is aware of a fairly recent contract for which the Engineer's estimate, compiled on the basis of a well-known database of construction prices, was some 30% lower than the *lowest* tender. In fairness to the parties concerned, however, it should be mentioned that the Works included a substantial proportion of Temporary Works, the exact nature of which was subject to tenderers' proposals, and that the contract was to be executed in a so-called 'hostile' (economic *and* weather-wise) environment. Whilst this particular example may be considered somewhat extreme, however, it does serve to illustrate the dangers of 'estimating by database'!)

Since the Employer's budget will have been based, to a large extent, on the Engineer's estimate, it may be necessary to recommend, as an infinitely more preferable option to the Employer's rejection of all tenders (which he is quite entitled to do – see section 6.1) followed by a repeat of the tendering exercise on the basis of a suitably revised contract content, the institution of negotiations with a view to reducing the contract price.

Such a reduction may be effected in one of three ways:

(a) the overall scope and content of the contract can be reduced, with the 'new' price being calculated on the basis of the originally tendered rates;

(b) the original scheme remains substantially unchanged with the original tender rates undergoing an 'across-the-board' reduction to fit the Employer's budget;

(c) a combination of (a) and (b) in which the scope of the contract is reduced, by a smaller amount than above, along with certain item rates;

Whichever approach is adopted, however, it goes almost without saying that a satisfactory outcome to the process depends virtually entirely on there being a considerable amount of cooperation between all parties concerned.

In such an event, negotiations are undertaken with the

'recommended Contractor', in the first instance, on the understanding that, if they fail or break down, negotiations will be opened with the tenderer who has submitted the next most favourable tender – and so on.

A particular point to bear in mind when contemplating this course of action is that if the negotiations become excessively protracted, the subsequent award of the contract may be delayed to the extent that the Contractor becomes fairly entitled to a revision of his tendered rates. Since this would serve largely to defeat the whole purpose of entering into negotiations in the first place, it is essential to put a mutually acceptable time limit on individual negotiations – particularly since the Contractor in question will be well aware of the strength of his position and will, not unreasonably, seek to capitalise to the maximum possible extent on such! Furthermore, it is noteworthy that, in the event that the chosen option entails some reduction of the scope of the contract, care should be taken not to reduce the original proposal to the point where the Contractor is entitled to a rate revision on the grounds that the revised project is substantially different from that for which he submitted his original tender – again defeating the object of the exercise.

Finally, it should be noted that *all* negotiations *must* be recorded with the resultant agreements suitably endorsed and appended to the contract documents.

7.3.5 Award of the Contract

Following receipt of the Engineer's report, and subject to the successful outcome of any subsequent negotiations which may have been deemed necessary, the Employer appropriately deliberates on the issue and makes his final choice of Contractor.

The first step in the 'award of contract' process is, fairly obviously, to notify the 'chosen one' that his tender has been successful. This is done by sending him a *Letter of Acceptance* containing words to that effect and informing him that all subsequent instructions will be issued by the Engineer to whom all subsequent communications, concerning the contract, should be addressed. Whilst such a letter will generally emanate from the Employer, since *he* is the party to whom the selected Contractor will henceforth be under contract, it may, on occasions, be issued by the Engineer on behalf of the Employer. In such an event, it is a legal requisite that the Engineer be in possession of *express* (or *reasonably implied*) authority to act thus.

In view of the legal importance of this document (it forms the 'acceptance' part of the 'offer and acceptance' aspect of contract legality), it should make reference to any agreements which have been reached as a result of previous negotiation. In wording this letter, great care should be taken not to include any terms or conditions which might be construed as being 'new', whether or not that is the case, since such inclusions will effectively constitute a 'counter offer' which requires subsequent unconditional acceptance by the Contractor and (more importantly) destroys the original offer.

In the light of the aforesaid, the contract is sealed immediately the Letter of Acceptance is sent out. Consequently, the 'unsuccessful' tenderers should be notified accordingly. Although not strictly necessary in any legal sense of the word, it is customary to provide tenderers (including the successful one) with an indication of their tendering performances for possible use in connection with their future tendering policies. By way of maintaining tendering confidentiality, such details are usually supplied in the form of two lists, one of which contains the names of tenderers in alphabetical order with the other containing a list of the tender sums, with the successful tender sum being suitably annotated, in ascending order of magnitude. Tenderers are thus able to observe their own ranking without being privy to the ranking of their competitors. By way of *further* ensuring the maintenance of tendering confidentiality, presumably bearing in mind the extremely efficient nature of the industry 'grapevine', it is generally the case that tenderers only be provided with 'performance' information in the event that the total number of tenders submitted exceeds four.

Where it is not immediately possible, for a variety of reasons, to issue a Letter of Acceptance, such a document may be preceded by a *Letter of Intent* informing the tenderer that his tender has been successful and that the Employer intends placing the contract with him. Since such a letter will generally contain no express reference to 'acceptance' of the tender, it does not constitute a legal document contract–wise. Consequently, it would not be wise for the Contractor in question to commence work on the basis of a Letter of Intent since any costs incurred by him prior to his receipt of the official 'acceptance' of his tender might subsequently transpire to be irrecoverable. In cases where urgency for commencement of the Works is paramount, the Letter of Intent may be used to instruct the Contractor to commence, in advance of the official acceptance of his tender, his preparations for carrying out the Works. In such circumstances, the Letter of Intent may be legally construed as constituting a binding contract *in itself* and should, therefore, be very carefully and precisely worded.

Following issuance of the Letter of Acceptance, the *Form of Agreement* (see section 6.7), constituting the legal contract between the Employer and the Contractor, is signed and sealed by both parties.

It may well be that, for a number of reasons, the signing of the 'Form of Agreement' is delayed. It is worth noting that this does not, in any way, affect the legality of the contract or the interests of the contracting parties in the interim period. Apart from the 'offer and acceptance' aspect of contract legality having previously been satisfied, thereby giving rise to a legal contract of the 'simple' variety, it bears repeating that the Form of Tender (see section 6.2), used in the case of a contract to be carried out in accordance with the I.C.E. Conditions of Contract, contains a statement to the effect that, in the absence of a formal agreement between the parties, the Letter of Acceptance is deemed to be sufficient for the purpose of forming a binding contract between the two parties.

An important effect of the formal Agreement, which should not be overlooked, is that, by virtue of it requiring the *seals* of the contracting parties to be affixed thereto, its completion adds a further six years to the contract limitation period (see section 3.6.7).

A further document which is generally completed at this stage is the *Form of Bond* (see section 6.8) which requires to be duly 'signed and sealed' by the Contractor and his Surety and forwarded to the Engineer. In some instances, this may be required as a pre-requisite to the issue of the Letter of Acceptance in which case the Contractor will be issued with a Letter of Intent including a statement to this effect.

Following completion of the aforementioned formalities, the Engineer sends the Contractor the formal 'instruction to proceed'. The importance of this document should not be overlooked since it will contain the official starting date of the contract which, when used in conjunction with the stated contract completion period (see section 6.2), forms the basis on which liquidated damages will be calculated in the event of contract completion being delayed.

Finally, it should be noted that neither contract law nor the I.C.E. Conditions of Contract make reference to a *specific* figure in respect of the maximum allowable elapse of time between submission of tenders and commencement of the Works. Clause 41 of the I.C.E. Conditions (5th edition), however, requires that the '... the Date for

Commencement of the Works ...' be '... within a *reasonable* (author's italics) time after the date of acceptance of the Tender ...'. Whilst the obvious implication of such wording is that the Contractor be notified of such date, in the Engineer's 'instruction to proceed', within a *reasonable* time following the acceptance of his tender, no specific remedy is afforded the Contractor in the event that such notification is *unreasonably* delayed. Whilst an aggrieved Contractor always has recourse to the law in the event of an *unreasonable* delay, the matter is generally dealt with in a more straightforward fashion *under the contract*: if the Engineer anticipates that the commencement of the Works will be delayed, for reasons outwith the control of the Contractor and beyond a *reasonable* time following the acceptance of the tender, he will generally issue an order to commence on time and treat any subsequent delay as a delay in the Contractor's being given possession of the site. Under the terms of Clause 42 of the Conditions, such a delay would fairly entitle the Contractor to an extension of the contract period *and* payment of any costs incurred by him as a result of the delay.

Unfortunately, tenderers have little such protection against an Employer procrastinating *unreasonably* in the matter of reaching a decision as to which tender to accept. Whilst a tender may legally be withdrawn at any time prior to its unconditional acceptance, most tenderers are reluctant, for commercial reasons, to do so even though, by way of avoiding the possibility of their becoming over-committed in the event of their tenders being successful, they may be inhibited from tendering for further contracts in the meantime. In cases where the terms of a tender include a specified 'validity period', that tender *automatically* lapses upon expiry of the relevant period and is not thereafter open to acceptance unless it is renewed. In the absence of such, however, a tender will be deemed to have lapsed after a *reasonable* time.

In both cases, the outcome depends on the opinion of the adjudging party (the Engineer in the first case and a Court of Law in the second) of what constitutes a *reasonable* (or *unreasonable*) delay. Little more can be said in that respect other than to suggest that the subsequent decision will depend upon the prevailing circumstances and the adjudging party's perception of 'reasonableness'.

These, then, are the procedural steps to be undertaken by way of *competitively* letting a contract for which the design and documentation are *substantially complete* at the time of tender.

In cases where, for a variety of reasons, the design (and, hence, the documentation) is *incomplete* at tender time, such an approach is (by definition) precluded and the Employer may be left with little alternative other than to award the subsequent contract *non–competitively* following negotiations with a suitable Contractor. Whilst there are a number of advantages (see Chapter 5) to be gained from this approach, the major flaw in such is that, by virtue of the absence of any *competitive* element in the tendering process, the negotiated contract price might well be in excess of that which might be described as 'truly commercial'. If this is considered to constitute a *significant* drawback, an alternative approach would be to let the contract on the basis of *two–stage tendering*, details of which are described hereunder.

7.4 Two–Stage Tendering Procedures

Although the opening paragraphs of this Chapter would suggest the following description to be a contradiction in terms, *two–stage tendering* might best be described as: a *competitive* form of *negotiated* tendering.

Used in situations where the project design is incomplete, or where there exists only a basic or provisional project Specification, at the time of tender, two–stage tendering comes in a number of forms, the two most commonly–used of which are as follows:

Version 1.

First Stage:

Selected contractors are supplied with the Employer's broad requirements concerning the project. They are invited to tender by submitting individual proposals, containing sufficient details to facilitate a meaningful appraisal of the proposals *and* to give clear indications of their intentions. Tenderers are additionally required to price a *notional* Bill of Quantities containing the *approximate* quantities of those items likely to be involved in the construction.

Second Stage:

Following an appropriate appraisal exercise, the Engineer selects the most suitable proposal. He and the Contractor in question then work together to produce a final economic design, based on the original proposal, and construction programme. The contract price is determined on the basis of the item rates contained in the tendered Bill.

Version 2.

First Stage:

An initial proposal is drawn up by the Engineer and is sent to selected contractors for their examination within a specified time. At the end of that time, the contractors meet with the Engineer to comment on, criticise or provide alternatives to, his proposal.

Second Stage:

The initial proposal is suitably revised in the light of the contractors' comments. The contractors are invited to submit competitive tenders for the revised scheme, such tenders to be appraised in the same fashion as that discussed in connection with single-stage tendering.

Although two-stage tendering is a considerably more protracted exercise than that of single-stage tendering, a clear advantage of the former over the latter is that the early involvement of the Contractor enables an earlier start to be made on the construction and, hence, an earlier completion date to be achieved.

That aside, however, it should be appreciated that, where single-stage tendering is envisaged, the Engineer has to rely totally on his own expertise and experience in completing the project design prior to tenders being invited. In cases where the Works are of a *general* nature, this poses no particular problem. In the case of a *specialised* project, however, or one in which the design is heavily dependent upon construction methods and general 'buildability', the Engineer may well not have the experience or expertise to carry out the design unaided since, by and large, he will not be as conversant with the problems of construction as would be a contractor. In such circumstances, therefore, and by way of ensuring that the end product was at least 'buildable', he might derive considerable benefit from enlisting the help of a suitably experienced contractor in carrying out the final design of the project.

Whilst a *package* contract (see section 5.4.3) would be a possibility in many such circumstances, it should be remembered that such an arrangement severely restricts the extent to which the Engineer may exercise control over the design – which may subsequently transpire to be more favourable to the Contractor than to the Employer – or, on account of his having little or no powers of veto in respect of the Contractor's selection of subcontractors, the construction. In the event that the Employer or Engineer considers the risks engendered by these restrictions to be unacceptable, a more 'conventional' contract let on the basis of two–stage tendering would constitute a less 'risk–fraught' alternative.

It might also be mentioned that, irrespective of a prospective Contractor's experience and expertise in the relevant field, an obvious pre–requisite for his submission of a *package* tender is his ability to command the considerable resources necessary to put together the appropriate package. This being the case, the number of tender submissions might often be less than that which would be considered ideal for the project in question (see section 7.2), a factor which might give rise to serious concern regarding the integrity of the tendering process and the 'competitiveness' of the subsequent contract price. In the case of a contract let on the basis of two–stage tendering, the design process will be a *joint* Engineer/Contractor effort and it is unlikely that, between them, they will be unable to command all the resources necessary. Consequently, an increased tendering response might reasonably be supposed, over that to an invitation to submit *package* tenders, with a corresponding increase in the level of 'competition–related' benefit to the Employer.

Notwithstanding the aforesaid, however, a number of disadvantages should be noted in connection with the use of the cited procedures:

In the case of the *first* variant thereof, the initial compilation of individual proposals for subsequent appraisal involves the expenditure of considerable effort and cost on the part of each tenderer. Whilst the *successful* tenderer will be able to recoup his incurred costs under the subsequent contract, the *unsuccessful* tenderers are left with little alternative other than to write–off their costs against *future* contracts – with obvious long term financial consequences for Employers in general.

The most noteworthy deficiency of the *second* variant of the procedure stems from the fact that, whilst the final proposal drawn up by the Engineer may well incorporate features suggested by a number of tenderers, it must be formulated in such a way as to avoid providing any individual tenderer with a second–stage tendering advantage over the others. Since this will preclude the final proposal containing features which require the use of 'state of the art' construction methods developed by individual tenderers, or which can only be completed using specialist items of plant of which a given tenderer is the sole possessor, the danger here is that of the project Specification ending up as a compromise which, although satisfactory from the point of view of tendering parity, may not constitute the 'optimum solution', in terms of cost and/or subsequent performance, to the Employer's problem.

Finally, it should be noted that, whilst *all* the tenderers will contribute, in some measure, towards the efficacy of the final proposal, only *one* will be awarded the contract for its construction. With this in mind, and by way of avoiding the possibility of their efforts serving only to fill the pockets of one of their competitors, tenderers may be somewhat less than forthcoming in suggesting improvements or alternatives to the Engineer's proposal. In this event, the Engineer might not derive as much benefit from the initial 'consultation process' as originally envisaged.

7.5 Selection of Subcontractors

In the case of a contract executed in accordance with the I.C.E. Conditions of Contract, subcontractors are classified as either *nominated* or *domestic*, depending upon whether they are selected by the Engineer or by the Main Contractor.

7.5.1 Nominated Subcontractors

In the course of carrying out the project design and drawing up the Specification, the Engineer may see the need for certain aspects of the Works to be completed by *specialist* contractors of *his* choice. Similarly, he may consider that certain materials, for incorporation into the *permanent* Works, should be supplied by *specialist* suppliers of *his* choice.

In this event, and prior to the completion of the main contract documentation, the Engineer will approach a number of suitable firms and, having established their willingness to undertake the relevant work or to provide the requisite materials, will supply them with the necessary information and documentation to enable them to submit competitive tenders for his appraisal. Whilst the method of appraisal will vary according to circumstance, it might reasonably be expected to be along similar lines to that described under 'single–stage tendering'.

In a case where there is only a *single* contender for the subcontract, or where the Engineer has a definite preference for a *particular* firm, competitive tendering is obviously out of the question and the eventual price will be arrived at by negotiation between the Engineer and the party in question.

Following the award of the main contract, the Main Contractor will be instructed, via the contract documents, to enter into a contract with the chosen party (the *Nominated Subcontractor*) for the provision of the specified goods or services. The appropriate price is entered in the main contract Bill of Quantities as a 'Prime Cost' lump sum (see section 8.2.4) and will be payable to the subcontractor, via the Main Contractor, upon satisfactory discharge of the former's obligations.

Although the subcontractor in question is chosen by the *Engineer*, he will subsequently be employed by the *Main Contractor* under a contract to which neither the Engineer nor the Employer will be party. Consequently, and by way of reflecting his assumption of *full* responsibility for the actions or omissions of the subcontractor, the Main Contractor is entitled to be paid a percentage of the Prime Cost sum. Furthermore, Clause 59A of the I.C.E. Conditions provides him with the right of veto in the event of his *reasonable* objection (e.g. on the grounds of the subcontractor's dubious reliability or financial status) to the Engineer's choice. Should that veto be exercised, the Engineer is required to nominate a mutually acceptable *alternative* subcontractor.

Finally, a particular point should be noted, namely that: in accordance with the I.C.E. Conditions, a 'Prime Cost Item' is that which '... contains a sum ... which will be used for the execution of work, or for the supply of goods, materials or services for the Works ...'. The key phrase here is '*will* be used' which implies that the work content of the item may only be reduced or omitted by the issuance of a *variation order* – in which case, the Main Contractor is entitled to claim for any resultant disruption *and* for the *full* (i.e. based on the *original* Prime Cost sum) amount of his 'risk percentage'. Consequently, nominated subcontract work should *only* be covered by Prime Cost items in the Bill if the work can be *fully* defined and costed at the outset. In circumstances where this is *not* possible, the *estimated* cost of the work should be entered as a 'Provisional' sum (see section 8.2.4) which '... is designated for the execution of work, or the supply of goods, materials or services ...(and)... may be used in whole, or in part, or not at all, at the direction and discretion of the Engineer ...'. At a stage where the relevant work *is* fully definable, the nominated subcontractor can be appointed and the 'Provisional' and 'Prime Cost' items amended accordingly.

7.5.2 Domestic Subcontractors

With the exception of those items of work which the Engineer requires to be executed by *nominated subcontractors*, the main contract envisages that the 'Works' will be executed by the Main Contractor. For a variety of reasons, however, the Main Contractor may elect to sub-let certain portions of the Works to other organisations. By virtue of the choice of such resting with the Contractor (as opposed to the Engineer), these organisations are referred to as *domestic* subcontractors.

The first step in the procedure for selecting domestic subcontractors is for the Main Contractor to identify, at the 'tender-compilation' stage, work suitable for execution by such.

There are many factors which influence the Main Contractor's decision to sub-let portions of the Works, not the least of which is the 'speciality' of the work involved. In the case of *building* Works, this usually means items such as lift installation and commissioning, air conditioning, electrical work, plumbing, glazing etc. etc. Whilst the use of subcontractors in *civil engineering* Works is not quite as extensive in terms of the *variety* of suitable work, subcontract work can be quite extensive in terms of its proportion of the overall contract value. For example, it is not uncommon for the bulk earthworks, which will constitute a *substantial* proportion of the contract value, to be sub-let in the case of a major roadworks contract. In both cases, firms will generally establish by practice the type of work they prefer to sub-let and, in view of the risks associated with sub-letting, will restrict the use of sub-contractors to those of 'tried and tested' performance.

The decision to sub-let may also be influenced by the *size* of the contract. If the contract in question is significantly larger than that which would normally be undertaken by the Contractor, he may wish to sub-let portions of such to impose less strain upon his existing resources or to offset some of the overall financial risk.

Irrespective of the underlying reasons for sub-letting, however, it should be noted that, for a contract executed in accordance with the I.C.E. Conditions, the Main Contractor requires the *written* permission of the Engineer, in his capacity as the Employer's *agent* (see sections 3.7.2 and 4.2.3) so to do. Furthermore, he is expressly prohibited from sub-letting the *whole* of the Works.

Following identification of the relevant aspects of the Works, enquiries are sent to suitable subcontractors (i.e. those on the

Contractor's 'approved' list), giving full details of the Contractor's requirements in terms of work content and timing, inviting them to tender for the work. The method of appraisal and selection will vary according to circumstances and will depend on the Contractor's attitude towards such. On the one hand, if time permits (it usually *doesn't*!), a formal 'single–stage' tendering procedure may be used. On the other hand, the procedure may be little better than an exercise in 'horse–trading'! Either way, it is important to note that the chosen subcontractor must meet with the Engineer's approval. It might also be mentioned that, although not compulsory by any means, the subcontractor's price should advisedly be *less* than that tendered by the Main Contractor for the work in question!

7.6 Selection of Management Contractors

Irrespective of the exact type of Management contract envisaged (see section 5.4.8), it is of considerable importance to realise that, in selecting a Management Contractor, the Employer is adding a new member to his project team. Consequently, the selection procedure is (advisedly) somewhat more involved than either the 'single–stage' or 'two–stage' tendering procedures described hitherto.

Whilst certain aspects of the selection procedure will vary according to the nature of the project, the attitude of the Employer and the envisaged role of the Management Contractor, it might reasonably be expected to proceed along the following lines:

7.6.1 The Invitation to Tender

The first task required of the Employer and his advisers is to compile, at the earliest possible opportunity, a *preliminary list* of around ten to fifteen suitable contenders who might be interested in managing the project through its various stages. Whilst the list will obviously concentrate on 'management' contractors *per se*, it should not exclude other more 'conventional' contractors who have a record of good contract management and may have suitable qualities for the position in question.

By means of informal enquiries, and in consultation with existing members of the project team, the preliminary list should be reduced to a *short list* of about four or five contractors who:

(a)   are acceptable to all parties involved in the project;

(b)   have expressed a definite and specific interest in the project;

(c)   have the neccessary resources, particularly in respect of personnel, available for immediate mobilisation;

In cases where the project is multi–disciplinary, particular attention should be paid to the extent of the contractors' expertise, both in design and construction, in the appropriate disciplines.

A formal *invitation to tender* is sent to the short–listed contractors and should include the following information:

(a)   full details of the roles and responsibilities of all members of the project team, including those of the Employer;

(b)   a full description of the project insofar as that is possible at the time;

(c)   full details of any constraints to which the project is, or might become, subject;

(d)   sufficient qualitative and quantitative information to enable the compilation of a budget estimate if such is required;

(e)   any critical dates which might apply to various stages of the project;

(f)   details of the duties of the Management Contractor, including the envisaged extent (if any) of his involvement in the design process or other aspects of the pre–construction stage of the project;

(g)   the Conditions of Contract;

(h)   full and precise details of the Employer's requirements regarding the submission of tenders;

Although not essential to the efficacy of the selection procedure, considerable benefit may be derived from a meeting held with the tenderers at the time of, or shortly after, the issuance of the formal invitations to tender. This will serve to resolve any points of doubt which the tenderers might have with regard to the project, or to the submission of tenders, and might avoid the possible expenditure of considerable time and effort at a later (and, possibly, more critical) stage in the project.

7.6.2 Submission and Evaluation of Tenders

As with the previously discussed tendering procedures, a primary requirement concerning the submission of tenders is that they should comply *exactly*, in terms of both content and format, with the Employer's stated requirements concerning such.

It is generally recognised that the success, or failure, of a management–type contract places substantially more dependence upon the calibre of the personnel employed on the project, and on their ability to work together as a team, than would a more 'conventional' contractual arrangement. To enable the Employer and his advisors to take such matters into account in evaluating tenders, the key personnel (in terms of their standing within the tenderer's organisation *and* their prospective involvement in the project under consideration) of each tenderer should be required to make a presentation of their company's submission. Apart from providing the Employer *et al* with the facility to closely question tenderers on various aspects of their individual proposals, an oral presentation provides the ideal opportunity for the former parties to assess the *personal* qualities of potential additions to their team.

Whilst tender–assessment procedures and techniques will vary according to circumstance, it is recommended that a system be employed whereby both the *quantitative* and *qualitative* aspects of individual submissions may be taken into account.

The quantifiable aspects will include:

(a)   the tenderer's fee for managing the construction process;

(b)   the tenderer's fee for providing any pre–construction services which may be required, including details of any cost–reimbursable elements thereof;

(c)   the budget estimate of construction costs (if required to be provided by the tenderer);

(d)   the estimated cost of providing site facilities (if required);

(e)   full details and costs of the staff which the tenderer proposes to provide;

The qualitative factors to be taken into account will include the tenderer's:

(a) appreciation and grasp of the project;

(b) approach towards the role of Management Contractor;

(c) proposed management strategy;

(d) management capabilities, in terms of the experience and expertise of the key personnel he proposes to employ on the project;

(e) expertise and experience in planning and programming projects of comparable size and complexity to the one in question;

(f) proposed methods of maintaining quality control and providing quality assurance;

(g) design capabilities (if relevant), in terms of expertise, experience and resource levels;

(h) compatibility with the other members of the project team;

Whilst an appraisal of tenders on the basis of their financially quantifiable aspects presents no great problems, the principal difficulty with the *qualitative* appraisal is that of instituting a procedure which facilitates an *objective* comparison of tenders in terms of *subjectively*-assessed variables. In this respect, it is often helpful to carry out the qualitative appraisal on the basis of a points system with each variable being weighted to reflect its relative importance in the prevailing circumstances. Whilst the financial aspects of tenders may be included, with appropriate weighting, in the points system, it is most commonly the case that, for reasons discussed previously, the subsequent contract will be awarded *primarily* on the basis of the qualitative appraisal of tenders but will be subject to the proviso that the selected tender's costs be within the Employer's budget.

CHAPTER 8

THE BILL OF QUANTITIES

As described previously (see sections 5.1.2 and 6.6), the Bill of Quantities forms a part of the contract documentation drawn up by the Engineer. In essence, the Bill is an itemised breakdown of the Works with each item being briefly described and approximately quantified. At the time of tender, the tenderer is required to insert a *unit rate* against each item (that being his price for carrying out each unit of the relevant work) and an *extended total* (the product of item quantity and unit rate). The *Tender Total*, on which the tender will *primarily* be appraised, is the aggregate sum of the *extended totals*.

An extract from a typical *unpriced* Bill of Quantities is shown in Figure 8.1.

The principal purposes of the Bill of Quantities are:

(a) to provide sufficient information to enable tenders to be compiled efficiently and accurately;

(b) to provide a sufficiently detailed breakdown of the Works to facilitate an itemised comparison and appraisal of tenders by the Engineer;

(c) to provide a basis for evaluating variations or additions to the Works which may be ordered in the course of construction;

(d) to facilitate the progressive valuation of work executed during the contract;

Number	Item description	Unit	Quantity	Rate	Amount £	p
	IN SITU CONCRETE					
	Provision of concrete; ordinary prescribed mix.					
F143	Grade C20; cement to BS 12; 20mm aggregate to BS 882.	m³	275			
F163	Grade C30; cement to BS 12; 20mm aggregate to BS 882.	m³	1500			
	Placing of concrete (Mass)					
F411	Blinding; thickness: 50mm.	m³	95			
	Placing of concrete (Reinforced)					
F521	Bases and ground slabs; thickness: not exceeding 150mm.	m³	220			
F522	Bases and ground slabs; thickness: 150 – 300mm.	m³	125			
F542	Walls; thickness: 150 – 300mm	m³	250			

Figure 8.1 Extract from a typical Bill of Quantities

With these aims in mind, the Bill should be *clear, easily interpretable* and should contain an itemised breakdown of the Works which is *sufficient* in detail to enable distinctions to be drawn between:

(a) *different* types of work;

(b) *similar* types of work which are carried out in *different* circumstances which give rise to considerations of cost;

To ensure that these aims are met, and to provide the users thereof with all the benefits of dealing with a standardised format, the majority of Bills of Quantities for contracts of any substance are compiled in accordance with one or other of the recognised standards (see section 5.1.2). In the case of Works of civil engineering construction, the most commonly used standard is: *The Civil Engineering Standard Method of Measurement.*

8.1 The Civil Engineering Standard Method of Measurement

Published by the Institution of Civil Engineers in conjunction with the Federation of Civil Engineering Contractors, the Civil Engineering Standard Method of Measurement is currently in its second edition and, in accordance with popular usage, will be referred to hereafter as CESMM2.

8.1.1 General Principles

CESMM2 was formulated with the stated intent that it be used in conjunction with the I.C.E. Conditions of Contract (5th Edition) and, to that end, contains a number of express references to Clause numbers in such. That said, however, it is generally recognised that the standard may be used with *other* Conditions of Contract *provided* that those Conditions invest the Bill of Quantities and the method of measurement with the same status and functions as do the I.C.E. Conditions *and* that the functions and responsibilities of the parties involved in the contract are broadly the same, under the relevant Conditions, as they would be under the I.C.E. Conditions. In this event, it is obviously necessary to coordinate the provisions of CESMM2 with the prevailing Conditions and to inform tenderers of any consequent amendments to the former by including full details of such in the Preamble to the Bill of Quantities (see section 8.2.1).

It is further stated in CESMM2 that it is intended for use '... only in connection with Works of civil engineering construction ...' and that it '... does not deal with the preparation of Bills of Quantities for, or for the measurement of, mechanical or electrical engineering work, building work or work which is seldom encountered in civil engineering contracts ...'. This statement should *not* be taken to imply that the standard should be used *only* in contracts where the work is *exclusively* civil engineering in character since few such contracts exist. Indeed, CESMM2 recognises this fact by stating measurement and Billing principles which should be followed in cases where a contract includes work elements appropriate to another discipline or in cases where aspects of the Works are not sufficiently common to justify their measurement being standardised in CESMM2. Broadly speaking, the method of dealing with such items is left largely up to the Bill compiler and is governed only by the *general* requirement that the items be Billed in *sufficient* detail to *identify the relevant work components* and to *enable adequate pricing thereof by tenderers.*

If the work is *measurable*, CESMM2 requires that full details be provided, in the Preamble to the Bill, of any 'non-standard' convention which is to be used for calculating the associated quantities.

Whilst this stated requirement provides *tacit* approval for the use of non-standard measurement conventions if necessary, it should *not* be construed as providing a licence for the indiscriminate use of such. Indeed, it is generally recommended that their use be avoided *if at all possible* since they can often become a source of contention in the preparation of Final Accounts.

In many cases, such as that of, say, a self-contained component of the Works which does not have a quantity as a relevant parameter of its cost, there may be no need to use a measurement convention at all. Since the principal function of a Bill item is to identify the work and to enable it to be priced, a component of the type cited may be entered in the Bill, to the satisfaction of all concerned, as a lump-sum item and described therein with no more detail than: 'The object shown in Detail A of Drawing 123/45B and covered by Clause 67.8 of the Specification.'

Where the component *is* associated with a quantity which forms the principal basis on which the price will be calculated, there might still be no need for a measurement convention *per se* to be used since CESMM2 states the *general* principle that quantities '... shall be computed net from the drawings ...'. If this general principle is insufficient for the item concerned, then a non-standard measurement convention may well be the only alternative. In such an event, the 'golden rules' are: keep it *simple*, keep it *straightforward* and ensure that it can be *easily understood and applied* by *all* concerned.

The quoted aim of the Standard is: '... to set forth the procedure according to which the Bill of Quantities should be prepared and priced and the quantities of work expressed and measured ...'. It achieves these aims by specifying a layout for the Bill (see section 8.2) and by making use of a *Work Classification* which defines:

(a) the breakdown of the Works into its constituent work items for entry in the Bill;

(b)   the information to be given in item descriptions;

(c)   the units to be used in expressing item quantities;

(d)   how the work is to be measured for quantification;

8.1.2 The Work Classification

The Work Classification divides work commonly encountered in civil engineering into 25 main classes as shown in Figure 8.2.

Class A:	General Items;
Class B:	Ground Investigation;
Class C:	Geotechnical and other specialist processes;
Class D:	Demolition and Site Clearance;
Class E:	Earthworks;
Class F:	In situ Concrete;
Class G:	Concrete Ancillaries;
Class H:	Precast Concrete;
Class I:	Pipework – Pipes;
Class J:	Pipework – Fittings and Valves;
Class K:	Pipework – Manholes and Pipework Ancillaries;
Class L:	Pipework – Supports and Protection; Ancillaries to Laying and Excavation;
Class M:	Structural Metalwork;
Class N:	Miscellaneous Metalwork;
Class O:	Timber;
Class P:	Piles;
Class Q:	Piling Ancillaries;
Class R:	Roads and Pavings;
Class S:	Rail Track;
Class T:	Tunnels;
Class U:	Brickwork, Blockwork and Masonry;
Class V:	Painting;
Class W:	Waterproofing;
Class X:	Miscellaneous Work;
Class Y:	Sewer Renovation and Ancillary Work;

Figure 8.2 CESMM2 Work Classification

Each class of work is subdivided into *three* divisions which classify the work at successive levels of detail, with each division containing a *maximum* of *eight* distinct features of the work. For example, Class I contains three divisions with Pipework being classified according to the pipe *material*, the *nominal bore* of the pipes and the *depth* at which the pipes are laid. The relevant extract from CESMM2 is shown in Figure 8.3.

CLASS I: PIPEWORK — PIPES

Includes: Provision, laying and jointing of pipes
           Excavating and backfilling pipe trenches

Excludes: Work included in classes J, K, L and Y

FIRST DIVISION		SECOND DIVISION	THIRD DIVISION
1 Clay pipes	m	1 Nominal bore: not exceeding 200 mm	1 Not in trenches
2 Prestressed concrete pipes	m	2 200–300 mm	2 In trenches, depth: not exceeding 1·5 m
3 Other concrete pipes	m	3 300–600 mm	3 1·5–2 m
4 Cast or spun iron pipes	m	4 600–900 mm	4 2–2·5 m
5 Steel pipes	m	5 900–1200 mm	5 2·5–3 m
6 Plastics pipes	m	6 1200–1500 mm	6 3–3·5 m
7 Asbestos cement pipes	m	7 1500–1800 mm	7 3·5–4 m
8 Pitch fibre pipes	m	8 exceeding 1800 mm	8 exceeding 4 m

Figure 8.3 Classification Table from CESMM2

The entries in the divisions are known as 'descriptive features' and, when linked together, with *not more than one* feature from *each* division being used, form a reasonably comprehensive description of a billed item.

Whilst the Work Classification breaks down the work into a considerable number of parts (the divisional classification illustrated gives rise to 512 possible combinations of pipe material, nominal bore and depth ranging from 'Clay pipes, nominal bore not exceeding 200mm, not in trenches' through to 'Pitch Fibre pipes, nominal bore exceeding 1800mm, in trenches, depth exceeding 4m'), it does not subdivide to the finest level of detail at which a distinction might need to be drawn, For example, the pipework subdivision does not take account of different specifications for pipe quality nor does it differentiate according to the various types of joint which can be used. Fairly obviously, if an attempt were made to account for every possible eventuality, the standard document would be considerably more complex (and more expensive!) than it is at present and, more importantly, something would inevitably be discovered to have been left out! Thus, it can be said that CESMM2 provides a *broad* system of work classification which can, if necessary, be extended by the Bill compiler to cover particular circumstances.

Each Class within CESMM2 contains a set of *Measurement*, *Definition*, *Coverage* and *Additional Description* rules which are intended to amplify the *basic* information given in the 'classification' section.

The *Measurement* rules contain information relating either to how the relevant item quantities will be calculated or to circumstances under which particular work will, or will not, be measured. Using the previously cited example concerning Class I, Measurement Rule M3 states that pipe lengths will be measured along the centre−lines of pipes − thus avoiding possible confusion in cases where pipes enter manholes at angles. The rule further states that, for pipes *in trenches*, the measured pipe length will *include* lengths occupied by valves or fittings but will *exclude* lengths occupied by backdrops into manholes.

The purpose of the *Definition* rules is to clarify words or expressions, used in the 'classification' section, which might give rise to ambiguities in the Bill. In the case of Class I, a prime example of such is Definition Rule D3 which states that the 'depth' referred to in the third division of the pipe classification (see Figure 8.3) is the depth of the *invert* of the pipe below the *commencing surface* (see below).

The *Coverage* rules are intended to draw attention to any work which will be deemed to be included in the relevant items. For example, Coverage Rule C2 for pipes states that items for pipes *in trenches* are deemed to include for excavation, backfilling, disposal of excess spoil and the like. It should be particularly noted that the Coverage rules for a particular item (or group of items) do *not* specify *all* the work covered by the item but refer *only* to *specific* elements of cost which are deemed to be included in the item cost.

The function of the *Additional Description* rules is exactly that which may be inferred from their title – namely that they specify any details which particular item descriptions should contain *over and above* that deriving from the divisional classification. For example, again citing the case of Class I, Additional Description Rule A1 requires the location of pipe runs to be included in the descriptions for pipework items in order that the items may be identified by reference to the drawings. Apart from the obvious benefits accruing to the tenderer's estimator, with regard to his pricing of the Works (it should be appreciated that the depth stated in the decription will *not* be the average depth upon which excavation costs and the like will be based), this particular rule requires, by implication, *separate* itemisation of pipework which is similar in type, nominal bore and depth but which is laid in different terrains – an obvious requirement in the light of possible cost differentials. A similar implication is inherent in Rule A2 of the same Class which requires the item descriptions to include details of the pipe material (which will state the specified quality thereof), the joint types and the lining requirements.

Before moving on to consider the layout and contents of the Bill of Quantities, it is necessary to clarify the meanings of several terms which are used in the Work Classification in respect of work involving excavation.

For example, *Additional Description* Rule A3 for Class E (Earthworks) states that:

'The Commencing Surface shall be identified in the description of each item for work involving excavation for which the Commencing Surface is not the Original Surface. The Excavated Surface shall be identified in the description of each item for work involving excavation for which the Excavated Surface is not the Final Surface.'

These various 'surfaces' may best be defined by reference to, say, a roadworks contract which involves, *inter alia*, excavation to the road formation level and subsequent excavation below this level to accomodate catchpits or the like.

When the Contractor first walks on to the site, the surface which he sees (and is possibly standing on) is the *Original Surface*. It is also the *Commencing Surface* for the Bill item concerning excavation to formation level. The *Excavated Surface* for this item is the road formation which also constitutes the *Commencing Surface* for the catchpit excavation items. The *Excavated Surface* for each of the latter items is, fairly obviously, the bottom of each catchpit hole. The *Final Surface* is the level to which excavation must be carried out to receive the Permanent Works. In this case, therefore, the *Final Surface* comprises the road formation, *excluding* those areas occupied by catchpits, and the bases of each of the lower pockets of excavation.

8.2 Contents of the Bill of Quantities

In accordance with the recommendations of CESMM2, the Bill of Quantities should comprise the following sections:

(a) The List of Principal Quantities;

(b) The Preamble;

(c) The Dayworks Schedule;

(d) Work Items;

(e) The Grand Summary;

These will now be discussed in turn.

8.2.1 The List of Principal Quantities

The List of Principal Quantities is an itemised breakdown of the *major* components of the Works, with their approximate quantities, and is, in effect, a tabulated summary of the 20% or so of the Billed items which generally contribute to about 80% of the contract value. It is included in the Bill as a means of enabling tenderers to make a rapid appraisal of the *general* scale and character of the Works, prior to their examination of the contract documents in *detail*, and allows them to assess immediately whether the job is suited to their particular skills and existing resource levels.

With a view to avoiding possible claims related to divergences between the List of Principal Quantities and the detailed contents of the Bill, CESMM2 *expressly* states that such a list is *solely* provided for the abovementioned purpose.

A typical List of Principal Quantities is shown in Figure 8.4.

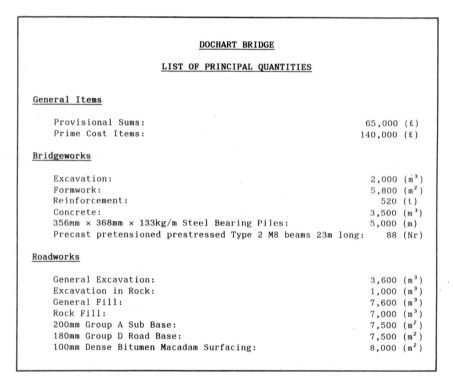

```
                          DOCHART BRIDGE

                    LIST OF PRINCIPAL QUANTITIES


General Items

        Provisional Sums:                                    65,000 (£)
        Prime Cost Items:                                   140,000 (£)

Bridgeworks

        Excavation:                                          2,000 (m³)
        Formwork:                                            5,800 (m²)
        Reinforcement:                                         520 (t)
        Concrete:                                            3,500 (m³)
        356mm × 368mm × 133kg/m Steel Bearing Piles:         5,000 (m)
        Precast pretensioned prestressed Type 2 M8 beams 23m long:   88 (Nr)

Roadworks

        General Excavation:                                  3,600 (m³)
        Excavation in Rock:                                  1,000 (m³)
        General Fill:                                        7,600 (m³)
        Rock Fill:                                           7,000 (m³)
        200mm Group A Sub Base:                              7,500 (m²)
        180mm Group D Road Base:                             7,500 (m²)
        100mm Dense Bitumen Macadam Surfacing:               8,000 (m²)
```

Figure 8.4 Typical List of Principal Quantities

8.2.2 The Preamble

The purpose of the *Preamble* to the Bill of Quantities is to provide tenderers with *general* information regarding the compilation of the Bill and to draw their attention to matters of *particular* importance in the context of pricing and subsequent payment.

Specifically, the Preamble should contain a statement indicating which, if any, standard method of measurement has been adopted for the purpose of compiling the Bill. In the case of a contract executed under the I.C.E. Conditions of Contract, such a statement is not strictly necessary since the Bill of Quantities for such will be assumed (see section 9.1.10) to have been prepared in accordance with CESMM2. In the event that non–standard methods of measurement, or amendments to the quoted (or assumed) standard method of measurement, have been adopted for certain components of the Works, *full* details of such, along with details of the extent of work covered thereby, should be given in the Preamble.

In contracts which involve excavation, piling or tunnelling, 'differences of opinion' sometimes arise between the Contractor and the Engineer as to what, for payment purposes, constitutes 'rock'. Whilst it might be suggested that some Contractors tend to consider anything slightly harder than normal as 'rock', it might equally be suggested that some Engineers are inclined to regard as 'normal material' any naturally occurring material which does not actually require a thermo–nuclear explosion to remove it! To eliminate (or, at least, to *minimise*) the possibility of confusion arising in this respect, CESMM2 requires a definition of 'rock' to be included in the Preamble. Whilst it is possible to define 'rock' in terms of the plant which, *in the Engineer's opinion*, is capable of removing it, it should be appreciated that this is a *subjective* definition which may serve only to 'fuel the flames' of confusion. This being the case, it is generally recommended that, wherever possible, the definition of 'rock' should be made with reference to geological strata shown in the borehole logs made available to tenderers.

Whilst Class E (Earthworks) and Class T (Tunnels) of the CESMM2 Work Classification each state a minimum size in respect of boulders being measured, and paid for, as 'isolated volumes of rock', no such definition appears in any other Class involving excavation. If such a definition is intended to be applied to such classes, a statement to that effect should be included in the abovementioned definition of 'rock'.

In view of the obvious cost implications of excavating under, or immediately adjacent to, a body of *open* water, the Preamble should contain full details of any such areas within, or on the boundaries of, the Site. This is best achieved by reference to drawings indicating the extents, and surface levels, of such. Where the boundary and surface level of any body of open water is subject to fluctuation, the drawings should state the anticipated range of fluctuation.

Whilst such details tend to refer to *permanent* bodies of open water (rivers, lakes, the sea etc.), mention should be made in the Preamble of the presence of any *temporary* bodies of open water (seasonal flooding etc.) which may affect work on the site. To minimise the potential for subsequent dispute over, for example, the difference between a large puddle and a 'temporary body of open water', the extent of areas likely to be thus affected should be indicated.

Finally, the Preamble should contain full details of the method of payment of *Method-Related Charges* (see section 8.2.4) and the *Adjustment Item* (see section 8.2.5).

8.2.3 The Daywork Schedule

As discussed previously, the Engineer may deem it necessary, in the course of the contract, to order variations or additions to the Works as specified at the time of tender. In the majority of cases, such work will be valued for payment on the basis of the Billed rates for items of a *similar* nature. On occasions, however, there may be additions or variations which cannot be easily valued in this manner. In such cases, the Engineer may consider it preferable to value the work on a quasi cost-reimbursement basis whereby the Contractor is paid *directly* for the resources used to carry out the work, along with an allowance to cover overheads and profit.

Since the Contractor is required to submit his accounts for such on a *daily* basis, the work is referred to as *Daywork* and payment for such is in accordance with the *Daywork Schedule*.

CESMM2 makes *express* provision for *two* alternative forms of Daywork Schedule to be used:

(a) In the course of compiling the Bill, the Engineer draws up a 'customised' Daywork Schedule comprising a list of the various classes of labour, plant and materials which are likely to be used in the execution of Daywork, and the conditions under which the Contractor will be paid for such. At tender time, the Contractor is required to insert a rate against each item on the list.

(b) The Daywork Schedule contains a statement to the effect that Daywork will be payable on the basis of the rates contained within the *current* (i.e. applicable at the time of execution of the work) edition of the *Schedules of Daywork carried out incidental to Contract Work* (published on a regular basis by the Federation of

Civil Engineering Contractors), such rates being adjustable (up *or* down) by percentages inserted by the Contractor at tender time. Since Contractors will, in the main, be familiar with the well–established arrangement of, and conditions attached to, the FCEC Schedules, this method of approach is generally preferred to the 'customised' approach detailed above.

Although not *expressly* provided for within CESMM2, there is a *third* option for contracts executed under the I.C.E. Conditions of Contract and that is not to include a Daywork Schedule in the Bill at all. Clause 52(3) of the Conditions states that, failing the provision of a Daywork Schedule, the Contractor will be paid for Daywork at the rates and prices contained within the FCEC Schedules with *no* adjustment.

If it is anticipated that Daywork payments will comprise a *significant* proportion of the eventual contract price, it is recommended that the Engineer take advantage of the facility provided by CESMM2 for *Provisional Sums* (see section 8.2.4) to be entered in the *General Items* section of the Bill (see also section 8.2.4) in respect of the estimated expenditure on each of Daywork labour, plant, materials and supplementary charges (on–costs and the like). If a Daywork Schedule in the form of option (b) above is used, each of the abovementioned provisional items is accompanied by an 'adjustment' item against which the Contractor is required to insert the percentage adjustment referred to previously. The price of each such item will be calculated by applying the inserted percentage to the associated provisional sum.

Whether or not the provisional sums are fully utilised in the course of the contract, the 'adjustment' sums will directly influence the *Tender Total*, on which basis tenders will primarily be compared, thereby making Daywork as much subject to the rigours of competition as any other item.

8.2.4 Work Items

This section constitutes the bulk of the Bill and comprises those items which are to be *priced* and which contribute to the *tender total*.

In accordance with CESMM2, the Work Items section of the Bill *may* be subdivided and arranged into *numbered* (to avoid confusion with the *lettered* system of the Work Classification) parts. Although such subdivision – which will mainly be determined by such factors as the nature of the work, locational considerations and limitations on the timing and sequencing of aspects of the Works – is not *compulsory*, it is generally considered to be *advantageous* in that it enables the Bill compiler to distinguish between parts of the Works which may involve different construction methods or which may have cost implications. For similar reasons, such subdivision is also considered helpful for estimating purposes.

If such a system *is* used, the only constraint imposed thereon by CESMM2 is that the items within each subdivision or numbered part be arranged in the general order of the Work Classification.

In dividing the Bill into parts, it is *customary* (but not *obligatory*) to group separately those items belonging to Class A (General Items) on account of the essential difference in character between such items and those belonging to other Classes: whereas items in Classes B – Y tend generally to relate to specific components of the Permanent Works (i.e. that which will remain for posterity after the Contractor and his entourage have moved on to bigger and better things), those in Class A tend either *not* to relate to the Permanent Works or are such that they may be conveniently grouped under a 'general' heading.

The items in Class A may be summarised as follows:

(a) Contractual Items

These items cover the costs incurred by the Contractor in respect of his discharge of certain obligations specified under the terms of the contract. They include his costs in respect of providing the specified Performance Bond and the cost to him of insuring the Employer against anything and everything (with the usual exceptions of sonic booms, civil war, invasion by a foreign power etc.). Also included are the costs incurred by him in insuring *himself* against damage to third parties, or the property thereof, in the course of his execution of the contract. Items in this section are invariably priced as *sums*.

(b) Specified Requirements

The situation here is best explained by reference to Definition Rule D1 of Class A which states that:

> 'All work, other than the Permanent Works, which is expressly stated in the Contract to be carried out by the Contractor, and of which the nature and extent is expressly stated in the Contract, shall be classed as *specified requirements*.'

In this context, 'in the Contract' may be read as: 'in the Specification or any other contract document'.

Specifically, such items cover the provision of the Engineer's site staff with the necessary facilities and services (e.g. offices and associated equipment, laboratories and associated staff/equipment, transport, surveying equipment, chainmen and the like) to enable them to discharge their functions during the construction of the Works, the testing of materials and/or the Works, where such testing is not itemised in other Classes, and the provision of *specified* Temporary Works.

The classification of such items as 'specified requirements' serves two purposes: it serves to draw the tenderer's attention to requirements which, in other circumstances, might be a matter for his own decision *and* it enables a price to be obtained for items which, although specified in character and extent at the outset of the Contract, *could* be subject to subsequent variation at the behest of the Engineer.

The items contained in this section are priced as sums *except* where their values will be determined from admeasurement.

(c) Method-Related Charges

The principal deficiencies of the 'traditional' (i.e. pre-CESMM) Bill of Quantities were that it was essentially a quantified and priced list of the various components of the *Permanent* Works and that its format implied a direct relationship between the prices (and, hence, costs) of items and their quantities. There was also the. *implied* assumption that the Contractor's costs for any given item, on which the item price would be based, were *direct* costs in that they were *exclusively* incurred in connection with that item.

The fact is, however, that, with the exception of those relating to the provision of materials for the Permanent Works, the bulk of the Contractor's costs are *not* quantity-proportional but are either based on the time taken to complete particular items of work, or are 'fixed' in the sense that they depend neither upon the duration of the work nor upon its quantity. Furthermore, a significant proportion of those costs will be *indirect* in that they will be incurred in connection with the completion of a *number* of Permanent Works items. For example, a 'traditional' Bill would require the costs of providing a tower crane for a 'high-rise' construction project to be distributed amongst all the Permanent Works items above the ground floor. The same can be said of the costs incurred in providing such *Temporary* Works (defined as anything which is not classified as *Permanent* Works) as are required to enable the Permanent Works to be completed. For example, the cost of providing the necessary falsework support systems for the construction of suspended floor slabs in the 'high-rise' example would require to be offset against the formwork, reinforcement and concrete-placement items for the slabs since none of those operations could be completed without the falsework being in place.

The upshot of all this was that the valuation of additions and variations ordered in the course of the contract often fell outside the realm of simple re-measurement since an adjustment of the *value* of a particular item, based simply upon an adjustment in its quantity, rarely reflected the actual variation in the Contractor's *costs* (after all, it costs no less to provide a tower crane to place 250 m^3 of high-level concrete than it does to place 300 m^3!). Consequently, the valuation of additional or varied work generally involved a complicated and protracted claims procedure which not only introduced a climate of contention to the proceedings but also tended to make a nonsense of any financial control of the project based on cash-flow forecasts. Since no mention was made in a 'traditional' Bill of either *time* or *resources*, a similar procedure was generally required to evaluate the cost of delays incurred in the course of the contract.

It might also be mentioned that the pattern of contractual payments to the Contractor, based as it was on the progressive (i.e. quantity-wise) completion of the *Permanent* Works, rarely (if ever) bore any relationship to the pattern of his expenditure. This often led to an unbalanced cash-flow situation which benefitted neither the Contractor nor the Employer.

To alleviate such problems, the CESMM (the original 1976 version) instituted the system of *Method-Related Charges* whereby the Contractor, if he so desired, could enter Bill items, which related to his proposed methods of carrying out the Works, against each of which he could enter a *lump-sum* to cover costs which were *not* quantity-proportional.

No constraints are placed upon the Contractor as to what he may, or may not, include in this section of the Bill – indeed, he is under no obligation to enter any method-related charges at all if he prefers otherwise. If he *does* avail himself of this facility, however, it is encumbent upon him to enter the items in the order suggested in Class A of the CESMM2 Work Classification, to provide a clear and precise description of the work and, since the method of payment of such differs accordingly, to indicate whether the charges are 'fixed' or 'time-related'.

An example of such a Bill entry, concerning the provision of site-batched concrete, is shown in Figure 8.5.

Number	Item description	Unit	Quantity	Rate	Amount £	p
	METHOD-RELATED CHARGES					
	Plant					
	Provision of Concrete Batching Facilities					
A335.1	Erect batching plant; (Fixed)	sum	–	–	5,500	00
A335.2	Operate for duration of concreting operations; (Time-related)	sum	–	–	50,700	00
A335.3	Dismantle batching plant and remove from site on completion of concreting operations; Reinstate batching area; (Fixed)	sum	–	–	1,000	00
A336	Transport concrete for duration of concreting operations; (Time-related)	sum	–	–	34,000	00

Figure 8.5 Example of Billing of Method-Related Charges

Although not shown, this entry would be accompanied by a quantity-proportional item, in the section of the Bill containing items in Class F (In Situ Concrete) of the Work Classification, covering the unit *material* cost of providing concrete.

In accordance with CESMM2, method-related charges '... shall be certified and paid pursuant to Clauses 60(1)(d) and 60(2)(a) ...' of the Conditions of Contract (see section 9.1.10). In other words, a method-related charge is treated, for payment purposes, as would be any other item price in that the proportion of such to which the Engineer considers the Contractor entitled is included in the relevant interim valuation. A statement to this effect should appear in the Preamble together with a statement to the effect that method-related charges will be subject to Baxter-formula adjustment in the event of the CPF Clause (see section 5.2) being applicable to the contract.

Whilst 'fixed' charges will generally be payable in full upon satisfactory completion of the relevant work or, if necessary in the light of protracted completion, in such interim proportions as the Engineer deems appropriate, 'time-related' charges will be payable according to the proportion of the *overall time* (generally not specified but easily determinable from the Contractor's construction programme) of the operation which has elapsed at the time of assessment.

For example, referring to the case illustrated in Figure 8.5, the 'fixed' charge for erecting the batching plant (Item A335.1) would be certified for full payment as soon as the Engineer was satisfied that the operation had been completed *and* that the plant was capable of producing concrete. Similarly, the dismantling charge (Item A335.3) would become payable as soon as the batcher was dismantled and removed from the site and the batching area was restored to its original pristine condition.

Assuming the Contractor's construction programme indicated a duration of 18 months for the concreting operations, the operation and transport charges (Items A335.2/3) would be payable, during the currency of the concreting activities, at the rate of 1/18 of the total 'time-related' charge per month. Thus, at 5 months into the concreting operation, the Contractor would be entitled to a *cumulative* payment of 5/18 of the relevant charges even though, as a result of the irregularity of concrete operations, only 10% (say) of the total volume of concrete required might have been produced at that stage.

It is important to note that the payment of the above-stated proportion would be dependent upon the work progressing according to schedule. If, for example, concreting was found to be 1 month *ahead* of schedule at the end of month 5, the *overall* operation could reasonably be expected to be completed in 17 months thereby entitling the Contractor to a cumulative payment, at that stage, of 5/17 of the total charge – and *not* 6/18 as might be suggested by the fact that he had effectively completed 6 month's work in 5 months. Using similar reasoning, a payment of 5/19 of the total charge would be appropriate if the work was found to be 1 month *behind* schedule at that point.

In the event that variations instructed in the course of the contract render the Billed 'time-related' charges inappropriate, those charges may be adjusted accordingly. If, therefore, in the cited example, the Engineer instructed additional work to the extent that the batching plant was required to remain on site for an additional month, the 'time-related' charges would be increased to 19/18 of their original value and, for interim payment purposes, the overall concreting duration would be taken as 19 months.

In the context of payment of method-related charges generally, it should be noted that such monies constitute part of the Tender Total which is, in effect, the price which the Contractor proposes to charge for the completion of the Permanent Works. It may consequently be stated that, provided such Works are completed to the extent originally tendered for, the Contractor is entitled to the *full* amount of his method-related charges *irrespective of whether he did fewer, or even none, of the things which were covered by such charges.*

This is of *particular* importance in respect of those items which may be regarded as *contingencies* in that the subsequent extent of such is dependent upon circumstance and, as a consequence, is largely indeterminate at the time of tender.

By way of illustration, consider the case where a Contractor has entered a method-related charge to cover the possible cost of de-watering foundations in the vicinity of a high water table, the amount of such charge being based on the extent to which he considers de-watering likely. The fact that such de-watering subsequently transpires to be unnecessary, on account of a severe drought lowering the water table to a level well below that of the foundation excavations, does not diminish the Contractor's entitlement to be paid the relevant method-related charge *in full.* The point here is that the Contractor has carried the risk, placed upon him by the contract, of having to do an indeterminate amount of de-watering (had the extent of such *exceeded* the amount allowed for in the charge, he could have expected very little sympathy – '... reasonably foreseen by an experienced Contractor etc. etc.') and, as a consequence, is entitled to be paid accordingly.

If nothing else, it should be appreciated that, had the Contractor chosen *not* to avail himself of the method-related charge facility, but had chosen instead to cover the de-watering contingency cost in the prices for the appropriate *quantity-proportional* items, the Employer/Engineer would be none the wiser!

Finally, it should be noted that, although such a state of affairs may reasonably be inferred from the preceding paragraphs, CESMM2 expressly states that the Contractor's insertion of an item as a method−related charge: '... shall not bind him to adopt the method stated in the item description ...'. Thus, a Contractor who has entered method−related charges for the provision of a concrete batching plant, as illustrated in Figure 8.5, is entitled to be paid the *full* amount of such charges even though he might subsequently elect to use ready−mixed concrete in preference to site−batched concrete. Interim payments of the charges will, however, be certified *in proportion to the volume of concrete provided* since to do otherwise would be contractually unreasonable.

(d) Provisional Sums

CESMM2 establishes the general rule that the quantities billed for measureable items should be those which are *expected* to be required. Consequently, and based on the reasonable principle that a tenderer should not be required to price unquantifiable work, any work which is *known to be required*, but is *uncertain in character and extent at the time of tender*, should be covered by a *Provisional Sum* in the Bill of Quantities.

For example, consider the case of a factory construction project where the pre−contract borehole survey has indicated the presence of 'bad ground' in certain areas of the Site, within which the factory foundations will require to be deeper than normal. Since the borehole survey provides only a *general* indication of the soil strata underlying the site, the *exact* extent of the extra foundation work is unlikely to be known until after the excavation for such has actually commenced. This being the case, the additional work is entered in the Bill as a *single* item, suitably described, against which a *provisional sum* is inserted to cover the estimated cost thereof.

Although the estimated value of 'provisional' work is entered as a *sum*, it is subject to admeasurement in that the *actual* work carried out is valued as would be any other variation instructed by the Engineer, namely on the basis of billed rates (if appropriate) or by negotiation. Since the extent of the work is unknown *a−priori*, as is the basis on which it will subsequently be valued, the *actual* value may deviate significantly from its *estimated* value. In this respect, Clause 58(1) of the I.C.E. Conditions of Contract states that a provisional sum '... may be used in whole, or in part, or not at all, at the direction and discretion of the Engineer ...'

Thus, the Engineer is given complete freedom, under the contract, to instruct as much of the relevant work as he considers necessary *up to the value of the Provisional Sum*. It is most important to note that he is given no *express* permission to *exceed* the sum. That said, however, it is generally recognised that the Engineer is no more accomplished at the art of clairvoyance than is anybody else and that, as a consequence, exceedance of such a sum is always a distinct possibility. If it *is* exceeded, however, there is a distict *probability* that the Engineer will be required to justify the fact. This being the case, and without wishing to suggest he adopt a 'multiply by 2 and add £10,000 just in case' approach to his estimation, the Engineer should advisedly bear in mind that it is preferable, and certainly less problematical, to *over-estimate* and *under-spend* than it is to do the converse!

(e) Prime Cost Items

As discussed previously (see section 7.5.1), *Prime Cost Items* are those items which cover work performed by *Nominated Subcontractors*, with each item being priced, by the Engineer, as a *lump sum* known as a *Prime Cost*.

Since there are obvious cost implications, CESMM2 requires separate itemisation of work carried out by Nominated Subcontractors according to whether it includes, or does not include, work on the Site.

During the performance of the relevant items of work, the Engineer certifies for interim payment such proportion of the Prime Cost as he deems appropriate in the light of his assessment of the progress of the work. Despite the fact that the price for such work is negotiated by the *Engineer* prior to the award of the main contract, it should be remembered that the relevant subcontract is between the *Main Contractor* and the *Nominated Subcontractor*. This being the case, interim payments for completion of the work are made to the Main Contractor who then pays the Nominated Subcontractor. Although no maximum permissable time interval is stated in this respect, the Main Contractor is generally required to pay the Nominated Subcontractor within a *reasonable* time following receipt of *his* payment. If the payment is *unreasonably* delayed, the Employer is empowered to pay the Nominated Subcontractor *directly* and deduct, via the Engineer, the appropriate amount from the next interim valuation (see section 9.1.6). Indeed, the Engineer is entitled to demand proof, prior to certifying *any* interim payment to the Main Contractor for such work, that the latter's account with the Nominated Subcontractor is up to date. Failure to provide such proof will result in the Engineer taking the previously mentioned course of action.

In the context of payment of Prime Costs, it should be mentioned at this point that, for fairly obvious reasons, Prime Costs are not subject to Baxter–formula adjustment.

As a consequence of the Nominated Subcontractor being employed, under contract, by the Main Contractor, the latter is responsible, under the terms of the main contract, for all the actions of the former. Not unreasonably, therefore, he is entitled to payment for his assumption of this responsibility (and its attendant element of risk) and, in this respect, each Prime Cost Item is accompanied by a 'profit and other charges' item which the Main Contractor is required to price as a percentage of the associated Prime Cost.

The Main Contractor is similarly entitled to charge a lump sum for the supply of any 'labours' in connection with his attendance upon the Nominated Subcontractor. In accordance with CESMM2, 'general' and 'special' labours require separate itemisation:

For nominated subcontracts involving work *on* the site, CESMM2 defines 'general' labours as those supplied by the Main Contractor *only* in connection with the Nominated Subcontractor's use of any '... temporary roads, scaffolding, hoists, messrooms, sanitary accomodation and welfare facilities ...' which the former has provided principally for his *own* use. Also included within this category are labours supplied '... for providing space for ... (the Nominated Subcontractor's) ... office accomodation and storage of ... (the Nominated Subcontractor's) ... plant and materials, for disposing of ... (the Nominated Subcontractor's generally vast quantities of) ... rubbish, and for providing light and water for the work of the Nominated Subcontractor ...'.

For nominated subcontracts *not* involving work on the site, 'general' labours are defined as those involved *only* in '... unloading, storing and hoisting materials supplied by ... (the Nominated Subcontractor) ... and returning ... (the Nominated Subcontractor's) ... packing materials ...'.

In both cases, 'special' labours are defined as those which involve anything not included in the above definition or, indeed, those which *specifically exclude* anything in that definition.

A typical Bill entry for Prime Cost Items is shown in Figure 8.6

Number	Item description	Unit	Quantity	Rate	Amount	
					£	p
	NOMINATED SUBCONTRACTS WHICH INCLUDE WORK ON THE SITE					
A510	Supply and Installation of Electrical Control Gear.	sum			27,500	00
A520	Labours.	sum				
A530	Special labours: attendance on testing.	sum				
A540	Other charges and profit.	%				
	NOMINATED SUBCONTRACTS WHICH DO NOT INCLUDE WORK ON THE SITE					
A610	Supply of precast prestressed pretensioned concrete bridge beams.	sum			88,000	00
A620	Labours.	sum				
A640	Other charges and profit.	%				

Figure 8.6 Example of Billing of Prime Cost Items

These, then, are the various items which are included in Class A – General Items.

Although the remainder of the items in the 'Work Items' section generally make up the bulk of the Bill of Quantities, little more can be said about them other than that they are, in the main, *measurable* items belonging to Classes B to Y of the Work Classification and should be entered in the order suggested by that classification.

There are, however, a number of rules which apply to the manner in which they are Billed but these will be dealt with in section 8.3.3

8.2.5 The Grand Summary

This is the final section of the Bill of Quantities and, in essence, is just that which is suggested by its title, namely: a summary of all the prices entered in the preceding sections of the Bill.

In accordance with CESMM2, the Grand Summary should firstly contain a tabular presentation of the totals of the various parts into which the Bill has been divided (see section 8.2.4).

This is followed, if required, by a provisional sum, inserted by the Engineer, known as the *General Contingency Allowance*. As discussed previously, *specific* contingencies are catered for by the inclusion of Provisional Sums in the General Items section of the Bill. There is, however, the possibility of something occurring which was *completely* unforeseen at the time the Bill was compiled and which could not, therefore, be *specifically* catered for. Should something of this nature occur, any extra costs incurred as a result can be met from this allowance. If deemed appropriate, such a sum rarely exceeds 5% of the Engineer's pre-contract estimate of the total contract price.

It is sometimes the case that the priced Bill of Quantities requires last-minute adjustment before it is finally submitted.

There are two *main* reasons why the need for such may arise, the first of which is to accomodate for the arrival, right at the end of the tender period, of revised quotations from subcontractors or suppliers. The second reason for such relates to a process – sometimes known as the 'final adjuducation' of the tender – whereby the tenderer's senior staff adjudge the *commercial viability* of the tender immediately prior to its submission. In view of the inverse relationship between the 'profitability potential' of a tender and its 'acceptance potential', the pricing of such is something of a 'balancing act' in that the general level of prices must be *high* enough to minimise the risk of the tenderer subsequently losing money, if awarded the contract, but *low* enough to ensure that he has a *realistic* chance of being awarded the contract in the first place. If the 'final adjuducation' indicates the need for a general pricing level above or below that adopted by the estimator in pricing the Bill, the total of the priced Bill will require to be adjusted accordingly.

Whichever is the case, it is unlikely that there will be sufficient time to effect such an adjustment by altering the relevant rates, prices and sub-totals in the Bill. As a convenience to tenderers, therefore, CESMM2 allows for the insertion of a lump-sum *'Adjustment Item'*, immediately before the Bill total, as a means by which the tenderer can raise or lower such total without altering individual rates and prices in the preceding sections of the Bill.

Payment (or *deduction*, if appropriate) of such will be made pro-rata to the *gross* value (i.e. *before* the deduction of any retention) of work items included in interim valuations until such time as the sum is fully realised. In this respect, it is of considerable importance to note that the 'proportionality' calculations will be based on the *original* (i.e. *tendered*) total of the priced Bill, including Prime Cost and Provisional items, *before* the addition or deduction of the Adjustment Item. Thus, it is quite possible, as a consequence of variations subsequently reducing the overall value of the contract, for the sum to be only *partially* realised by the date of issuance of the Completion Certificate. In such a case, the outstanding balance becomes payable immediately following such. Conversely, an *increase* in the overall contract value will probably result in the sum being fully realised in advance of substantial completion of the Works. In this event, it is sufficient to state that the *aggregate* adjustment should not exceed the sum specified in the Bill.

As discussed previously (see section 8.2.2), details of the method of payment (or deduction) of the Adjustment Item should appear in the Preamble to the Bill. A statement should also be included therein to the effect that the Adjustment Item will be taken into account in the calculation of any contract price fluctuations (see section 5.2).

The final insertion in the Grand Summary is the *Total of the Priced Bill of Quantities* (sometimes referred to, rather grandly, as the *Grand Total*) which, rather unsurprisingly, is the total of everything included in the Grand Summary.

8.3 Quantification and Billing of Measured Items

This process is carried out in three stages:

(a)   Taking-off;

(b)   Abstraction;

(c)   Billing;

Each of these aspects are now discussed in turn with the intent of providing an overview of the general procedures involved.

8.3.1 Taking-off

In essence, this exercise is the calculation of the item quantities from the dimensions given in the drawings.

Although the items are generally considered in the order in which they appear in the CESMM2 Work Classification, there is no strict rule in this regard. Indeed, it may often be inconvenient to do so. For example, it will often be easier to calculate the quantity of backfill required round a foundation (Class E: Earthworks) by simply substracting the volume occupied by the foundation (previously calculated in connection with Class F: In Situ Concrete) from the gross volume of the foundation excavation (previously calculated under Class E: Earthworks) than it would be to perform a separate calculation based on the backfill dimensions.

For obvious reasons, particular attention should be paid to the divisional classification in CESMM2, and to the appropriate Measurement, Definition, Coverage and Additional Description rules therein, since these will determine:

(a) how the work is to be itemised;

(b) the units in which the quantities are to be expressed;

(c) how each item is to be measured;

(d) the nature and extent of work deemed to be included in each item;

Whilst the CESMM2 Work Classification will determine the bulk of the itemisation, it should be appreciated that it is not the 'last word' on the subject in that it does not cover *every* eventuality. It should particularly be remembered that separate itemisation is required not only for *different* types of work but also for *similar* types of work carried out under *different circumstances giving rise to considerations of cost*. In this respect, a certain amount of engineering judgement, based on a knowlege of possible construction methods and approaches, should be exercised.

Whilst it may be argued that, as long as the resulting quantity is correct, the means of achieving such is irrelevant, it should be appreciated that, in the majority of cases, the taking-off calculations will be checked with a view to determining whether or not they *are* correct. To minimise the possibility of subsequent confusion arising, therefore, the quantification should be carried out in accordance with the standard procedure detailed hereunder:

All mathematical calculations should be entered on 'dimension sheets' conforming to the requirements of BS 3327 *Stationary for Quantity Surveying*. As shown in Figure 8.7, each sheet is divided into two identically ruled sections with each section containing four columns:

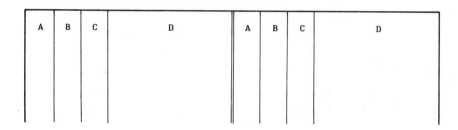

Figure 8.7 Dimension Sheet Layout

Column A is termed the 'timesing' column and is used to enter the number of times an item occurs if more than one such is required to be measured. It is also used to enter additional multiplication factors if the item in question is anything other than a rectangle or rectangular prism. For example, a calculation to determine the volume of a sphere of given diameter would require the entry of $\pi/6$ in this column. It is customary for each entry in this column to be followed by a 'slash' which, in terms of the calculation, may be regarded as the (less confusing) equivalent of a multiplication sign.

Column B is referred to as the 'dimension' column in which the dimensions of the item, as abstracted or calculated from the drawings, are entered. Such dimensions are *always* entered one under the other in order of length, width and depth/thickness. Depending upon whether the required quantity is linear, areal or volumetric, there may be one, two or three lines of dimensions.

Column C is called the 'squaring' or 'cubing' column and is used to enter the calculated quantity of the item.

Column D, the 'remarks' or 'description' column, is that in which the description of the item is entered. Any sketch used for clarification purposes may be entered in this column, as may any 'waste' calculations (see below). If the item quantity in Column C is anything other than a *volume*, its type should advisedly be specified in Column D to avoid possible confusion.

Before moving on to consider a worked example of taking–off, a number of particular points should be noted, the first of which concerns the entry of dimensions in the appropriate column of the dimension sheet.

Although the CESMM2 Work Classification specifies the units in which item quantities should be expressed in the Bill, it appears to be almost *de rigeur* for the dimensions on the drawings to be expressed in *different* units. To avoid confusion, not to mention subsequent embarrassment (e.g. 12.5m^3 of concrete being billed as 125m^3 owing to the Bill compiler's ignorance as to how many cubic millimetres make up a cubic metre!), dimensions entered in the 'dimension' column should *always* be expressed in the units specified in CESMM2.

For information purposes, the units used by CESMM2 are as follows:

(a) **Linear measure (m)**: items of constant cross–section which are more conveniently priced on a linear basis – e.g. pipes, kerbs;

(b) **Area (m^2, ha)**: items which are areal in character or are essentially two–dimensional layers – e.g. site clearance, formwork, road bases; the most notable exceptions to this rule are blinding and concrete placed in slabs which, despite being two–dimensional layers, are measured volumetrically to be consistent with other concrete placement items;

(c) **Volume (m^3)**: bulk items priceable on a volumetric basis – e.g. earthworks, concrete;

(d) **Mass (kg, t)**: items which are more conveniently priced on the basis of their mass – e.g. reinforcement, structural steelwork;

(e) **Time (hr, wk)**: items whose cost depends upon the time for which they are required – e.g. pumping;

(f) **Number (nr)**: complex fabricated items which cannot be broken down into individually priceable components – e.g. traffic signs, pre–cast units;

(g) **Lump sum (sum)**: specified lump sum charges – e.g. all of the items in Class A (General Items) of the Work Classification;

As with the units in which they are expressed, the dimensions given on the drawings will rarely be those which can be *directly* abstracted from the drawings for measurement purposes. In such cases, intermediate calculations – 'wastes' – will be required to 'build up' the appropriate dimension for entry in the 'dimension' column. No matter how trivial or elementary such calculations appear, they should be entered in the 'description' column by way of preventing any subsequent doubt arising as to how the 'measurement' dimensions were arrived at. Since the drawings will invariably be in the form of prints, which are rarely (if ever) true to scale, the practice of scaling dimensions from such, for measurement purposes, should be avoided wherever possible.

Whilst there are no set rules governing the general format and layout of entries on the dimension sheets, other than those mentioned previously in connection with column entries, the following guidelines are useful:

(a)   The *first* dimension sheet for each *main* section of the Works should be suitably headed and should contain a list of the drawings from which the relevant item quantities have been derived.

(b)   In order to facilitate a *systematic* transfer of quantities from the dimension sheets to the Bill, and to ensure the early discovery of missing sheets, each sheet should be suitably headed and numbered.

(c)   Each item should be described and numbered in accordance with the convention adopted for the Bill (see section 8.3.3). In order to save space and time, many of the words used in such descriptions may be abbreviated *except* where the use of abbreviations might, on account of possible misinterpretation thereof, give rise to subsequent confusion. A selection of some of the more commonly–used abbreviations is given in Figure 8.8.

(d)   For purposes of clarity, and to avoid confusion arising over the tranference of data to abstract sheets (see section 8.3.2) or the Bill, each item should be demarcated by a line drawn across the full width of columns A to D of the dimension sheet. With particular regard to the avoidance of confusion, each item on the dimension sheet should be 'scored out' following its transference to the Bill (or abstract sheet, if appropriate).

(e)   In order to minimise the need for 'volume = length × cross–sectional area' type calculations, complex–shaped items should

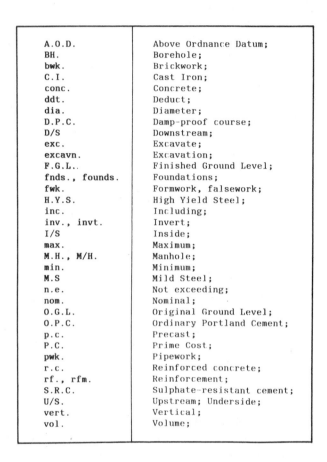

A.O.D.	Above Ordnance Datum;
BH.	Borehole;
bwk.	Brickwork;
C.I.	Cast Iron;
conc.	Concrete;
ddt.	Deduct;
dia.	Diameter;
D.P.C.	Damp-proof course;
D/S	Downstream;
exc.	Excavate;
excavn.	Excavation;
F.G.L..	Finished Ground Level;
fnds., founds.	Foundations;
fwk.	Formwork, falsework;
H.Y.S.	High Yield Steel;
inc.	Including;
inv., invt.	Invert;
I/S	Inside;
max.	Maximum;
M.H., M/H.	Manhole;
min.	Minimum;
M.S	Mild Steel;
n.e.	Not exceeding;
nom.	Nominal;
O.G.L.	Original Ground Level;
O.P.C.	Ordinary Portland Cement;
p.c.	Precast;
P.C.	Prime Cost;
pwk.	Pipework;
r.c.	Reinforced concrete;
rf., rfm.	Reinforcement;
S.R.C.	Sulphate-resistant cement;
U/S.	Upstream; Underside;
vert.	Vertical;
vol.	Volume;

Figure 8.8 Examples of Commonly–used Abbreviations

be subdivided into simple–shaped components with the individual quantities of such being summed to give the overall item quantity. A sketch should be provided showing the componental breakdown. The calculations for each component quantity should be demarcated by a line drawn across columns A and B of the dimension sheet. Under the last component quantity, a line should be drawn across the width of columns A to C, under which the overall item quantity should be entered. In the event of a component quantity requiring to be deducted, such as that for an opening in a suspended slab, a note to this effect (i.e. 'ddt') should be made in column D.

By way of illustrating the taking–off procedure, consider the construction of foundation slabs for 20 identical houses. A 'typical detail' of the foundations is given in Figure 8.9, giving sufficient information for construction purposes;

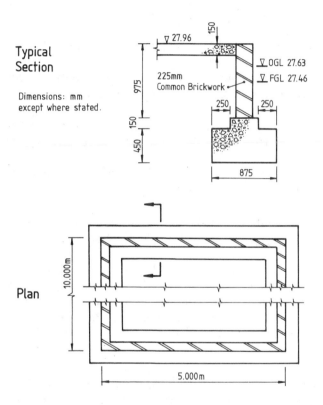

Figure 8.9 Detail for Worked Example of Taking–Off

The following information should also be noted:

1. Topsoil may be assumed to have been previously stripped in connection with another contract;

2. Footings to be constructed from C30/20 mass concrete;

3. Brickwork to be laid centrally on footing upstand;

4. Filling to be carried out to Finished Ground Level (FGL) or

underside of slab as appropriate;

5.   Ground slab to be constructed from C30/20 concrete reinforced with 2 layers of H.Y.S Mesh Reinforcement to B.S. 4483 (Type A252: 3.95 kg/m^2).

6.   50mm cover required to all reinforcement;

7.   Top surface of ground slab to have a wood float finish;

8.   Earthworks bulking and compaction factors, with respect to 'virgin' soil, should be taken as 1.2 and 0.9 respectively;

The dimension sheets, containing the taking-off calculations, are shown in Figures 8.10a to 8.10d. Although only one set of dimension sheet columns has been used in the examples shown, to facilitate the inclusion of an 'explanation' column, it should be noted that both sets of columns would be used in practice.

A particular technique which is not covered in the worked example, but which should be noted, is that of 'dotting-on'.

By way of illustration of this technique, consider the taking-off calculations for a basement car park. If the drawings showed there to be 3 rows of 5 identical columns supporting the roof slab, the calculation for (say) placement of concrete in the 1.5m × 1.2m × 0.45m column bases would be entered on the dimension sheet as shown in Figure 8.11a.

If, subsequently to the calculations being carried out, the design of the car park was changed such that 4 rows of 6 columns were now required, the *additional* quantities could be 'dotted-on' as shown in Figure 8.11b. Note that the 'dot' is equivalent to a 'plus' sign.

8.3.2 Abstraction

In circumstances where items on dimension sheets cannot conveniently be transferred *directly* to the relevant sections of the Bill, they may be grouped together in abstract — generally using A3 size *abstract* (or *analysis*) *paper* ruled vertically into 25mm (or thereabouts) columns — where they will be suitably classified and collated for entry into the Bill.

Abstraction is most commonly used where individual quantities of

			Exc. for founds. Max. depth 1.25m (E324) 450　27.96 150　27.63 975　0.33 1575 330 1245 Mean perimeter 5000　10000 225　　225 4775　9775 　　4775 14550×2 = 29100	Mean perimeter used since this will save further calculation when dealing with the brickwork, the footings etc.
20/	29.1 0.875 1.245	634.0		
20/	29.1 0.875	509.25	Preparation of exc. Surfaces (E522.1) (Area) U/s of footing	Rule M11 : for surfaces which are to receive permanent work – whether expressly required or not.
20/2/	29.1 0.45	523.8	Preparation of vert. exc. Surfaces (E522.2) (Area) Sides of footing.	Rule A7: separate items for vert. and horiz. Surfaces. Where no 'orientation' included in description, Surface is assumed to be horizontal.
20/	29.1 0.875 0.45	229.16	Placing of mass conc. in footings: thickness 450 mm (F423) Lower section	Requires separate itemisation on account of differing concrete thicknesses.
20/	29.1 0.375 0.15	32.74	Placing of mass conc. in footings: thickness 150 mm (F421)　875 　　　　　-250 　　　　　-250 Upstand　375	Would still be the case if same thickness since separate operations giving rise to considerations of cost.

Figure 8.10a Worked Example: Dimension Sheet No. 1

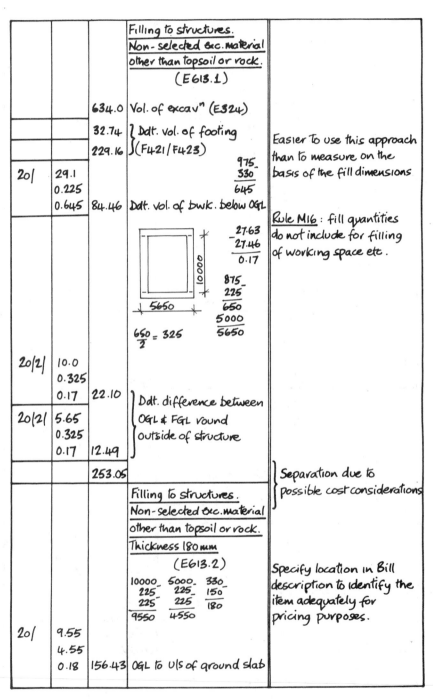

			Filling to structures. <u>Non-selected exc.material</u> <u>other than topsoil or rock.</u> (E613.1)	
		634.0	Vol. of excavⁿ (E324)	
		32.74	} Ddt. vol. of footing	Easier to use this approach
		229.16	} (F421/F423)	than to measure on the
20/	29.1		975 330 ‾‾‾ 645	basis of the fill dimensions
	0.225			
	0.645	84.46	Ddt. vol. of bwk. below OGL	Rule M16 : fill quantities do not include for filling of working space etc.
			27·63 27·46 ‾‾‾ 0·17	
			875 225 ‾‾‾ 650 5000 ‾‾‾ 5650	
20/2/	10.0		$\frac{650}{2}$ = 325	
	0.325			
	0.17	22.10	} Ddt. difference between	
20/2/	5.65		} OGL & FGL round	
	0.325		} outside of structure	
	0.17	12.49	}	
		253.05		} Separation due to } possible cost considerations
			Filling to structures. Non-selected exc.material other than topsoil or rock. Thickness 180 mm (E613.2)	
			10000 5000 330 225 225 150 225 225 ‾‾‾ ‾‾‾‾ ‾‾‾‾ 180 9550 4550	Specify location in Bill description to identify the item adequately for pricing purposes.
20/	9.55			
	4.55			
	0.18	156.43	OGL to U/s of ground slab	

Figure 8.10b Worked Example: Dimension Sheet No. 2

			Trimming of filled surfaces (E712)	Rule M22: for surfaces which are to receive no permanent work — whether expressly required or not.
				Generally taken to refer to surfaces which will remain on view after completion of the permanent work.
20|2|	10.0 0.325	130.0	⎤ o/s of structure.	
20|2|	5.65 0.325	73.45	⎬ See above and diag. ⎦ for E613.1	
		203.45	(Area)	
			Preparation of filled surfaces (E722)	Rule M23: for surfaces which are to receive permanent work — whether expressly required or not.
20|	9.55 4.55	869.05	(Area) u/s of ground slab	
			Disposal of exc. material (E532)	
		634.0	Excav" (E324)	Rule M12: net exc - net fill
		253.05	Ddt. fill (E613.1)	
		156.43	Ddt. fill (E613.2)	
		224.52		
			Double handle excavated material (E542) 1.2/0.9 = 1.33	Rule M13: effectively the volume of the 'hole' made in the temporary stockpile. i.e. the bulk equivalent of the fill volume.
			253.05 156.43 ——— 409.48	
1.33|	409.48	544.61		
			Placing of reinf. conc. in ground slab. Thickness 150 mm (F521)	
20|	9.55 4.55 0.15	130.36		

Figure 8.10c Worked Example: Dimension Sheet No. 3

			Provision of C30/20 conc. Ordinary prescribed mix. (F163)	
		229.16	Item F423	
		32.74	Item F421	
		130.36	Item F521	
		392.26		
20/	9.55 4.55	869.05	Wood float finish to ground slabs (G811) (Area)	
20/2/	9.45 4.45	1682.1	Reinforcement. HYS Fabric to BS 4483 : Type A252 (G563) $\frac{4550}{100}$ $\frac{9550}{4450}$ $\frac{9550}{100}$ $\frac{9450}{9450}$ (Area) 2 layers	Deduct for cover round outside of ground slab
20/	29.1 0.975	567.45	Common brickwork Thickness 225 mm (U121) (Area)	Rule M4 : mean perimeter used to calculate area
20/2/	29.1 0.15	174.6	Plane vert. fwk. Rough finish. Width 150 mm (G142) (Area) Sides of found. Upstand.	Easiest approach is to use previously calculated mean perimeter.

Figure 8.10d Worked Example: Dimension Sheet No. 4

			Placement of reinf. conc. In col. bases; Thickness 450mm (F523)	
3/5/	1.5 1.2 0.45	12.15		

Figure 8.11a Example of 'Dotting–On': Original Calculations

			Placement of reinf. conc. In col. bases; Thickness 450mm (F523)	
3/5/ 1/2/	1.5 1.2 0.45	22.68		Extra quantities 'dotted-on'.

Figure 8.11b Example of 'Dotting–On': Amended Calculations

particular items require to be collated, from the dimension sheets, to determine the overall quantities of such. An example, concerning the collation of manhole earthworks quantities, is shown in Figure 8.12a. In such a case, each item in question is suitably described and occupies a single vertical column of the abstract sheet. For ease of transference of the quantities to the Bill, items should preferably (i.e. unless there are sound reasons indicating otherwise) be entered on the abstract sheet in the order in which they will appear in the Bill.

The process of abstraction is also used in circumstances where the quantities of particular items are most conveniently taken–off, in the first instance, in units other than those which are used to define the

		EARTHWORKS TO MANHOLES				
	Exc.for founds. Max.depth 1-2m (E424)	Exc.for founds. Max.depth 2-5m (E425)	Prepare exc. Surfaces (E522)	Disposal of exc. material (E532)	Double handle exc.material (E542)	Filling to structures (E613)
MH 1	8.13 (5)	—	10.75 (5)	7.31 (5)	1.09 (6)	0.82 (6)
MH 2	10.46 (8)	—	10.75 (8)	9.33 (9)	1.50 (9)	1.13 (10)
MH 3	—	21.62 (12)	14.40 (14)	19.50 (13)	2.82 (14)	2.12 (14)
MH 4	—	32.44 (15)	14.40 (15)	28.76 (15)	4.89 (15)	3.68 (15)
MH 5	—	55.13 (17)	18.55 (17)	49.66 (18)	7.28 (18)	5.47 (18)
MH 6	16.26 (20)	—	14.40 (20)	14.58 (20)	2.23 (20)	1.68 (21)
	34.85	109.19	83.25	129.14	19.80	14.93

Figure 8.12a Example of Abstraction (1)

quantities in the Bill and must, therefore, be subsequently reduced to the appropriate units. A notable example of such concerns the calculation of steel reinforcement quantities which, although expressed in the Bill in terms of mass, are most conveniently taken-off, from the drawings, in terms of length and bar diameter. The Bill quantities, for each particular bar size, may be derived by grouping in abstract (in the same fashion as shown in Figure 8.12a) the individual lengths of such and multiplying each column total by the appropriate 'mass per unit length' conversion factor.

A similar need for an intermediate abstraction process arises in circumstances where the method of calculating particular item quantities is not ideally suited to the 'dimension sheet calculation' format. For example, bulk earthworks quantities are generally calculated by the application of a suitable 'prismoidal formula' to the areas of cross-sections taken at regular intervals over the extent of the area in question. Figure 8.12b illustrates the use of abstraction to calculate roadworks cut and fill quantities by applying Simpson's Rule to the areas of cross-sections – extracted from dimension sheets – taken at 100m intervals along the road centre-line.

For similar reasons as those described previously in connection with taking-off, abstract sheets, like dimension sheets, should be suitably headed and numbered. For purposes of facilitating a subsequent check on the calculations, suitable reference should be made in the abstract sheets to the dimension sheets from which individual quantities have been extracted. In the examples shown, each quantity which has been extracted from a dimension sheet is accompanied by a bracketed number which represents the number of the appropriate dimension sheet from which it originated.

8.3.3 Billing

Following the item quantification exercise, each item is numbered, described and entered in the Bill along with its quantity.

A number of rules and guidelines are given by CESMM2 in this respect and may be summarised under the following heads:

(a) General Format

Work items should be entered on A4 size paper ruled and headed as shown in Figure 8.1. The 'number' column should be sufficiently wide to take a 7-digit item number whilst the 'quantity', 'rate' and '£' columns should each have a capacity of ten million less one. Each page should be numbered and should contain provision for the entry of a page total at the bottom of the 'amount' columns.

(b) Arrangement of Items

If the Bill is divided into parts, as discussed previously (see section 8.2.4), each part should be suitably numbered and headed and the items therein entered in the order of the Work Classification. If deemed appropriate, items within each part of the Bill may be grouped under suitable headings and sub-headings which should be taken as comprising part of the description of items to which they apply, and which should be repeated at the start of each page on which the relevant items are listed. For purposes of clarity, each group or sub-group should be demarcated by a line drawn across the 'item description' column immediately underneath the last item to which the heading or sub-heading applies.

CROSS SECTION	CUT AREA (m^2)	FILL AREA (m^2)	WEIGHTING	WEIGHTED CUT (m^2)	WEIGHTED FILL (m^2)
EARTHWORKS FOR ACCESS ROAD					
5+000	7.25 (12)	14.74 (12)	1	7.25	14.74
5+100	29.35 (12)	4.13 (12)	4	117.40	16.52
5+200	30.78 (12)	7.03 (13)	2	61.56	14.06
5+300	35.83 (13)	5.32 (13)	4	143.32	21.28
5+400	26.62 (13)	8.64 (14)	2	53.24	17.28
5+500	15.80 (14)	12.25 (15)	4	63.20	49.00
5+600	1.68 (15)	17.36 (15)	2	3.36	34.72
5+700	10.53 (15)	14.19 (15)	4	42.12	56.76
5+800	12.68 (15)	13.56 (16)	2	25.36	27.12
5+900	18.36 (16)	11.32 (16)	4	73.44	45.28
6+000	26.42 (17)	8.58 (17)	1	26.42	8.58
				616.67	305.34

Exc. for cuttings
Max. depth 0.5-1 m
(E223)

$$\frac{100}{3}(616.67)$$

$$= 20,556$$

Filling to embkmts
(E623)

$$\frac{100}{3}(305.34)$$

$$= 10,178$$

Disposal of exc. material
(E532)

$$20,556 - 10,178$$

$$= 10,378$$

Figure 8.12b Example of Abstraction (2)

(c) Item Numbering

Although not compulsory, it is generally considered advantageous for items to be numbered in accordance with the Works Classification coding system. Under this system, each item is assigned a code number consisting of a *letter* – corresponding to the Class in which the item occurs – followed by a *maximum* of *three digits* corresponding to the numbers of the relevant descriptive features contained in the first, second and third divisions in the Class.

For example, a code number of I236 (see Figure 8.3) would identify an item as:

Class	I	Pipework – Pipes
First Division	2	Prestressed concrete pipes
Second Division	3	Nominal Bore: 300 – 600mm
Third Division	6	In trenches, depth 3 – 3.5m

In the case of an item for which *fewer* than *three* divisions of classification are given, or for which a divisional classification does not apply, the digit 0 is to be inserted in the appropriate position in the item number. An example of such may be seen in Figure 8.6.

Similarly, the digit 9 may be inserted in the appropriate position where a descriptive feature of the item is not listed in the Work Classification.

Finally, additional suffix numbers may be used to avoid duplication of item numbers. For example, separate items for 375mm and 450mm diameter concrete pipes in trenches 3 – 3.5m deep may be numbered I336.1 and I336.2 respectively. A further example of such may be seen in Figure 8.5.

Irrespective of the item numbering system adopted for the preparation of the Bill, it is most important to note that, as opposed to item *descriptions*, item *numbers* in the Bill have no contractual significance. It may thus be inferred that the description of an item prevails in the case of conflict with the item number. For example, an item numbered as I436 and described as: 'Concrete pipes; Nominal Bore: 600 – 900mm; In trenches, depth 3.5 – 4m;' should be taken as just that and not, as might be inferred from the number, as: 'Cast Iron pipes; Nominal Bore: 300 – 600mm; In trenches, depth 3 – 3.5m;'.

(d) Item Descriptions

CESMM2 states that each item: '... shall be described in accordance with the Work Classification ...'. Broadly speaking, this may be taken as requiring each item description to comprise not more than *one* descriptive feature from each division *and* any *additional* description required by the 'Additional Description' rules. In the event that clarity and conciseness would suffer as a result of *all* relevant descriptive features being included in an item description, CESMM2 permits any descriptive feature to be omitted *providing* a reference is given in its place as to where, in the other contract documents, the omitted information may be found.

Further description may be added if the description compiled solely in accordance with the Work Classification is considered insufficient to identify *precisely* the work covered by the item, or if any *other* characteristic of the work (such as its location) is thought to have possible cost implications.

In the interests of brevity, descriptions of items for the Permanent Works should be limited to an identification of the *component* of the Works and should not extend to a description of the *tasks* required of the Contractor. For example, 'Mild steel bar reinforcement to BS 4449, nominal size 20mm' is preferred to 'Supply, deliver, cut, bend and fix mild steel bar reinforcement to BS 4449, nominal size 20mm' since the Contractor will (or, at least, *should*) know that he is required to supply, deliver, cut, bend and fix the reinforcement and that the cost of such should be included in his rates. That said, however, specific mention should be made in the item description where the tasks required of the Contractor are specifically *limited* in extent. Thus, if the Contractor in the above example was *not* required to supply and deliver the reinforcement, the item should be described as: 'Mild steel bar reinforcement to BS 4449, nominal size 20mm, excluding supply and delivery to site.'.

Where the Works components covered by a particular item are of a *single* dimension, that dimension may be stated in the item description in place of the range of dimensions given in the Work Classification. Thus, 'Excavation for foundations; maximum depth 1.8m' may be used in preference to the somewhat less precise: 'Excavation for foundations; maximum depth 1 – 2m'.

Finally, and despite the absence of any *express* statement to this effect in CESMM2, the use of abbreviations in item descriptions is, for obvious reasons relating to possible ambiguity, not advisable!

(e) Item Quantities

Bearing in mind that the Billed quantities are, for contractual purposes, regarded as 'approximate' and, in many cases, will be subsequently amended .as a result of variation orders issued in the course of the contract, phenomenally accurate quantification of items is not required at this stage. It is generally acceptable in the majority of cases, therefore, for calculated quantities to be rounded up or down prior to their entry in the Bill. Where fractional quantities *are* significant, however, those who are inclined to express numbers to ten decimal places – 'because that is what my calculator says' – should note that *one decimal place* will usually suffice! It is, of course, a different matter when remeasuring for the Final Account – in which case the quantities will undoubtedly be calculated with sufficient accuracy to produce a price to the nearest £0.01!

As an illustration of the above principles, a sample Bill, based on the dimension sheets shown in Figures 8.10a to 8.10d, is shown in Figures 8.13a and 8.13b.

8.3.4 Taking–off and Billing Exercises

The following exercises may be undertaken as a means of gaining practice in the use of CESMM2 for taking–off and billing. No abstraction will be necessary in connection with these exercises.

General Notes

1. Taking–off must be done, using the standardised approach, in accordance with the CESMM2 Measurement, Definition and Coverage rules;

2. Sufficient wastes must be included to facilitate subsequent checking;

3. Billing must be completed in accordance with the general principles given in CESMM2;

4. Bill descriptions must include any additional requirements specified in CESMM2 and must be sufficiently detailed to permit unambiguous pricing of each item;

Number	Item description	Unit	Quantity	Rate	Amount	
					£	p
	EARTHWORKS					
	Excavation					
E324	Excavation for foundations; Maximum depth: 1.25m.	m³	634			
	Excavation ancillaries					
E522.1	Preparation of excavated surfaces.	m²	509			
E522.2	Preparation of vertical excavated surfaces.	m²	524			
E532	Disposal of excavated material.	m³	225			
E542	Double handling of excavated material.	m³	545			
	Filling					
E613.1	Filling to structures.	m³	253			
E613.2	Filling to structures; Original Ground Level to underside of ground slab.	m³	156			
	Filling ancillaries					
E712	Trimming of filled surfaces.	m²	203			
E722	Preparation of filled surfaces.	m²	869			
	IN SITU CONCRETE					
	Provision of concrete					
F163	Ordinary prescribed mix; Grade C30; 20mm aggregate; Cement to BS 12.	m³	392			
	Placing of concrete					
F421	Mass; in footings; thickness: 150mm.	m³	33			
				PAGE TOTAL:		

Figure 8.13a Bill of Quantities for Worked Example (1)

Number	Item description	Unit	Quantity	Rate	Amount £	p
	Placing of concrete					
F423	Mass; in footings; thickness: 450mm.	m³	229			
F521	Reinforced; in ground slabs; Thickness: 150mm.	m³	130			
	CONCRETE ANCILLARIES					
	Formwork					
G142	Rough finish; plane vertical; Width: 150mm.	m²	175			
	Reinforcement					
G563	High Yield Steel fabric to BS 4483; Type A252: 3.95 kg/m².	m²	1682			
	Concrete accessories					
G811	Wood float finish to top surface of ground slabs.	m²	869			
	BRICKWORK, BLOCKWORK AND MASONRY					
U121	225mm common brickwork to foundations; vertical.	m²	567			
				PAGE TOTAL:		

Figure 8.13b Bill of Quantities for Worked Example (2)

5. For convenience, topsoil stripping quantities may be ignored in all cases. Reinforcement quantities may also be ignored unless details of such are given in the drawings, or in the accompanying notes.

Exercise 1.

The footings for a multi–storey car park consist of two rows of five reinforced concrete column bases and a 200mm thick concrete slab as shown in Figure 8.14. All dimensions are given in mm. The rectangular slab is 40m wide by 100m long, and is reinforced with 2 layers of High Yield Steel Fabric to B.S.4483. (Type A393: 6.16kg/m^2).

Notes:

(a) Original ground level is equal to finished slab level;

(b) Earthworks bulking and compaction factors may be taken as 1.2 and 0.9 respectively;

(c) Fair finish formwork is to be used for all *exposed* concrete surfaces as appropriate; otherwise rough finish formwork must be used;

(d) All structural concrete is to be underlaid by 50mm blinding;

(e) A minimum of 40mm cover is required for all reinforcement;

Exercise 2.

The plan and typical section for a reinforced concrete settlement tank are shown in Figure 8.15 with all dimensions being given in mm.

Notes:

(a) Original ground level is equal to finished ground level as shown;

(b) All structural concrete is to be underlaid by 50mm blinding;

(c) Formwork shall be rough–finish to the outside of the structure and fair–finish to the inside of the structure;

(d)   Brickwork for vertical walls must be fair–faced both sides;

(e)   Earthworks bulking and compaction factors may be taken as 1.25 and 0.85 respectively.

Exercise 3.

The plan and typical carriageway section of a road junction are shown in Figure 8.16.

Notes:

(a)   Original ground level is equal to finished road level;

(b)   The top paving layer is to be C40/20 paving–quality concrete reinforced with 2 layers of HYS Fabric to B.S.4483 (Type A252: 3.95kg/m^2);

(c)   Minimum cover to reinforcement is 50mm;

(d)   Earthworks bulking and compaction factors may be taken as 1.3 and 0.9 respectively;

(e)   CBGB to be cast tight against excavated surface;

(f)   Rough finish formwork to be used to support lean concrete and paving layer;

Exercise 4.

Figure 8.17 shows the plan and section of a brick manhole.

Notes:

(a)   Original Ground Level = 28.95m AOD;

(b)   Finished Ground Level = 29.54m AOD;

(c)   Top slab to be cast in situ and reinforced with 2 layers of HYS Fabric to BS 4483 (Type A193: 3.02kg/m^2);

(d)   Minimum cover to reinforcement is 40mm;

(e)   Vertical concrete faces to be cast, where appropriate, tight against excavated surfaces;

(f)   All brickwork to be engineering brickwork and fair-faced on exposed surfaces;

(g)   Earthworks bulking and compaction factors are 1.2 and 0.85 respectively;

(h)   All structural concrete to be underlaid, where appropriate, by 50mm blinding;

(i)   Remember that these examples are given to enable practice to be gained in taking-off and the use of CESMM2. Do not be tempted to Bill the item under Class K!

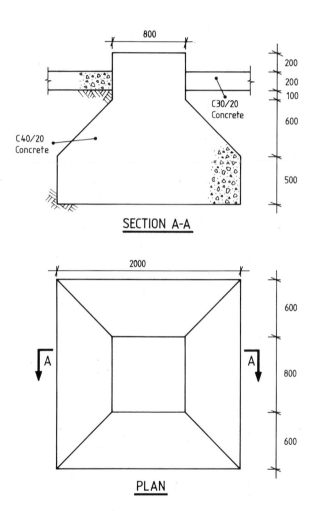

Figure 8.14 Detail for Exercise 1.

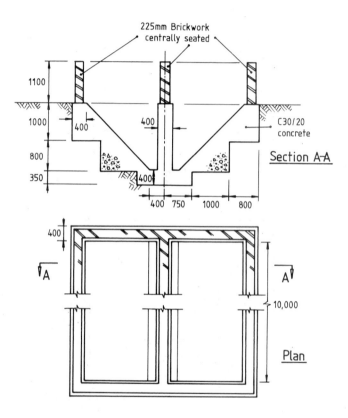

Figure 8.15 Detail for Exercise 2

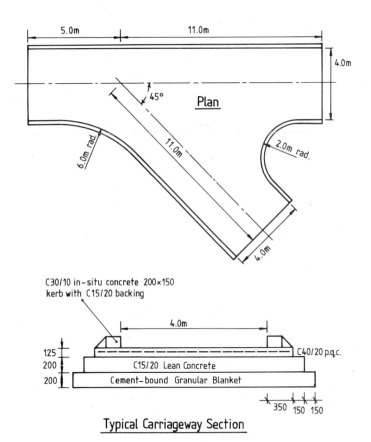

Figure 8.16 Detail for Exercise 3

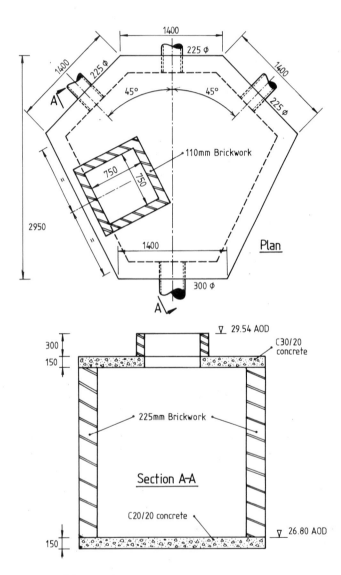

Figure 8.17 Detail for Exercise 4

CHAPTER 9

THE I.C.E. CONDITIONS OF CONTRACT

The I.C.E. Conditions of Contract, formally entitiled: *The Institution of Civil Engineers Conditions of Contract and Forms of Tender, Agreement and Bond for use in connection with Works of Civil Engineering Construction*, is reputedly the most commonly used standard for civil engineering Works executed in the UK. Issued jointly by the *Institution of Civil Engineers*, the *Association of Consulting Engineers* and the *Federation of Civil Engineering Contractors*, the standard first appeared on the scene in December 1945 and has undergone successive revision as follows:

Second Edition:	January	1950
Third Edition:	March	1951
Fourth Edition:	January	1955
Fifth Edition:	June	1973
Sixth Edition:	January	1991

Whilst the Conditions, as may be implied from the full title thereof, are specifically formulated to provide a comprehensive contractual code for the execution of Works of *civil engineering* construction, they may be used for *building* Works or, indeed, for *electrical* or *mechanical* Works if such are included in a contract which is substantially civil engineering in character.

As can be seen from the rate of revision, the standard document has had a rather chequered history in that it was revised three times in the first nine years of its life. Although the fourth edition lasted somewhat longer than its predecessors, it is generally recognised that its eighteen year life span owed more to the inability of the sponsoring bodies to reach agreement on a completely new and radically altered document than it did to its inherent adequacy. Finally, a compromise was reached and a further revision – the fifth edition – was issued which was broadly similar in wording and format to the fourth edition but contained a number of radical changes. (It might be suggested, as an 'aside', that one radical change which might beneficially have been included, but was not, would have been the use of the occasional comma in the wording and structure of clauses – if not for purposes of avoiding confusion over interpretation, then at least for the purpose of allowing the average user the opportunity to take the occasional 'breath' when reading clauses!)

This edition survived virtually unchanged – save for a few minor revisions in January 1979 – for eighteen years and, although not without its defects and imperfections, was generally acceptable to Employers, Engineers and Contractors alike. That said, however, it is sometimes suggested in certain quarters that this was largely due to a 'better the devil you know ...' attitude being adopted on the part of its users!

During the late 1980s, moves were emerging towards developing a standard Conditions of Contract more suited to alternatives to the traditional form of contract which were, at that time, coming into vogue. Whilst there was a general recognition of the need for such, opposition was being expressed, by *inter alia* the Association of Consulting Engineers and the Federation of Civil Engineering Contractors, to the radical changes proposed by the Institution of Civil Engineers. The end result was, once again, a compromise and the three sponsoring bodies issued yet another revision – the sixth edition – in January 1991.

The intent of this Chapter is to provide the reader with an introduction to the basic working provisions of the Conditions and, in this respect, discusses the content and effects of a selection of the more important clauses contained therein. Following this discussion, and based on the philosophy that the best way of 'getting to grips with' the standard document is to actually *use* it 'in anger', a selection of typical contractual problems is given to which the reader is invited to apply the Conditions.

At the time of writing, the use of the sixth edition of the Conditions of Contract is still in its infancy and it will be some time before the effects (if any) of the changes become known. Consequently, the content of this chapter will be based largely on the fifth edition but will make reference, where applicable to the discussion, to the major differences between the fifth and sixth editions.

9.1 Contents of the Conditions

The fifth edition of the I.C.E. Conditions contains a total of 198 clauses and sub–clauses (not to mention a plethora of 'sub–sub–clauses' and various parts thereof) grouped under 25 headings. For the purpose of providing a *general* understanding of the workings of the document, the coverage hereunder will be limited to those aspects of such which are deemed to have particular significance. For a *comprehensive* commentary on the document, reference should be made to one or other of the several available texts specifically dedicated to such.

For discussion purposes, the relevant clauses may be considered under the following heads:

9.1.1 Definitions

Clauses: 1(1)(h), 1(1)(i), 1(1)(n), 1(5);

To avoid any confusion or misunderstanding arising over the use of particular terms in the Conditions, all necessary definitions are given in Clause 1.

Whilst most of such are fairly straightforward, the following points are worthy of particular note:

The *Tender Total*, as defined in Clause 1(1)(h), is the total of the priced Bill of Quantities at the date of acceptance of the Contractor's tender. It serves as the principal basis on which tenders are compared but, with the notable exception of its subsequent uses in connection with calculating the amounts of the Performance Bond and the cumulative retention (see section 9.1.10) to be deducted from contractual payments made to the Contractor, has little or no significance thereafter.

In accordance with Clause 1(1)(i), the *Contract Price* is the sum of money which the Employer pays to the Contractor in return for the latter's complete and satisfactory discharge of *all* his contractual obligations. It comprises all contractual payments made on the basis of measurement, including any variations instructed in the course of the Contract, any price fluctuations where appropriate and the settlement of all claims. In short, it is the 'bottom line' price which the Employer pays for the Works.

The *Site* is defined by Clause 1(1)(n) as the lands and other places on, under, in or through which the Works are to be executed and any other lands or places provided by the Employer for the purposes of the Contract. It thus encompasses all possible locations where the Works may have to be executed. It should be noted that, depending upon whether the phrase 'provided by the Employer' is deemed to refer to the whole of the sentence preceding it, or merely to the phrase 'and any other lands or places', a certain amount of confusion has been known to arise in respect of the use of the definition, as it appears (comma−free, naturally!) in the Conditions, in determining the Contractor's entitlement to payment for materials on site in accordance with Clause 60(1).

In accordance with Clause 1(5), the word *'Cost'*, when used in the Conditions, is taken to mean the cost of labour, plant, materials *and*, unless elsewhere stated to the contrary, overheads. By implication, 'cost' does not include profit − although, it has to be said, such exclusion gives rise to a number of anomolies in the application of the definition of 'cost' and is a bone of considerable contention in the contracting community!

6th Edition Changes

The definition of the Tender Total *is as in the 5th Edition but contains the proviso that, in the absence of a Bill of Quantities, the Tender Total will be taken as the agreed estimated total value of the Works at the time of tender.*

The definition of the Site *has been extended to cover additional areas subsequently considered by the Engineer as forming part of the site.*

The fact that 'cost' does not include for profit (unless otherwise indicated) is now clearly stated.

9.1.2 The Contract

Clauses: 1(1)(e), 5, 7(1), 9, 13(3), 44;

As referred to from time to time in the Conditions, the 'Contract' is defined by Clause (1)(1)(e) as comprising: '... the Conditions of

Contract Specification Drawings Priced Bill of Quantities the Tender the written acceptance thereof and the Contract Agreement (if completed) ...'.

Subsequent to the award of the contract, Clause 9 of the Conditions empowers the Engineer, on behalf of the Employer if the latter so desires, to instruct the Contractor to '... execute a Contract Agreement ...' with the Employer. If such is the case, the completed Form of Agreement constitutes the 'legal contract' between the two parties whereby the Contractor '... covenants ... to construct and complete the Works and maintain the Permanent Works ...' in return for the Employer's covenant to pay him '... the Contract Price ... in the manner prescribed by the Contract ...'. In the absence of such an Agreement, the 'legal contract' is deemed to comprise the Contractor's tender and the Employer's written acceptance thereof.

It should be noted that, in the former case, the contract formed is 'under seal' and has a twelve-year limitation period whilst the latter course of action results in the formation of a 'simple' contract with a six-year limitation period (see Chapter 3).

In accordance with Clause 5 of the Conditions: '... the several documents forming the Contract are to be taken as mutually explanatory of one another ...'. In the event that discrepancies or ambiguities in such become apparent in the course of constructing the Works, the clause empowers the Engineer to explain and adjust such and requires him, in such an event, to accordingly notify the Contractor *in writing*. If any such adjustments result in the Contractor suffering delay or disruption to his execution of the Contract, or cause him to incur costs '... beyond that reasonably to have been foreseen by an experienced contractor at the time of tender ...', Clause 13(3) (see section 9.1.11) entitles the Contractor to recompense for the additional cost and requires the Engineer to take the delay into account when determining any extension of the contract completion time to which the Contractor is entitled under Clause 44 (see section 9.1.8).

A further clause which should be noted in respect of the 'adjustment' of documentation in the course of the contract is Clause 7(1) which empowers the Engineer to issue to the Contractor '... such modified or further drawings and instructions as shall in the Engineer's opinion be necessary for ... the adequate construction completion and maintenance of the Works ...' and to bind him contractually by such.

In the context of 'document adjustment' generally, it should be particularly noted that the Engineer's wide powers of such are limited to the 'technical' documentation and do not extend to his amendment or alteration of the fundamental terms and conditions of the Contract – simply because the latter is the sole province of the parties thereto.

6th Edition Changes

Clause 7 contains additional sub–clauses to cover the supply of documents by the Contractor in the event that he is involved in the design of parts of the Permanent Works.

9.1.3 Control of the Works

Clauses: 1(1)(d), 2, 15, 37;

The Conditions envisage that the execution of the Works will be given full–time supervision on behalf of the Engineer and the Contractor.

In respect of the former, the key personage is the *Engineer's Representative*, commonly known as the *Resident Engineer*, who, in accordance with Clause 1(1)(d), is appointed by the Engineer or the Employer '... to perform the functions set forth in Clause 2(1) ...', namely: '... to watch and supervise the construction completion and maintenance of the Works ...'.

As a convenience, and to eliminate delays in the day–to–day administration of the contract, Clause 2(3) permits the Engineer to '... authorise the Engineer's Representative ... to act on behalf of the Engineer either generally in respect of the Contract or specifically in respect of particular clauses of these Conditions of Contract ...' and states that any act of the Engineer's Representative '... within the scope of his authority shall for the purposes of the contract constitute an act of the Engineer ...'.

Fairly obviously, the phrase 'within the scope of his authority' may be taken to imply that any actions of the Engineer's Representative outwith such will, for contractual purposes, be regarded as *ultra vires*. To clarify the position and, perhaps more importantly, to avoid the possibility of the Contractor's unknowing acquiescence in any such acts compromising his associated rights under the contract, Clause 2(3) requires that any delegation of authority to the Engineer's Representative should be *in writing* and that the Engineer should provide the Contractor with *prior written notification* of such.

In exercising the powers delegated to him by the Engineer, particularly those which are discretionary in nature, the Engineer's Representative should endeavour to act as he thinks the Engineer would act and should not impose his own views if he knows that they conflict with those of the Engineer. In particular, he should appreciate that Clause 2(4) provides the Contractor, in the event of his being dissatisfied with any decision of the Engineer's Representative, with the right to refer the matter to the Engineer for *his* decision.

Whilst the extent of vicarious authority invested in the Engineer's Representative will vary according to circumstance, it should be particularly noted that the Engineer is expressly prohibited from delegating any authority in respect of his powers under:

Clause 12(3)   —   Delay and additional cost due to unforeseeable physical and artificial obstructions;

Clause 44   —   Extension of time for completion;

Clause 48   —   Certificate of Completion of the Works;

Clause 60(3)   —   Final Account;

Clause 63   —   Forfeiture;

Clause 66   —   Settlement of Disputes;

In order to aid the Engineer's Representative in the exercise of his function, Clause 2(2) provides for the appointment, by the Engineer or the Engineer's Representative, of any number of assistants subject to the proviso that the Contractor be notified in writing of the names and functions of such. It should be particularly noted that such assistants have no power to issue instructions to the Contractor save those which are deemed necessary to enable them to discharge the functions accorded them by the Engineer or the Engineer's Representative, and those related to their acceptance of workmanship and materials. For contractual purposes, the Contractor is required to regard any instructions from such assistants as emanating from the Engineer's Representative. In the event of his dissatisfaction with any such instructions, Clause 2(4) provides the Contractor with the right to refer the matter to the Engineer's Representative who may confirm, reverse or vary such.

Finally, and by way of enabling the Engineer, or any person

authorised thereby, to discharge his contractual responsibilities *vis–a–vis* supervision, Clause 37 requires that such persons be afforded unlimited access to the site and to any other place in which work is being carried out in connection with the Contract.

Clauses 15 and 16 cover the supervision of the Works from the Contractor's standpoint.

First and foremost, Clause 15(2) requires the Contractor to provide a '... competent and authorised agent or representative ...' to take '... full charge of the Works ...'. Such a person requires the *written* approval (which may be withdrawn at any time) of the Engineer, must be '... constantly on the Works ...' (certain readers with experience of site work will note this requirement with wry amusement!) and must '... give his whole time to the supervision of the same ...' (ditto). All instructions given to the Contractor in the course of the contract are to be given *via* the Contractor's Agent and are deemed to take effect immediately upon such.

In respect of supervision of the Works generally, Clause 15(1) requires the Contractor to provide '... sufficient persons ... as may be requisite for the satisfactory construction of the Works ...' and further requires that those persons should have '... adequate knowledge ...' of the proposed construction operations *and* the safety aspects thereof. By way of ensuring that these criteria are complied with, Clause 16 empowers the Engineer to object to, and require the removal from site of, any such person (supervisory or otherwise) who, in *his* (!!) opinion, '... misconducts himself ...', or is '... incompetent or negligent in the performance of his duties ...', or who persistently disregards any specified safety provisions or general safety requirements.

6th Edition Changes

Clause 2 is now entitled 'Engineer and Engineer's Representative' and, unlike the 5th Edition version of the clause, deals with the duties and responsibilities of the Engineer as well as those of the Engineer's Representative. The clause has been re–written and substantially extended to contain eight sub–clauses instead of four. In particular, Clause 2(2) introduces (through the 'back door'!) the requirement that the person acting as the Engineer should be a Chartered Engineer (although not necessarily a Chartered Civil Engineer). Additionally, new Clause 2(8) requires that the Engineer act impartially in all matters save those in which he is required to obtain the Employer's consent – thus expressly stating a requirement which could, previously, only be implied.

9.1.4 General Obligations of the Contractor

Clauses: 1(1)(l), 1(1)(o), 7(3), 8, 11, 13(1), 19–26, 30, 44, 52(4);

In accordance with Clause 8(1), the Contractor assumes responsibility for *all* aspects of the construction, completion and maintenance of the Works (defined by Clause 1(1)(l) as comprising the Permanent *and* Temporary Works) in accordance with the provisions of the Contract. This includes the provision of all labour, plant, materials and temporary Works whether the need for such is *specified* or whether it may merely be *reasonably inferred* from the Contract.

This general requirement is repeated in Clause 13(1) which further states that the Works are to be constructed, completed and maintained '... to the satisfaction of the Engineer ...', to whose instructions the Contractor should strictly adhere, *unless* it is legally or physically impossible so to do.

With regard to the 'caveat' concerning physical impossibility, it is generally held that the Contractor is absolved from further performance if it transpires to be impossible, from a commercial or practical standpoint, to construct the Works as designed. Potential invokers of this 'loophole' should, of course, note that there is a considerable difference between work which is impossible and that which is found to be substantially more difficult and involved than originally anticipated!

In constructing, completing and maintaining the Works, Clause 8(2) requires that the Contractor take full responsibility for the adequacy, stability and safety of all site operations and construction methods. Not unreasonably, he is expressly exempted from any responsibility in respect of the design or specification of the Permanent Works – unless provisions to the contrary are made elsewhere in the contract documentation – and in respect of any Temporary Works designed by the Engineer.

The 'safety' aspect is also covered by Clause 19(1) which requires the Contractor to '... have full regard for the safety of all persons entitled to be upon the site ...', to keep the Site and the Works '... in an orderly state appropriate to the avoidance of danger to such persons ...' and to take such measures as are deemed appropriate to minimise any danger to the general public. Clause 19(2) places a similar 'duty of care' upon the Employer in the event that he carries out work on the Site with his own workforce.

To a certain extent, this Clause is redundant since the Occupiers Liability Act 1984 (see section 2.2) imposes a *statutory* duty of care on the Contractor (or the Employer, if present on site in any 'controlling' capacity) to ensure that authorised persons entering the site are reasonably safe whilst there. It should also be mentioned that the Act imposes upon the Contractor a limited duty of care to *trespassers* in that he is required to take all reasonable steps to prevent their injury.

In accordance with Clause 20(1), the Contractor is responsible for the care of the Works (Temporary *and* Permanent) from the Date of Commencement of such until 14 days after the issuance of the Certificate of Completion. A similar provision is made for parts of the Works for which Completion Certificates have been issued prior to the entire Works having been completed. Should the Works suffer any damage, *from any cause whatsoever* (including, it would seem, the Employer's negligence!), during that period, Clause 20(2) requires the Contractor to '... repair and make good ...' such *at his own expense*. Some relief is, however, provided by Clause 20(3) which absolves the Contractor from any responsibility in respect of so-called 'excepted risks'. It is of particular interest to note that, apart from the usual battery of 'civil war, invasion, revolution, nuclear contamination, sonic boom' type events, the list of such risks includes any '... fault defect error or omission in the design of the Works ...' except, of course, where such design has been supplied by the Contractor as part and parcel of his contractual obligations.

Clauses 21 to 25 contain the contractual requirements regarding insurance against, and liability for, any losses:

Clause 21 requires the Contractor to insure, in the joint names of himself *and* the Employer, the Permanent and Temporary Works, including any materials or suchlike which have been delivered to the site for incorporation in the Works, and the Constructional Plant (deemed, according to Clause 1(1)(o), to comprise '... all appliances or things of whatsoever nature ...' used in connection with the construction, completion and maintenance of the Works) against all loss or damage.

Since this excludes the 'excepted risks', it is of particular importance for the Employer to note that he will not be insured (under *this* particular policy, anyway) against losses arising from any defects in the Engineer's design!

Clause 21 (not unreasonably) further *expressly* requires that the terms of such insurance meet the Engineer's approval and, by empowering the Engineer to examine receipts and suchlike from time to time, includes the *implied* requirement that the Contractor's premium payments be kept up to date at all times.

Under Clause 22, the Contractor is required to indemnify the Employer against liability for damage to any person or property whatsoever (other than the Works, since such is covered by Clause 21) arising out of the construction of the Works. Similarly, the Employer is required to indemnify the Contractor against liability for damages in respect of matters outwith the latter's field of responsibility. The clause also contains details of the apportionment of contributory liability in either of the abovementioned cases.

Clause 23 requires the Contractor to insure against the third party claims for which he is, according to Clause 22, liable. The Contractor is also required to insure the Employer in connection with instances where he (the Employer) is required to indemnify the Contractor in accordance with Clause 22. In addition to reiterating the requirements of Clause 21, namely that the terms of the insurance should meet the Engineer's approval and that the Engineer is entitled to request proof of payment of the associated premiums, the clause includes the requirement that the insurance be effected '... for at least the amount stated in the Appendix to the Form of Tender ...'.

Clause 24 ensures that the Employer has no liability in respect of accidents or injuries to the Contractor's workmen *unless* he (or his servants) have contributed toward such.

Finally, in respect of insurances, Clause 25 comprises a statement to the effect that: if the Contractor fails to take out such insurances as are required under the contract, the Employer will do it for him and will charge him accordingly – either by treating the premium payments as a straightforward debt or by deducting such from subsequent contractual payments made to him.

One of the fundamental tenets of contract law is that the courts will refuse to enforce any contract which is tainted by an element of illegality (see section 3.4.4). In this respect, Clause 26 of the Conditions is of particular relevance.

In accordance with Clause 26(1), the Contractor is required to give all notices, and to pay all fees, as may be required by any '... Act of

Parliament or any Regulation or Bye–law of any local or other statutory authority ...' in connection with the execution of the Works or by '... the rules and regulations of all public bodies and companies whose property or rights are or may be affected in any way by the Works ...'. In addition to reimbursing the Contractor, *via* interim certification, for any fees paid by him in this respect, the Employer is required to reimburse him for any and all rates and taxes paid by him in connection with the Site or any other places where temporary structures, the use of which is *exclusively* in connection with the Works, have been erected.

Clause 26(2) requires the Contractor to '... ascertain and conform in all respects with ...' any of the provisions, rules and regulations described in Clause 26(1) and to indemnify the Employer against all penalties and liabilities for breach of such. The indemnification requirement is subject to the notable exception of a breach of such regulations being the unavoidable consequence of the Contractor's compliance with the '... Drawings Specification or instructions of the Engineer ...'. In such a case, the Contractor is required to draw the Engineer's attention to the matter whereupon the latter is required to issue such instructions as are necessary to ensure conformity with the said regulations. In the event of the Contractor suffering delay or incurring cost in this respect, Clause 7(3) entitles him to claim, under Clause 44 (see section 9.1.8), for an extension of time and, under Clause 52(4), for the additional costs incurred (see section 9.1.11). A noteworthy exception to the Contractor's responsibility to 'ascertain' the applicable rules and regulations is that concerning planning permission for the '... Permanent Works or any Temporary Works specified or designed by the Engineer ...'. Under part (c) of this sub–clause, the responsibility for obtaining such is placed firmly on the shoulders of the Employer who warrants that all the necessary permissions '... have been or will in due time be obtained ...'. The Contractor is expressly absolved of all responsibility in this respect.

In the context of the Contractor's *general* obligations, the contents of Clause 30, concerning the avoidance of damage to public highways in the course of transporting materials or plant to or from the site, should be noted.

Clause 30(1) requires the Contractor to take all reasonable steps to reduce the likelihood of public highways in the vicinity of the site being subject to extraordinary traffic either in terms of volume or loads carried.

In accordance with Clause 30(2), the Contractor is responsible for, and is required to meet the costs of, any bridge—strengthening or highway improvement if such becomes necessary to facilitate the movement of *Constructional Plant*, *Equipment* or *Temporary Works* required in connection with the execution of the Contract. He is also required to indemnify the Employer against any claims, made by the highway authorities, for the repair of damage incurred as a result of such movements.

Providing the Contractor has complied with the requirements of Clause 30(1), Clause 30(3) requires the Employer to meet the cost of any highway damage incurred in connection with the transportation of *materials* and *manufactured or fabricated articles* required for the construction of the Works. If such damage is deemed to have resulted from the Contractor's failure to observe the provisions of Clause 30(1) then, of course, the Contractor is required to foot the bill for the necessary repairs.

Perhaps the most important of the so—called 'general obligation' clauses in the Conditions, and often the source of considerable contention, is Clause 11.

In accordance with Clause 11(1), the Contractor is deemed to have satisfied himself, prior to the submission of his tender, as to *inter alia* the nature of the site, the nature of its ground and sub—soil, the extent and nature of the work involved in the construction and completion of the Works and, most importantly, the nature and extent of the risks which he is required to assume under the contract.

Having taken all the above factors into account, the Contractor is deemed, under Clause 11(2), to be satisfied that his tender is sufficient to cover *all* his contractual obligations.

It is of particular importance to note that the Contractor is required to satisfy himself as to the nature of the ground and sub—soil only '... as far as is practicable ...' and, in this respect, is expressly permitted to take into account '... any information in connection therewith which may have been provided by or on behalf of the Employer ...'.

Although somewhat imprecise and subjective, the phrase: 'as far as is practicable' is generally interpreted as: 'to an extent which would *reasonably* be expected of an experienced Contractor *in the limited time available*'. Any interpretation should, of course, take into account the amount and detail of any '... information in connection therewith

...' which has been supplied to tenderers since this will have a direct bearing on what could reasonably be considered 'practicable'

Without wishing to suggest that tenderers should totally mistrust any such information unless it is accompanied by an express warranty (included in the Contract) as to its accuracy and correctness, or unless it is the subject of a separate collateral agreement between the tenderer and the Employer, it should be particularly noted that the Contractor has no contractual remedy in the event that the information transpires to be inaccurate – simply because such information (usually data from the pre–contract site investigation carried out by the Engineer) does not, and is not intended to, comprise part of the Contract.

The fact that the Contractor is expressly permitted to make use of such information in adjudging the sufficiency of his tender, however, implies his right, in such an event, to instigate legal proceedings (against the Employer) under the Misrepresentation Act 1967 (see section 3.4.2). For such an action to succeed, however, the aggrieved Contractor will be required to demonstrate, to the Court's satisfaction, either that the information was misleading or that it had been compiled with less care than would reasonably have been expected in light of the importance placed on such by tenderers. In adjudging the validity of the Contractor's complaint, the Court will take into account whether or not the Employer had reasonable grounds to believe (and *did* believe) the information to be true at the time it was supplied and, bearing in mind the inherent uncertainties of geotechnical investigations, whether or not it would have been *reasonable*, or *practicable*, for the Contractor to have ascertained the validity of the information prior to submitting his tender.

6th Edition Changes

Clause 8 is substantially unchanged but now contains three sub-clauses instead of two. A notable addition to the clause is that the Contractor is required to exercise all reasonable skill, care and diligence in designing (if required) any parts of the Permanent Works.

In accordance with Clause 20, the Contractor's responsibility for the care of the Works now ceases upon the issue of the Completion Certificate (now called, more correctly, the 'Certificate of Substantial Completion of the Works') instead of, as previously, 14 days thereafter. The clause also contains an additional provision for apportionment of liability for damage which is occasioned partly by an 'excepted risk' and partly from an occurrence which is the Contractor's responsibility.

Whereas the previous Clause 21 required the Contractor to effect insurance to the full *value of the Works and the Constructional Plant, the revised clause requires such insurance to be for the* full replacement value of the Works plus 10% for additional costs which may be incurred. *The requirement to insure the Contractor's plant and equipment is now omitted.*

Clause 23 now requires third–party insurance to be in the joint *names of the Employer and the Contractor and for the relevant policy to contain a cross–liability clause so that the insurance applies to the parties as 'separate insured'.*

Clause 25 now requires the Contractor, prior to the Works Commencement Date, to provide evidence of satisfactory insurance. An additional sub–clause is included which requires both the Contractor and the Employer to comply with all the conditions set out in the insurance policies, failing which each party is required to indemnify the other against all losses and claims arising out of the said failure.

Clause 11(1) is a new sub–clause under which the Employer is deemed to have made available to tenderers all existing relevant ground investigation data with the Contractor being deemed to be responsible for the interpretation thereof. Clause 11(3) deems that the Contractor's tender will be based on such information and *his own inspection and examination.*

9.1.5 Execution of the Works

Clauses: 13, 14(3)–14(7), 17, 29, 33, 36, 38–39, 44;

The 'blanket' clause, in this respect, is Clause 13 which, in its first part (see section 9.1.4), requires the Works to be constructed, completed and maintained '... in strict accordance with the Contract to the satisfaction of the Engineer ...'. The clause further states, in its second part, that the plant, labour and materials used by the Contractor in the execution of the Works are subject to the Engineer's approval, as is the '... mode manner and speed of construction and maintenance of the Works ...'.

Clause 13(3) outlines the remedies available to the Contractor in the event that instructions issued by the Engineer, pursuant to the first two parts of the clause, have the effect of delaying or disrupting him to the extent that he incurs additional unforeseen costs (see sections 9.1.2 and 9.1.11).

With regard to the 'mode and manner' of executing the Works, the Engineer has a particular responsibility to ensure that it is such as to enable the Works to be '... executed in accordance with the Drawings and Specification ...' and that the Permanent Works, as and when completed, will not detrimentally be affected thereby. In this respect, Clause 14(3) requires the Contractor, upon the Engineer's request, to submit such details of his proposed methods of construction as are reasonably required to enable the latter to ensure that this will be the case.

If he *does* request such information, the Engineer is required under Clause 14(4) to notify the Contractor '... in writing within a reasonable period ...' either that the latter's proposed methods meet his approval or, failing that, in what respects they are considered by him to be unsatisfactory *vis-a-vis* the above requirements. In the latter event, the Contractor must then '... take such steps or make such changes in the said methods ...' as are necessary to obtain the Engineer's consent to their adoption and must not thereafter change the 'approved methods' without the Engineer's consent.

Although such may reasonably be implied from Clauses 14(3) and 14(4), Clause 14(5) expressly requires the Engineer to provide the Contractor with '... such design criteria relevant to the Permanent Works or any Temporary Works designed by the Engineer ...' as are adjudged necessary to enable his compliance with the requirements of the aforesaid sub-clauses.

In the event of the Engineer's approval of the Contractor's methods under Clause 14(4), or his supply of relevant design criteria to the latter under Clause 14(5), being unreasonably delayed, Clause 14(6) entitles the Contractor to claim for any delays or additional costs incurred as a consequence of such. Furthermore, a similar claim may be lodged in respect of any delays or additional costs which are incurred by the Contractor as a consequence of the said criteria, or the Engineer's requirements under Clause 14(4), imposing 'reasonably unforeseeable' limitations on his choice of construction methods.

With regard to the Engineer's appraisal of the Contractor's proposed methods, it should be particularly noted that he has no authority to relieve the Contractor of any of his responsibilities under the Contract.

Indeed, this is confirmed by Clause 14(7) which effectively states that, notwithstanding the Engineer's approval of such, the 'buck' stops with the Contractor in the matter of ensuring that his proposed methods meet the Contract requirements. It may thus be stated that the Engineer's powers in respect of controlling the Contractor's methods of executing the Works are strictly limited to those of veto. In other words, he is perfectly entitled to tell the Contractor what he should *not* do but is not empowered to tell him what he *should* do. To 'overstep the mark' in this respect may entitle the Contractor to a claim under Clause 13(3).

In the context of contractual responsibility, it should be noted that, in accordance with Clause 17, the overall responsibility for '... the true and proper setting–out of the Works ...' rests firmly on the shoulders of the Contractor *notwithstanding any checks carried out by the Engineer or his site representatives*. Thus, the Contractor is required to bear the cost of any reconstruction or remedial work which becomes necessary as a consequence of inaccurate setting–out even though such setting–out might well have been checked, and found to be 'correct', by the Engineer's site staff – setting–out engineers please note!

This is, not unreasonably, subject to the limitation that the Contractor be absolved of such responsibility in the event of setting–out errors being the result of incorrect data supplied *in writing* by the Engineer or the Engineer's Representative. In such a case, the cost of rectification will be borne by the Employer.

Clause 36 requires that the materials and workmanship used in the execution of the Works be of the kind described in the Contract or in the Engineer's instructions. In order to ascertain the Contractor's compliance with this requirement, the Engineer is empowered to direct, at his discretion, tests to be carried out on the Site or, if appropriate, at the place of manufacture or fabrication. He may order tests to be carried out in other places but only insofar as such places are specified in the Contract. The cost of such testing, including that of providing such samples as are required, or such assistance as is *normally* required in connection with the testing, is to be borne by the Contractor *providing* such is clearly intended under the Contract and is particularised in sufficient detail therein to be priced. Failing this, the cost is to be borne by the Employer – unless, of course, the testing shows the materials or workmanship in question not to comply with the provisions of the Contract.

In order to enable the Engineer to ascertain that the Works comply with the provisions of the Contract, Clause 38(1) prohibits the Contractor from covering up any work, or from placing any permanent work on foundations, until the Engineer has had sufficient opportunity to inspect and approve such. In this respect, the Contractor is required to give due notice to the Engineer when any such work is ready for examination, whereupon the Engineer is required to carry out such, unless he considers it to be unnecessary *and* advises the Contractor accordingly without unreasonable delay.

Clause 38(2) authorises the Engineer to order the Contractor to open up or uncover any parts of the Works for inspection and to subsequently carry out the necessary reinstatement to his (the Engineer's) satisfaction. If the work is found to be in accordance with the Contract, *and* has been covered up after the Contractor's compliance with Clause 38(1), the Employer will bear the cost of making the opening and carrying out the requisite reinstatement. In all other cases, the cost will be borne by the Contractor.

It should be particularly emphasised that, irrespective of whether the work subsequently transpires to be in accordance with the Contract, the Employer will not bear the abovementioned costs if the Contractor has not afforded the Engineer sufficient opportunity to examine the work prior to its having been covered up or put out of view. In this light therefore, and in the event of the Engineer delaying unreasonably in his examination of such work, or in notifying the Contractor that such examination is not necessary, the Contractor should (advisedly) *not* cover up such work prior to its approval (tempting though such a course of action might well be at the time!) but should wait until such time as the Engineer eventually 'performs' and then claim, under Clause 44 (see section 9.1.8), for an appropriate extension of completion time. It should be noted, however, that the subsequent success or otherwise of such a claim will depend upon whether the delay was considered *unreasonable* and whether or not the Engineer (or, more realistically, the Engineer's Representative) was given *reasonable* notice of his being required to examine the work in the first place.

In accordance with Clause 39, the Engineer is empowered, at any time during the progress of the works, to order *in writing* the removal of any materials which, in his opinion, are not in accordance with the Contract, the replacement of such with proper (i.e. contractually approved) materials and the removal and proper re-execution of any work which, in respect of materials or workmanship, is not, in his opinion, in accordance with the Contract.

In the event of the Contractor's default in complying with such an order, the Employer is entitled to employ another Contractor to do the work and to charge the defaulter for such by deducting the appropriate monies from subsequent certificates.

Worthy of particular note is the statement, contained in Clause 39(3), that the failure of the Engineer, or his representatives, to disapprove any work or materials does not prejudice their powers to *subsequently* disapprove such. Whilst this may be (and, indeed, often *is*) interpreted in some quarters as providing the Engineer and his representatives with a tacit licence to change their minds and disapprove certain work or materials which had previously been approved, it is more usually taken as providing the said parties with the reasonable right to presume that the Contractor knows what he is about and is aware of his contractual obligations. In the event that such a presumption proves to be ill-founded, the clause empowers the Engineer or his representatives to take the appropriate action and (implicitly) requires the Contractor to take the consequences.

With respect to the execution of the Works in general, Clause 29 is wothy of particular note since it deals with the effect of the construction work on surrounding properties and the general public.

In essence, the clause requires the Contractor to execute the Works in such a way as to minimise the detrimental effects thereof on the general public (specified, rather amusingly, as the 'public convenience'), and in a manner which does unnecessarily or improperly interfere with the use of, or access to, public or private roads, footpaths or properties. Additionally, the work is to be carried out without unreasonable noise or disturbance. In both cases, the Contractor is required to indemnify the Employer against any claims arising out of his liabilities in these respects.

In the general context of undue interference with the general public or adjoining properties, the Contractor should particularly note his *tortious* (as opposed to *contractual*) liabilities in respect of *private, public* and *statutory nuisance* (see section 2.3), *trespass* (see section 2.4) and *strict liability* (see section 2.5).

Finally, Clause 33 requires the Contractor, following the completion of the Works, to clear the Site of all Constructional Plant, surplus material, rubbish and Temporary Works and to leave it in a '... clean and ... workmanlike condition ...' to the satisfaction of the Engineer.

6th Edition Changes

The revision to Clause 13(3) is covered in section 9.1.11.

Clause 14(7) is substantially the same as the previous Clause 14(4) but now requires that the Engineer inform the Contractor, vis-a-vis *the acceptance or otherwise of his proposals, within 21 days as opposed to 'within a reasonable period'. Clause 14(8) replaces Clause 14(6) and, apart from indicating the addition of profit to cost in respect of additional Permanent or Temporary Works, is substantially the same thereas.*

Clauses 29(2), (3) and (4) are based on the original Clause 29(2) but are extended to cover 'other pollution'. The new Clause 29(4) contains the reservation that the Employer will indemnify the Contractor in the event of claims arising out of noise, disturbance or other pollution which is the unavoidable *consequence of carrying out the Works.*

9.1.6 Sub-contractors

Clauses: 4, 31, 58(4), 59A–59C, 63(1)(e);

In accordance with Clause 4, the Contractor is expressly prohibited from sub-letting the *whole* of the Works. He is, however, permitted to sub-let *parts* of the Works, to sub-contractors of *his* choosing, providing he has obtained the Engineer's prior consent so to do. The 'consent' requirement does not, of course, apply to *nominated* sub-contractors.

In the event that the Engineer consents (in his capacity as the Employer's *agent* – see section 3.7.2) to the Contractor's sub-letting of certain parts of the Works, it should be noted that the sub-contractors will be employed, under contract, by the *Contractor* and *not* by the *Employer*. Consequently, the Employer has no *direct* right of redress against such in the event of their default. He *does*, however, have an *indirect* right of action in such an event by virtue of Clause 4 making the Contractor *fully* responsible for the actions and omissions of any sub-contractor employed by him as if such actions and omissions were *his*. It should be emphasised that, in accordance with Clause 59A(4), this responsibility applies as much to *nominated* sub-contractors as it does to the *domestic* variety.

A noteworthy limitation, in respect of the Contractor's responsibility, is that, for the purpose of applying this clause, the use of 'labour only' subcontractors will not be regarded as 'sub-letting'.

In the event that the Contractor sub-lets *without* the Engineer's consent, it is of particular interest to note that he is not, strictly speaking, entitled to be paid by the Employer for the relevant work even though such work may have been executed satisfactorily by the sub-contractor. It should be further noted that unauthorised sub-letting may provide the Employer with the right to *forfeiture* (see section 9.1.13) of the Contract under Clause 63(1)(e).

In accordance with Clause 58(4), the Engineer is authorised to instruct the Contractor to enter into a contract with a *Nominated Sub-contractor* (see section 7.5.1) for the execution of work which is covered in the Bill of Quantities by a *Prime Cost Item* (see section 8.2.4). Additionally, the clause empowers the Engineer, if he so desires, to instruct the Contractor to carry out such work himself. Since the work involved may be of a specialist nature and outwith the field of expertise of the Contractor, however, the Contractor is given an unrestricted right to withhold his consent to such a course of action.

Notwithstanding the Engineer's powers under Clause 58(4), Clause 59A(1) entitles the Contractor to decline to enter into a contract with any Nominated Sub-contractor against whom he raises *reasonable* objection or who:

(a) refuses to undertake obligations towards the Contractor which are similar to those imposed on the Contractor under the main Contract;

(b) declines to safeguard the Contractor against any claims for damages and suchlike which arise out of the nominated sub-contractor's failure to perform;

(c) declines to provide a similar safeguard in respect of the Nominated Sub-contractor's negligence or misuse of plant, equipment or temporary works provided by the Contractor;

(d) declines to accept the provisions of a 'forfeiture' clause, similar to those of Clause 63 (see section 9.1.13), which the Contractor is required, under Clause 59B(1), to include in any sub-contract with a Nominated Sub-contractor;

In such an event, the Engineer is required, under Clause 59A(2), to take one of the following alternative courses of action:

(a) nominate an alternative sub–contractor, in which case the above procedure is repeated;

(b) vary the Works or, if necessary, omit the relevant Prime Cost item from the Bill so as to enable the work to be performed under a separate contract to be executed either concurrently with the main contract – in which case the provisions of Clause 31 (see section 9.1.11) apply – or at some later date;

(c) direct the Contractor, with the Employer's consent and providing the Contractor's objections are based on one or other of the above grounds, to enter into a sub–contract with the Nominated Sub–contractor on such other terms as may be specified by the Engineer;

(d) arrange for the Contractor to do the work – subject, of course, to his consent;

In the event that the Engineer selects the 'omission' option in alternative (b) above, it should be noted that this does not diminish the Contractor's entitlement to be paid the full 'other charges and profit' percentage (see section 8.2.4) associated with the Prime Cost item in the original Bill.

Clause 59B details the procedures which are to be followed in the event of the Contractor giving notice of his intention to exercise his right, under the aforementioned 'forfeiture' clause in the sub–contract, to treat the sub–contract as having been repudiated by the Nominated Sub–contractor.

In the event that such notice is given, the Engineer is required to do any one or more of the things described in (a), (b) or (d) of Clause 59A(2) above. The complicated procedure following this requirement guarantees that the Contractor does not suffer consequential loss and requires him to use his best endeavours to minimise the consequential losses of the Employer.

Clause 59C provides a 'safety net' for Nominated Sub–contractors in the event of the Contractor's failure to pay them the amounts paid to *him*, by the Employer, under the Contract. Broadly speaking, the clause entitles the Engineer to demand proof from the Contractor that the Nominated Sub–contractor *has* been paid. If the Engineer is not satisfied that payment has been made, he is empowered to certify direct payment of the Nominated Sub-contractor by the Employer and deduct the relevant monies from subsequent payments made to the Contractor under the main Contract.

6th Edition Changes

Clause 4 is revised to the extent that the Employer's consent is now no longer required for the sub-letting of parts of the Works. The Engineer is, however, required to be notified of such and has the power, under certain circumstances, to require the removal of sub-contractors from the Site.

Clauses 59A, 59B and 59C are now combined into a re-written Clause 59 which, although substantially the same in content as previously, contains a few additions and/or omissions. Most notably, the Contractor is provided with an additional ground on which he entitled to object to a Nominated Sub-contractor, namely the failure of the said sub-contractor to provide the Contractor with security of performance of the sub-contract. In the event of such objection, the previous provision for a Nominated Sub-Contractor to be employed on terms specified by the Engineer is now removed. The removal of this option is, however, made up for by the new power granted the Engineer, by Clause 59(2), to instruct the Contractor to secure a sub-contractor of his own choice – subject, of course, to the Engineer's approval thereof.

9.1.7 Completion and Maintenance

Clauses: 20-21, 48(1)-48(3), 49(1)-49(4), 50, 60(5), 61;

The Contractor's obligations fall into two separate periods: the period from the commencement of the Works up to 14 days following the issue of the Certificate of Completion – during which the work is carried out and during which, in accordance with Clause 20, he takes full responsibility for the care of the Works – and the Period of Maintenance which, in accordance with Clause 49(1), is the period specified in the Appendix to the Form of Tender (see section 6.2) during which he must repair any defects and complete any work outstanding at the time of issue of the Certificate of Completion.

In accordance with Clause 48(1), the initiative rests with the Contractor regarding the issuance of a Certificate of Completion. When *he* considers that the Works have been *substantially* completed – i.e. to an extent which, notwithstanding any minor deficiencies, will enable the Employer to take possession of the Works *and* use them for the purpose for which they were intended – and to have passed any required tests, he *may* give *written* notice of such to the Engineer and request the latter to issue a Certificate of Completion. Following the receipt of such notice, which should be accompanied by the Contractor's undertaking to complete any outstanding work during the specified Period of Maintenance, the Engineer shall *either* issue a Certificate of Completion *or* specify the work which, in *his* opinion, requires to be completed before a completion Certificate will be issued.

In the *former* case, the Engineer is required to issue the Certificate, stating the date on which, in his opinion, the Works were substantially completed in accordance with the Contract, within 21 days of his receipt of the Contractor's written request so to do. In the *latter* case, the Contractor is entitled to the Certificate within 21 days of his completion of the specified work to the Engineer's satisfaction.

Under Clause 48(2), a similar facility exists for any section of the Works for which a separate completion time is provided for in the Appendix to the Form of Tender and for '... any substantial part of the Works which has been both completed to the satisfaction of the Engineer and occupied or used by the Employer ...'.

It should be particularly noted that the Engineer is under no obligation to certify completion until a request to that effect has been received from the Contractor. Indeed, he has no right to do such, in respect of the *entire* Works, *until* the latter's request has been submitted. Whilst this latter constraint may appear to 'leave the door open' for a Contractor to delay the submission of such a request by way of maximising any compensation to which he is entitled (under various other clauses) as a result of the Employer's liability for disruption and delay, the fundamental tenets of contract law place a duty upon the Contractor to mitigate his own potential losses (see section 3.6.1) by requesting and obtaining a completion Certificate – thereby releasing his resources for use on other Works – as soon as possible. It should also be noted that the 'loss mitigation' duty imposed upon the Contractor *also* applies to the *Employer's* losses. Thus, it is reasonable to suggest that any attempt, on the part of the Contractor, to 'screw' the Employer in the aforesaid manner would be regarded somewhat unfavourably by the courts.

With this in mind, perhaps, Clause 48(3) provides the Engineer with the *discretionary* authority to issue completion Certificates, *without* reference to the Contractor and without relieving the Contractor of his obligations *vis-a-vis* the completion of any outstanding work during the Maintenance Period, for *any* sections or parts of the Works which are, in his opinion, substantially complete and have passed the requisite tests.

In view of the consequences of such a course of action to the Employer – a reduction in the extent of liquidated damages payable to him in the event of delayed completion of the Works and a refund, under Clause 60(5), of a proportion of the retention monies deducted from the Contractor (see section 9.1.10) – the Employer's consent to such would appear to be a reasonably implicit pre-requisite. Since the Employer has a fundamental duty to mitigate any damages payable by the Contractor in the event of delayed completion, it is *equally* implicit that, despite his possibly having no immediate need for the relevant part(s) of the Works, the Employer's consent to the Engineer's issuance of a Certificate of partial completion should not be unreasonably withheld.

In exercising such discretionary authority as that cited, the Engineer should bear in mind that, whilst the *agency* relationship between him and the Employer imposes upon him a duty of care towards the latter, his responsibilities under the Contract impose upon him the duty to act fairly as between *both* parties thereto. Indeed, if the Agreement between the Engineer and the Employer is drawn up in accordance with the *Association of Consulting Engineers' Model Form 'A'*, such a 'duty of fairness' will be an *express* requirement thereof.

In accordance with Clause 49(2), the Contractor is required to complete, as soon as is practicable following the issue of the Certificate of Completion, all work outstanding at the certified date of completion.

To avoid the financial impracticability of the Employer using another Contractor solely to carry out remedial work, the Contractor is required, under Clause 49(2), to remedy *all* defects to the Works (including those which are not due to his fault but excluding those attributable to fair wear and tear) which arise during the Period of Maintenance, or which are notified to him *in writing* by the Engineer, following the latter's inspection of the Works prior to the expiry of the Maintenance Period, within 14 days after the said expiry.

It is of importance to note that, since there is no *further* Maintenance Period (or retention) for work carried out under this clause, strict supervision of such, by the Engineer, is essential.

In accordance with Clause 49(3), the cost of such remedial work, including that of carrying out searches (if such is required by the Engineer under Clause 50) for the underlying cause of any defects arising during the Maintenance Period, is to be borne by the Contractor *only* where the work is necessitated by his use of materials or workmanship not in accordance with the Contract, or by his failure to comply with any other of his contractual obligations. In all other cases, the work will be valued and paid for as additional work (see section 9.1.9).

It should be particularly noted that, in accordance with the final sentences of Clauses 20(1) and 20(2), the Contractor is liable for the repair of any damage to the work outstanding at the date of completion (as specified upon issuance of the Certificate of Completion pursuant to Clause 48) *and* the repair of any damage to the Permanent Works resulting from the maintenance operations. If he has complied *fully* with the requirements of Clause 21, however, the associated costs will be borne by his insurers.

By way of providing a remedy in the event of the Contractor's failure to carry out any of the aforesaid remedial works, Clause 49(4) entitles the Employer to carry out the relevant work using his own workforce, or another Contractor, and to deduct the cost thereof from any monies owing to the defaulting Contractor.

Upon the expiry of the Maintenance Period, and upon the Contractor's completion of any outstanding work referred to in Clauses 48–50, the Engineer is required, under Clause 61(1), to issue, to the Employer with a copy to the Contractor, a Maintenance Certificate stating that, on the specified date, the Contractor has '... completed his obligations to construct complete and maintain the Works to the Engineer's satisfaction ...'. Since the Maintenance Certificate is, in essence, written evidence that the Contractor has fulfilled *all* his contractual obligations, it is important to note that it will not have any legal effect whatsoever unless it contains these precise words – or words to that effect. Equally important is the requirement to state the relevant date since this 'starts the clock running' in respect of the Contractor's submission of the Final Account (see section 9.1.10).

Where there is more than one Maintenance Period, on account of there being different completion dates for certain specified sections or parts of the Works, the '... expiration of the Maintenance Period ...', for purposes of issuing the Maintenance Certificate, is taken as the expiration of the last such period.

Finally, Clause 61(2) states that the issue of the Maintenance Certificate shall not be taken as relieving *either* party to the Contract of any of their liabilities to one another in connection with the performance of their respective obligations under the Contract. This is, of course, a reflection of the fact that, under contract law, the contracting parties remain liable to one another for the duration of the appropriate Limitation Period (see section 3.6.7) which, in the case of a contract *under seal* (the most probable case for a construction contract), is *twelve years*. Thus, an action for *breach of contract* (the only remedy open to the parties after the issue of the Maintenance Certificate) must be brought within twelve years of the *last relevant breach* being committed or within a similar period of the *time at which the breach should reasonably have been discovered.*

This latter option is of particular importance in the case of the Employer dicovering a latent defect in the Works caused, in *his* opinion, by the Contractor's failure to discharge his obligations under the Contract. In such a case, of course, the 'burden of proof', *vis-a-vis* the Contractor's alleged breach of contract, would rest with the Employer.

In the event of such a breach being proved, it is of value to note that the Employer is not entitled to demand that the remedial work be carried out by the defaulting Contractor. Equally, the defaulter has no right to demand that *he* carry out the relevant work. If the work *is* carried out by another Contractor, however, with the defaulter being liable for the cost thereof, the Employer has a duty (see section 3.6.1) to mitigate the defaulter's prospective losses by being *reasonably thrifty* in the matter.

6th Edition Changes

Clause 48 contains a new sub-clause – Clause 48(3): Premature Use by the Employer – which conveys more strongly the Contractor's right (previously contained in Clause 48(2)) to receive a completion Certificate in respect of substantial parts of the Works which, other than as specified elsewhere in the Contract, are occupied and in use by the Employer.

Clause 49 – which refers to the 'Defects Correction Period' instead of the previously used 'Period of Maintenance' – contains a new sub-clause which provides for the inclusion, in the undertaking given under Clause 48(1), of an agreed programme for the completion of any outstanding work. Failing such inclusion, the work is required to be completed 'as soon as practicable' during the Defects Correction Period.

In line with the above change in terminology, the 'Maintenance Certificate', as previously referred to in Clause 61, is now called the 'Defects Correction Certificate'.

9.1.8 Time

Clauses: 14(1)–14(2), 40–41, 42(1), 43–47, 51(1), 51(3), 52(4), 63;

In accordance with Clause 41, the Engineer is required to notify the Contractor, within a *reasonable* time (see section 7.3.5) following the acceptance of the latter's tender, of the Date for Commencement of the Works. The Contractor is required to commence the Works '... on or as soon as is reasonably possible after ...' this date and, thereafter, to proceed with the Works '... with due expedition and without delay in accordance with the Contract ...'.

Whilst the Employer (*via* the Engineer) has a number of contractual remedies – such as those provided by Clause 46 (see below) or, in more serious cases, by Clause 63 (see section 9.1.13) – open to him in the event of the Contractor delaying *unreasonably* in commencing the Works, the Contractor has little, if any, contractual protection in the event that the Engineer delays *unreasonably* in notifying him of the Date for Commencement of the Works (see section 7.3.5 for a discussion of this situation).

Clause 43 requires the Contractor to complete the Works within the Time for Completion, as stated in the Appendix to the Form of Tender (see section 6.2), and states that such time will be taken as running from the '... Date for Commencement of the Works notified under Clause 41 ...' and will include any extensions of time to which the Contractor is entitled under Clause 44 (see below). In view of the importance of the date of completion of the Works *vis-a-vis* the payment of liquidated damages (see below), it should be particularly noted that the 'clock' starts running on the *notified* Date for Commencement of the Works and *not* on the date thereafter on which, in accordance with Clause 41, commencement is '... reasonably possible...'.

In order to enable the Engineer to satisfy himself that the Contractor is able to discharge his contractual obligations within the specified completion time, Clause 14(1) requires the Contractor, '...

within 21 days after the acceptance of his Tender ...', to submit a construction programme to the Engineer for his approval. Since the Engineer's appraisal of the said programme would be largely meaningless in the absence of details of *how* the Contractor intends to discharge his obligations, the clause further requires the Contractor to provide, in writing, a '... general description of the arrangements and methods of construction ...' which he proposes to adopt, and thereafter to provide '... such further details and information as the Engineer may reasonably require ...'.

Whilst the requirement to provide such a programme is mandatory, the exact format in which it is presented is generally recognised to be a matter for the Contractor's discretion. That said, however, it is worth noting that the Engineer is entitled to withhold his approval thereof in the event that he considers the format to be inappropriate for the project in question. For example, whilst a standard bar-chart format might well be eminently suitable for a straightforward project involving a comparatively small number of large-scale self-contained operations, a detailed construction network might be considered more appropriate in the case of a project which involves a complex sequence of a large number of interlinking operations. In this respect, and whilst such a requirement is by no means mandatory, it might reasonably be suggested that much unnecessary 'hassle' may be avoided by the Engineer notifying the Contractor, in advance (i.e. *via* the Contract Documents), of any *specific* requirements concerning the format of the latter's 'Clause 14' programme.

In the event that the Engineer's approval of the Contractor's programme is withheld for any reason whatsoever, Clause 14(2) authorises the Engineer to require the Contractor to submit a revised programme containing such modifications as are adjudged, *by the Contractor*, to be necessary such that the end-result meets with the Engineer's approval. Indeed, this sub-clause authorises such a course of action at any time during the execution of the Works if the Engineer considers that the Contractor's *actual* progress does not conform with his *scheduled* progress. In invoking this clause, however, it should be emphasised (as previously stated in connection with the Engineer's approval of the Contractor's proposed methods of executing the Works – see section 9.1.5) that the Engineer's powers are strictly limited to those of veto in that, whilst he is entitled to demand that the Contractor's programme *be* revised, he is not entitled to specify exactly *how* it should be revised.

In the context of the Engineer's appraisal of the programme, it should be mentioned that, whilst there are a number of valid reasons why a Contractor might wish to submit an optimistic or ambitious 'Clause 14' programme, it is not unknown for a Contractor to submit an *excessively* optimistic programme with the specific intent of subsequently using it to support an exaggerated claim for delay and disruption, under Clause 44 (see below), should a suitable opportunity arise. Even though such nefarious intent may well be suspected, the Engineer has no *express* authority to reject a programme solely on the grounds of its over–optimism. He may, however, have the right to withhold his approval of such on the grounds that, in contravention of the provisions of Clause 14(1), it does not *truly* represent the Contractor's intended plan of action.

Under Clause 42(1), the Contractor must be given possession, on the Date for Commencement of the Works, of as much of the site as may be required to enable him to '... commence and proceed with the construction of the Works in accordance with the programme referred to in Clause 14 ...'. In the event that a timetable of staged handover of the Site is contained within the Contract, then *this* timetable (as opposed to the Contractor's 'Clause 14' programme) will govern the Contractor's entitlement to possession. Following initial possession of the Site having been given, the Contractor is entitled to subsequent possession of such portions of the Site as are necessary to enable him to proceed with the Works '... with due dispatch in accordance with the said programme ...'. There is, of course, the reasonable assumption that the said programme will have taken into account any staged handover of the Site which has been prescribed by the Contract.

In the event of possession being delayed, the Contractor is entitled to claim, under Clause 44 (see below), for the appropriate extension of completion time and, under Clause 52(4) (see section 9.1.11), for any additional costs which he incurs as a consequence of such delay.

Under Clause 46, the Contractor is expected to maintain his planned rate of progress in carrying out the Works. If, at any stage in the proceedings, the Engineer is of the opinion that the Contractor is behind schedule to an extent that the scheduled completion date (which will take into account any extensions of time granted the Contractor under Clause 44 – see below) is in jeopardy, he is required to notify the Contractor accordingly whereupon the Contractor is obliged, at his own cost, to take the necessary steps to remedy the situation. The Contractor's proposed course of action, in this event, will be subject to the Engineer's approval and will probably include the submission of an appropriately revised programme under Clause 14(2).

In appraising the Contractor's proposals, it should (once again!) be emphasised that, in order to avoid subsequent claims for instructed expedition, the Engineer should *strictly* limit his actions to approval or disapproval of such.

In accordance with Clause 45, and unless the Contract contains provisions to the contrary, the Contractor requires the Engineer's *written* permission to carry out work at night or on a Sunday (i.e. outside normal working hours) *except* where such work is *unavoidable*, or where it is *absolutely necessary* for *the saving of life or property* or for *the safety of the Works*, or where it is *customary* (as, for example, in tunnelling or tidal Works) *for the work to be carried out on a rotary- or double-shift basis*.

It should be noted, however, that where night or Sunday work is adjudged, *by the Contractor*, to be necessary to enable him to comply with the requirements of Clause 46, he is *still* required to seek the Engineer's permission for such but, even though the work might not fall into any of the above categories, such permission must not be unreasonably refused.

In the context of the 'time' element of the Contract performance, it is worth noting that Clause 40(1) empowers the Engineer to suspend the Contractor's operations anywhere on the Site and, if necessary, to bring the whole of the Works to a halt. Such suspension can be ordered at any time and for as long a period as the Engineer adjudges necessary. The Contractor is entitled to claim for any costs incurred by him as a result of the suspension *except* to the extent that such suspension is:

(a)   provided for in the Contract;

(b)   necessary as a result of weather conditions or as a result of some default on the part of the Contractor;

(c)   necessary for the proper execution of the work or for the safety of the Works insofar as this does not arise from any default on the part of the Engineer or the Employer;

Except where the suspension is provided for in the Contract, or where it arises out of some default on the part of the Contractor, the Engineer is required to take any consequential delay into account in determining any extension of time to which the Contractor is entitled under Clause 44 (see below).

Despite the sweeping powers invested in the Engineer in this repect, it is generally recognised that, unless such is provided for in the Contract, a suspension of the Works should only be considered in the most extreme circumstances and only as a last resort – i.e. in instances where none of the multitude of *other* contractual powers invested in the Engineer are likely to suffice.

There are two noteworthy circumstances in which suspension would not only be regarded as a *justifiable* course of action, but might be considered to be *advisable*, namely: where the Engineer considers that continuation of the work would pose a threat to the safety of site personnel or the general public, or where it would be seriously detrimental to the Works.

In the first case, it should be noted that the strict 'duty of care', imposed by the law on those involved in the work process, requires only that the Engineer be *reasonably* sure that a hazard exists or is likely to develop if the work continues. In the second case, however, the degree of certainty required of the Engineer, as to the consequences of his allowing the work to continue, is *substantially* greater. Suffice it to say that, in *this* case, suspension should only be contemplated if the Engineer is satisfied (as far as is humanly possible) that the consequences of continuing the work *would* be as stated and that he has *no other alternative* but to suspend the Works.

By way of emphasising the need for certainty on the part of the Engineer, it should be noted that the consequences of an invalid or ill–judged suspension are, to say the very least, undesirable. Not only will the Employer find himself meeting the cost of the Contractor's plant and labour standing idle during the suspension, he will also probably find himself footing the bill for almost every subsequent delay to the Works which, whether on the basis of fact or as a result of 'creative revisions' to the Contractor's programme, will be 'incontravertibly' found to be a direct consequence of the said suspension!

In the event that the suspension lasts for three months, Clause 40(2) empowers the Contractor to serve notice on the Engineer requesting the latter's permission to proceed. If such permission has not been received within 28 days following the Engineer's receipt of the notice, and if the Engineer is unable to demonstrate that the reason for continued suspension is some default of the Contractor, then the Contractor is entitled to treat the suspension as an omission under Clause 51(1) (see section 9.1.9) – where the suspension applies only to a *part* of the Works – or, where the suspension affects the Works in their entirety, as the Employer's abandonment of the Contract.

In accordance with Clause 44, an extension of the Contract completion time may be granted the Contractor in respect of:

(a) ordered variations under Clause 51(1) (see section 9.1.9);

(b) increased quantities under Clause 51(3) (see section 9.1.9);

(c) any other cause of delay referred to in the Conditions;

(d) *exceptional* adverse weather conditions;

(e) other special circumstances of any kind whatsoever;

In the event that the Contractor considers himself fairly entitled to an extension on one or other of the above grounds, he is required to provide the Engineer, '... within 28 days after the cause of the delay has arisen or as soon thereafter as is reasonable in all the circumstances ...', with full and detailed particulars of his claim. These details should, of course, provide evidence that the delay in question *will* have the claimed 'follow–on' effect on the contract completion time.

Following his receipt of the particulars, the Engineer is required to make an interim assessment of the extension of time (if any) to which he considers the Contractor entitled and to notify the Contractor accordingly.

On, or as soon as possible after, the due date for completion (which takes into account any extensions of time granted pursuant to this clause) of the Works (or relevant Section thereof), the Engineer is required to make an 'updated' assessment in the light of circumstances prevailing *at that time.* In the event that a *further* extension of time is granted, the Contractor is notified accordingly, the 'due date for completion' is amended accordingly and the process is repeated. If the Engineer considers that the Contractor is *not* entitled to any further extension, he is required to notify the Contractor *and* the Employer accordingly.

Upon the issue of the Certificate of Completion of the Works (or relevant Section thereof), the Engineer is required to review *all* the circumstances of each and every instance of delay and to certify the *overall* extension of time to which he considers the Contractor entitled.

In view of the 'progressive' nature of the assessment procedure outlined above, and in the interests of fairness to a Contractor completing in good faith of the Engineer's interim decisions, the clause expressly states that no review of the circumstances (either at the due date for completion or upon issue of the completion Certificate) shall '... result in a decrease in any extension of time already granted by the Engineer ...' pursuant to this clause.

Two further points should be noted in respect of the provisions of this clause, the first of which is that the Contractor's fair entitlement to an extension of time is not diminished by his failure to claim for one. Consequently, the Engineer is given the discretionary power, under this clause, to make an interim assessment of an extension of time in the absence of an appropriate claim from the Contractor.

Secondly, the granting of an extension of time under this clause does not entitle the Contractor to reimbursement of any additional costs incurred as a consequence of the delay. However, the underlying reason for the delay may entitle him to payment of such under another clause – e.g. Clause 13(3).

In the event that the Contractor's completion of the Works (or relevant Section thereof) is delayed beyond the due or extended date for completion, Clause 47 entitles the *Employer* (*not* the Engineer) to deduct liquidated damages from him to the extent specified in the Appendix to the Form of Tender (see section 6.2). In accordance with this clause, the specified liquidated damages are required to represent the Employer's '... genuine pre-estimate ...' of the damages likely to be incurred by him for each day or week (as appropriately specified in the Appendix to the Form of Tender) that the Contract completion is delayed beyond the due or extended date for such.

In accordance with Clause 47(3), and in compliance with the requirements of contract law, the damages payable to the Employer in the event of delay are expressly prohibited from being *punitive*. If the damages *are* considered by a Court, or an Arbitrator, to be punitive, the Employer is entitled only to recover damages to the extent *actually* incurred by him.

Finally, it should be noted that, under Clause 47, the Employer is *only* entitled to deduct liquidated damages from the Contractor if the Engineer has:

(a) determined and certified any extension of time to which he considers the Contractor fairly entitled;

(b) notified the Contractor *and* Employer, pursuant to Clause 44(3), that he is of the opinion that the Contractor is not entitled to any further extension of time;

6th Edition Changes

Clause 41 is now divided into two sub-clauses to identify separately the Date of Commencement of the Contract and the date that work actually starts.

The old Clause 14(1) is rewritten to include the requirement that, in the event of his 'Clause 14' programme being rejected, the Contractor must submit a revised programme within 21 days of the rejection. Clause 14(2) is a new sub-clause which requires the Engineer to respond, within 21 days, to the Contractor's submission of his programme, failing which such programme will be deemed to have been accepted and approved. Clause 14(3) – also a new sub-clause – requires the Contractor to respond, within 21 days, to a request by the Engineer to supply further information relating to the above programme. Failing a timely response, the programme will be deemed to have been rejected.

The provisions of Clause 42(1) have been clarified by subdividing the original clause into three sub-clauses.

Clause 46 contains a new sub-clause providing for agreed acceleration of completion (and appropriate payment therefor) if requested. It should be noted that the Engineer's authority in this respect may not be delegated to his Representative.

Clause 44 is substantially the same as its predecessor except that it now contains five sub-clauses instead of four. A notable addition thereto is the requirement that the Engineer, prior to making his interim assessment of the appropriate extension of time, make an initial assessment of any delay claimed by the Contractor and notify the latter accordingly. Whereas the previous version of the clause required the Engineer to update his assessment of any extension to the contract completion time '... on or as soon as possible after the due date for completion ...', Clause 44(4) now requires that such be made '... within 14 days ...' after the said date. Furthermore, Clause 44(5) now requires certification of the overall extension of completion time

'... within 14 days ...' after the issue of the completion Certificate and not, as previously required, '... upon ...' such issue.

Whilst the provisions of Clause 47 remain substantially unchanged, a new sub-clause has beeen added which outlines the procedure which applies following the issue of a variation order, or the acceptance of a 'Clause 12' situation, after liquidated damages have become payable. In essence, the Employer's entitlement to liquidated damages in respect of the affected parts of the Works is suspended pending the Engineer notifying the Employer and Contractor, in writing, *that the further delay (occasioned by the above) has come to an end.*

9.1.9 Alterations, Additions and Omissions

Clauses: 51, 52(1)–52(3), 52(4)(a);

Whilst the principles of contract law effectively prohibit the Engineer from altering or varying the *terms* of the Contract – that being a matter *solely* for the contracting parties, namely: the Employer and the Contractor – the Engineer is provided with substantial powers *under* the Contract to vary the Works involved in such.

In particular, Clause 51(1) authorises the Engineer, without vitiating or invalidating the Contract, to order *any* variation to the Works – not only in terms of quality and quantity but also in terms of the timing and sequencing thereof – that, in *his* opinion, is '... desirable for the satisfactory completion and functioning ...' of such and, in so doing, to bind both the Contractor (to carry out the necessary work) and the Employer (to pay for them).

Clause 51(2) prohibits the Contractor from effecting variations to the Works *unless* he has received the appropriate *written* order from the Engineer. In the event that, for reasons of urgency, an *oral* order is given in the first instance, such will suffice provided it is subsequently confirmed in writing. Where no such confirmation is provided, the Engineer's non-contradiction of the Contractor's written confirmation of such an order will be taken as implying the requisite confirmation.

In accordance with Clause 51(3), no specific order is required in connection with 'apparent' increases or decreases in work resulting *solely* from innacuracies in the item quantities expressed in the Bill of Quantities.

Under Clause 52(1), work ordered by the Engineer under Clause 51 shall be valued as follows:

(a) Where the work is of a *similar character*, and is executed under *similar conditions*, to that priced in the Bill of Quantities, it will be valued at such billed rates as may be applicable.

(b) Where the work is *not* of a similar character to that priced in the Bill, or is *not* executed under similar conditions thereto, it will be valued *on the basis* of the billed rates *so far as may be reasonable*.

(c) Where the work bears no similarity whatsoever to that which has been priced in the Bill, a *fair* valuation shall be made.

Whilst it is envisaged that such valuation be arrived at by consultation between the Engineer and the Contractor, it is not intended that prolonged negotiations should be entered into in this regard. This is emphasised by the express duty placed upon the Engineer, in the event that agreement is not reached within a reasonable time, to bring matters to a close by valuing the work *himself*, in accordance with the above principles, and by notifying the Contractor accordingly. This is not final, however, in that the Contractor is entitled to pursue the matter further through Clause 52(4)(a) (see section 9.1.11).

Clause 52(2) provides for the fixing, by the Engineer, of rates and prices for billed items in the event that either he *or* the Contractor consider that the Bill rates have been rendered *unreasonable* or *inapplicable* by any variation ordered by the Engineer. It should be emphasised that *no* consultation is required with the Contractor, it being the *sole* province of the Engineer to fix the rates and prices which he deems reasonable and proper in the circumstances. As previously, the Contractor is entitled to pursue the matter further through Clause 52(4)(a) if he is dissatisfied with the Engineer's decision.

Finally, it should be noted that Clause 52(3) entitles the Engineer, if he deems it preferable, to instruct that any additional or substituted work be carried out by the Contractor on a *Daywork* basis (see section 8.2.3).

6th Edition Changes

Clause 51(1) has been extended to include for variations ordered during the Defects Correction Period. The original Clause 51(2) has been replaced by Clauses 51(2) and 51(3) with the new Clause 51(3) containing a qualification excluding payment for variations necessitated by the Contractor's default.

9.1.10 Measurement, Certification and Payment

Clauses: 47, 52(1)-52(3), 52(4)(a), 55-57, 60(1)-60(6), 60(8);

In accordance with Clauses 56(1) and 57, the Contract is deemed to be of the *admeasurement* variety making use of a Bill of Quantities prepared, except where expressly stated to the contrary, in accordance with *The Civil Engineering Standard Method of Measurement* (see section 8.1).

Clause 55(1) confirms the generally accepted fact that the quantities expressed in the Bill are *estimates* of the quantities of work required of the Contractor in fulfilment of his obligations under the Contract and should not be taken as an *accurate* and *correct* representation thereof.

In the event that the *actual* quantities of work (following remeasurement) differ from the *Billed* quantities to an extent which, *in the opinion of the Engineer*, renders the Bill rates *inapplicable* or *unreasonable*, Clause 56(2) empowers the Engineer to determine, after consultation with the Contractor, an appropriate modification to the relevant Bill rates and to notify the Contractor accordingly. As previously discussed in connection with the Engineer's valuation of variations and additions, the Contractor is entitled to invoke the provisions of Clause 52(4)(a) in the event of his being dissatisfied with the Engineer's decision.

Particular note should be taken of the Engineer's powers, in this respect, by those tempted to indulge in the practice of 'spot' loading of Bill rates (see section 7.3.2).

Clause 55(2) details the procedures to be undertaken in the event that work detailed in the Drawings or Specification is omitted from the Bill, or in the event that item descriptions in the Bill either contain errors or, by implication from the assumed use of *CESMM* in accordance with Clause 57, are insufficiently detailed as to enable the work to be clearly identified, and accurately priced, by the Contractor. Broadly speaking, such errors or omissions are rectified by the Engineer's revaluation, in accordance with Clause 52 (see section 9.1.9) of the *actual* work done.

It should be noted, however, that, despite the provisions of this clause, such errors or omissions can be the source of considerable contention in view of the provisions of Clauses 5, 7, 11 and 13.

In respect of the progressive admeasurement of the work, it should be noted that, whilst Clause 56(1) places the duty for such on the shoulders of the Engineer, Clause 56(3) envisages that the process will be carried out in the presence of a representative of the Contractor − not unreasonably since the Contractor is responsible, under Clause 60 (see below), for the preparation of the monthly statement. Notwithstanding the *desirability* of joint measurement − from a practical standpoint − there will be times at which the Contractor's presence will be required in order to assist the Engineer in the process and to provide him with certain details and information regarding the work to be measured. In this event, Clause 56(3) requires the Engineer to give *reasonable* notice to the Contractor (or his representative) that his presence is required for measurement purposes, and places an obligation upon the Contractor to provide any details or information which may be required by the Engineer in connection with such. Failing the Contractor's co−operation in the matter, the Engineer is authorised to proceed on his own with the resulting measurement being taken as '... the correct measurement of the work ...' − subject, of course, to the findings of the Arbitrator in the event of a subsequent dispute (see section 9.1.12) over such.

In accordance with Clause 60(1), the Contractor is required, at the end of each month, to submit to the Engineer a statement showing:

(a) the estimated Contract value of the Permanent Works completed up to the end of that month;

(b) an evaluated list of materials *on site* for, but not yet incorporated in, the Permanent Works − the materials being valued in accordance with details contained within the Appendix to the Form of Tender;

(c) an evaluated list of materials *off site* which are vested in the Employer pursuant to Clause 54 (see section 9.1.13) − the materials being valued as in (b) above;

(d) the estimated amounts to which the Contractor considers himself entitled in respect of all *other* matters for which provision is made under the Contract;

Such a requirement is relaxed if, in the opinion of the *Contractor*, the net amount (see below) to which he is entitled is less than the 'minimum Certificate value' (if applicable) specified in the Appendix to the Form of Tender.

Clause 60(2) requires the Engineer to certify, and the Employer to pay, such amounts as the Engineer considers due to the Contractor on the basis of the latter's statement, such certification and payment to be effected within 28 days of the Engineer being in receipt of the Contractor's statement.

It should be noted that the Engineer is not bound to issue any interim Certificate for a sum less than that specified in this regard in the Appendix to the Form of Tender.

In accordance with Clause 60(4), the amounts payable to the Contractor, with the notable exeption of those in connection with materials on or off site, are subject to retention at 5% of the amount due until such times as the cumulative retention amounts to:

(a)   the *lesser* of *£1,500* or *5% of the Tender Total* where the Tender Total does not exceed £50,000;

(b)   *3% of the Tender Total* where the Tender Total exceeds £50,000;

Clause 60(5) requires that 50% of the overall retention be refunded to the Contractor within 14 days following the issue of the Certificate of Completion of the Works. In cases where specified Sections of the Works require the issuance of individual completion Certificates, 50% of the retention associated with each Section will be refunded within 14 days of the issue of the relevant completion Certificate – subject to the aggregate of such refunds not exceeding 50% of the *overall* amount of retention deducted under the Contract.

The balance of the retention, less the amount which, in the Engineer's opinion, represents the cost of any outstanding work, is refundable to the Contractor within 14 days following the expiry of the Period of Maintenance. In the event of specified Sections of the Works having separate Maintenance Periods, the 'expiry of the Period of Maintenance' shall, for the purposes of this clause, be taken as the expiry of the last such period.

By way of removing any temptation, on the part of the Employer, to hold back any of the above payments in order to improve his cash flow, Clause 60(6) requires the Employer to pay the Contractor interest, to the tune of 2% p.a. above the prevailing Minimum Lending Rate, on any payment delayed beyond the appropriate time limit.

In accordance with Clause 60(3), the Contractor is required to submit his Final Account, along with such supporting information as is reasonably required for verification purposes, within 3 months of the date of issue of the Maintenance Certificate. Within 3 months of the date of his receipt of the Final Account and the relevant supporting information, the Engineer is required to issue a Final Certificate stating the amount which, in his opinion, is finally due under the Contract. Realisation of the appropriate balance – which, in view of the fact that it is subject to the provisions of Clause 47 (see section 9.1.8), may not always be in the Contractor's favour – is due within 28 days of the date of issue of the Final Certificate.

Finally, it should be noted that, under Clause 60(8), *every* Certificate issued by the Engineer, pursuant to Clause 60, should be sent to the Employer with a copy thereof sent to the Contractor.

6th Edition Changes

Clause 60(1) is substantially the same as its predecessor except that the Contractor's statements are now required at monthly intervals *and not, as previously,* at the end of each month. *Clause 60(3) is an amplification of the reference to a minimum Certificate value contained in the original Clause 60(2). The minimum value of Certificates stated in the Appendix to the Form of Tender now applies only up until the issue of the Certificate of Substantial Completion. Clause 60(4) states that the rate and limit of retention may now be specified in the Appendix to the Form of Tender. In accordance with Clause 60(6), interest on overdue payments from the Employer will now be levied at a rate of 2% above the Base Lending Rate of the bank specified in the Appendix to the Form of Tender.*

9.1.11 Claims

Clauses: 5, 7(2)–7(3), 11–13, 14(3)–14(4), 19(1), 31, 40, 44, 51, 52(1)–52(2), 52(4), 56(2);

As used in this context, a 'claim' is a request by the Contractor, *under the Contract*, for recompense for a loss or expense (in terms of money and/or time) which he has suffered for reasons outwith his control, and which could not reasonably have been foreseen and accounted for in his tender.

Within the Conditions, there are 58 clauses and sub-clauses which contain provisions for payments to be made, and/or extensions of time to be granted, to the Contractor. It follows that these clauses *could* result in claims being made. Additionally, there are a considerable number of clauses which, by virtue of their imposing specific duties, or specific limitations thereto, on the Employer or the Engineer, may *also* give rise to claims.

The principal clauses under which such claims may be submitted are dealt with hereunder in the order in which they appear in the Conditions.

Although the responsibility for the supply of drawings, instructions and general information relating to the completion of the Works rests with the Engineer (see section 9.1.2), Clause 7(2) entitles the Contractor to request the supply of such further drawings and instructions as he (the Contractor) considers necessary for the execution of the Works. In the event that the Contractor suffers delay, or incurs costs, '... by reason of any failure or inability of the Engineer to issue at a time reasonable in all the circumstances drawings or instructions requested by the Contractor and considered necessary by the Engineer ...' for the '...proper and adequate construction completion and maintenance of the Works ...', then Clause 7(3) requires the Engineer to take such delay into account when determining the Contractor's entitlement to any extension of time pursuant to Clause 44, and entitles the Contractor to submit a claim for the costs incurred.

It should be particularly noted that, in strict accordance with the cited phrase, the Contractor has no remedy under this clause for the late supply of drawings and instructions *unless* such were *requested* from the Engineer, prior to their issue, *and* were adjudged by the latter to be *necessary* for the construction, completion and maintenance of the Works.

It should also be noted that the Engineer is required to supply the necessary documentation '... *at* a time reasonable in all the circumstances ...' and *not* '... *within* a time reasonable in all the circumstances ...'. It may thus be inferred that the Engineer may take as much time as he likes in complying with the Contractor's request as long as the requested information is provided in *reasonably* sufficient time to avoid any delay to the programmed completion of the Works.

One of the most frequently used, and, possibly, most contentious, of the 'claims' clauses in the Conditions is Clause 12 which entitles the Contractor to claim when, in *his* opinion, extra costs are incurred as a result of his encountering *adverse physical conditions* (with the notable exception of *weather conditions* or *'weather-consequent' conditions*) or *artificial obstructions* where such conditions or obstructions '... could not reasonably have been foreseen by an experienced contractor ...'.

In the event that the Contractor intends to submit a claim under this clause, he is required to give the Engineer *written* notice of his intentions in accordance with Clause 52(4)(b) – i.e. '... as soon as reasonably possible after the happening of the events giving rise to the claim ...' – and include, with such notice, a full description of the said conditions or obstructions. The Contractor is also required, '... with the notice if practicable or as soon as possible thereafter ...', to provide the Engineer with full details of:

(a) the effects which he anticipates the conditions or obstructions will have on the Works;

(b) the measures he is taking, or is proposing to take, to overcome the problem;

(c) the extent of any delays or difficulties which he anticipates in connection with his having to overcome the problem;

Following his receipt of the notice and the accompanying details, the Engineer may, if he thinks fit, *inter alia*:

(a) require the Contractor to provide an estimate of the cost of the measures he is undertaking or proposing to undertake;

(b) approve the Contractor's proposals;

(c) require the Contractor to modify his proposals;

(d) issue the Contractor with written instructions as to how the problem is to be dealt with;

(e) issue the Contractor with a variation order pursuant to Clause 51 (see section 9.1.9);

(f) order a suspension of the Works pursuant to Clause 40 (see section 9.1.8);

It should be suggested, at this point, that the Engineer act very cautiously in contemplating the issue of specific instructions as to how the Contractor should deal with the problem. Apart from the fact that he is putting himself in a potentially awkward situation should his proposed solution transpire not to be as straightforward as anticipated, there is always the possibility of such instructions 'opening the door' for subsequent claims, under Clause 13(3) (see below), in the event that they delay or disrupt the Contractor's execution of the Works.

For reasons which have been discussed previously (see section 9.1.8), *even more* caution should be exercised by the Engineer in contemplating a suspension of the Works or parts thereof!

In the event that the Engineer accepts the principle of the claim – namely that the conditions or obstructions, in part or in whole, could not '... reasonably have been foreseen by an experienced contractor ...' – he is required to assess, and certify for payment, any *reasonable* extra costs (including a *reasonable* allowance for profit) to which he considers the Contractor entitled. Similarly, he is required to take into account, when assessing any extension of time to which the Contractor is entitled under Clause 44, any delays which the Contractor has suffered as a consequence of his encountering the said conditions or obstructions.

If, on the other hand, he considers that the conditions or obstructions were, in part or in whole, reasonably foreseeable, he is required to notify the Contractor accordingly.

It should be particularly noted that the partial or total rejection of such a claim does not diminish the Contractor's entitlement to payment for work carried out in connection with any variation order issued under option (e) above.

Whilst the clause is fairly wide-ranging in its application, it is generally recognised that the vast majority of claims submitted thereunder concern ground conditions, the usual assertion being that the Engineer's Site Investigation data, as included in the Contract documents and upon which the Contractor is assumed to have relied in preparing his tender, is either inaccurate or insufficient.

In this respect, the conditions imposed upon the Contractor by Clause 11 (see section 9.1.4) appear, if viewed in isolation, to be somewhat onerous in that the Contractor is thereby deemed to have satisfied himself as to the nature of the ground and sub-soil of the Site and to have satisfied himself that his tender is sufficient to cover his contractual obligation to bear the burden of responsibility for any problems, whether foreseeable or not, which might arise in connection with such. Although Clause 12, by limiting his responsibility in this regard, provides a 'safety net' for the Contractor, it might also be mentioned (for the particular benefit of certain Employers who regard Clause 12 with something less than total enthusiasm!) that a tenderer's knowlege that such a 'safety net' exists removes the temptation for him to overload his tender in an attempt to cover himself against every conceivable eventuality.

Particular note should, however, be taken of the fact that, by virtue of the carefully worded phrase: '... could not reasonably have been foreseen by an experienced contractor ...', Clause 12 does *not* cover the Contractor against the *normal* (and considerable!) elements of risk inherent in most construction work of which an experienced Contractor would be aware or to which he would be alerted by the information available at the time of tender. The clause also excludes *calculated* risks taken by the Contractor.

Despite such careful wording on the part of the compilers of the Conditions, however, a considerable 'grey' area remains and, as long as Clause 12 exists, the question of what could or could not *reasonably* have been foreseen by an *experienced Contractor* will continue to be the subject of fierce debate.

Possibly the widest-ranging contractual provision for claims is under Clause 13(3) – which entitles the Contractor to claim, under Clause 44 (see section 9.1.8) and/or under Clause 52(4) (see below), in the event that '... instructions or directions ...', given by the Engineer '... in pursuance of Clause 5 or sub-clause (1) of this Clause ...', '... involve the Contractor in delay or disrupt his arrangements or methods of construction so as to cause him to incur cost beyond that reasonably to have been foreseen by an experienced contractor at the time of tender ...'.

Should such instructions or directions require any variation to any part of the Works, they will (effectively) be regarded as a variation order issued by the Engineer under Clause 51 (see section 9.1.9).

An instruction or direction '... in pursuance of Clause 5 ...' is one given to explain and adjust any ambiguity in the Contract documents (see section 9.1.2). It should be particularly noted, in this respect, that the Contractor is *only* entitled to claim in the event that the *resolution* of the ambiguity results in him suffering delay or incurring additional costs and *not* in the event that, following the said resolution, the work transpires to be more costly and time—consuming than he anticipated — unless, of course, the work amounts to a variation of the originally specified Works.

An instruction or direction given pursuant to '... sub—clause (1) of this Clause ...' is one in connection with the requirement of the Contractor to execute the Works in strict accordance with the Contract to the satisfaction of the Engineer (see sections 9.1.3 and 9.1.4).

A considerable amount of confusion surrounds the application of this clause, particularly regarding the *extent* to which the Contractor may rely upon it to claim recompense for 'disruptive' or 'delay—inducing' instructions or directions given by the Engineer.

For example, one school of thought holds that if the Engineer's powers to instruct or direct are *specifically* provided for in a particular clause — as in Clause 38(2) (see section 9.1.5), which empowers the Engineer to direct the Contractor to uncover specified parts of the Works to facilitate the former's inspection thereof — then any instructions or directives so issued should be regarded as having been issued pursuant to *that* clause and *not* pursuant to Clause 13(1). This interpretation is strengthened by the statement in Clause 5 that an instruction issued under such should be regarded as an instruction '... issued in accordance with Clause 13 ...', and by the corresponding reference, in Clause 13(1), to that statement. A spectre of doubt is cast upon the above interpretation, however, by the fact that Clause 71(2)(a) contains a similar statement yet no corresponding cross—reference appears in Clause 13.

The alternative viewpoint — strengthened by the requirement in Clause 13(1) that, in respect of his execution of the Works, the Contractor should '... adhere strictly to the Engineer's instructions and directions on any matter connected therewith (whether mentioned in the Contract or not) ...' — is that the provisions of Clause 13(1) and, hence, Clause 13(3), cover *any* instruction or direction issued by the Engineer in the course of the Contract. The validity of *this* interpretation, however, is called into question by the specific reference, in Clause 13(3), to Clause 5 which, if the provisions of the former *were* intended to cover *any* instruction from the Engineer, would be unnecessary.

The above confusion notwithstanding, it should be noted that the *wording* of Clause 13(3) gives rise to a number of curious situations which, it might reasonably be suggested, were not envisaged or intended by the compilers of the Conditions.

Consider, for example, a situation where the Engineer, under Clause 19(1), has instructed the Contractor to effect substantial repairs to a damaged safety fence surrounding the site. Because of his legitimate concern for the safety of the general public, the Engineer has instructed that the work be carried out *immediately*. In the event that the 'immediacy' requirement in the instruction disrupts the Contractor's arrangements, or delays other work, the *strict* wording of Clause 13(3) entitles him to recover the cost of the delay or disruption, if such *cost* is considered to be in excess of that which was *reasonably* foreseeable at the time of tender − *despite* the fact that Clause 19(1) stipulates that such work be carried out at his own expense!

Although the *spirit* of the clause is reasonably clear, the fact remains that, as a direct consequence of the aforesaid confusion, Clause 13(3) (as it stands) effectively provides an 'open invitation' for a certain type of Contractor whose *modus operandi* includes a significant element of 'claimsmanship' − i.e. subjecting the Engineer (or, more realistically, the Resident Engineer) to a continuous deluge of (mostly spurious) claims in the hope that some will creep through the 'rejection net' unnoticed.

Whilst the obvious solution to all of the abovementioned problems is a revision of the appropriate clauses, it should be noted that the Engineer has the power to effectively 'overwrite' potentially problematical clauses by writing 'customised' (and less ambiguously worded!) alternatives into the 'Special Conditions' under Clause 72. He should, of course, do this prior to tenders being invited (since he has no power to alter the Conditions after the Contract has been drawn up) and should draw tenderers' attention to the relevant changes by means of an appropriate inclusion in the tender documents.

Failing either of the above options, the Engineer has the power to substantially reduce the scope of Clause 13(3) by restricting the issue of instructions (without which the clause becomes inoperable!) to those situations where they are *absolutely unavoidable*. In particular, the Engineer and his site staff should avoid the issuance of gratuitous instructions − thereby opening the door to possible claims − which merely confirm obligations of the Contractor which are stated either in the Conditions of Contract or elsewhere in the Contract documentation.

Whilst the basic working provisions of Clause 14(6) have been discussed previously (see section 9.1.5) – namely that the Contractor is provided with remedies in the event that he suffers delay or disruption occasioned in respect of his being required to obtain the Engineer's approval of his working methods – some particular points therein merit further discussion.

In accordance with this clause, the Contractor is entitled to claim in the event that *inter alia* '... the Engineer's consent to ...(his)... proposed methods of construction ...(is)... unreasonably delayed ...'. It should be noted that not only does this cover the *original* consent, it *also* covers the Engineer's consent, pursuant to Clause 14(4), to any modifications to the original method which the Contractor has adjudged necessary to meet the Engineer's requirements *and* to any changes in methods which have been previously approved.

Whilst it is difficult to state precisely what would be adjudged as an 'unreasonable delay' on the part of the Engineer, since this will depend largely upon the prevailing circumstances, it should be noted that any adjudication thereof will not depend *solely* upon the convenience and financial interests of the Contractor and will take into account the urgency – in the context of the Contractor's programmed completion of the Works – with which the Engineer's response is required. It should, of course, be noted that the Engineer must be allowed a *reasonable* amount of time to appraise the Contractor's proposals – particularly in respect of their effect on the Permanent Works – and to make the appropriate response.

An examination of the grounds which entitle the Contractor to claim under this clause reveals the important fact that the possibility of such a claim *only* arises if the Engineer has 'started the ball rolling' by *requesting* (not *necessarily* in writing!) information, pursuant to Clause 14(3), pertaining to the Contractor's methods. It must, therefore, be suggested that, by way of minimising the extent of possible claims under Clause 14(6), the Engineer restrict his control of the Works, wherever possible, to exercising his powers of approval (or disapproval) of the Contractor's methods under Clause 13(2) (see section 9.1.5).

Clause 31(1) makes it clear that the Contractor is not entitled to exclusive possession of the Site in that it requires him to '...afford all reasonable facilities ...', in accordance with the Engineer's requirements, to other contractors and statutory bodies who are working on or near the Site. In the event that, as a consequence of his compliance with the above requirement, he suffers delay, or incurs cost, '... beyond that reasonably to have been foreseen by an experienced contractor at the time of tender ...', Clause 31(2) entitles him to claim recompense in the form of an extension of time and payment of '... such cost as may be reasonable ...'.

The phrase: '... afford all reasonable facilities ...' implies the requirement that the Contractor allow other contractors or statutory bodies to make use of facilities – such as, for example, a haul road – which he has provided for his own purpose. In the event that he is required to provide *special* facilities – i.e. those which he would not otherwise have been required to provide – then such a requirement would be adjudged as *unreasonable* and, failing the provision of such facilities being instructed as a variation (see section 9.1.9), would thereby relieve him of any obligation to comply with such.

It is worth noting that a similar *a–priori* judgement would be passed in the event of the Contractor being required to provide facilities which, although *reasonable* in themselves, would cause *unreasonable* disruption to the Contractor's programme – provided, of course, the programme had taken into account any information provided in the Contract documents in respect of the extent and timing of work to be carried out by other contractors or statutory bodies.

Finally, in respect of this clause, it should be noted that, in accordance with the *strict* wording of such, the 'unforeseeability' proviso in Clause 31(2) applies to the *consequences* of affording the required facilities and *not* to the facilities themselves. Thus, the Contractor would appear to be entitled to recompense for *unforeseen costs or delay* even though such may have been incurred as a result of his affording *reasonable and foreseeable facilities*.

Having dealt with the *principal* clauses in the Conditions under which claims may be submitted, it is relevant now to consider the procedure for the submission and appraisal of claims.

In this respect, it is necessary to make a distinction between claims for *time* and those for *cost* – even though the majority of 'claimable' occurrences will give rise to a combination of both – since the relevant procedures are covered by different clauses. In *all* cases, claims for *extensions of time* should be submitted and dealt with in accordance with the procedure specified in Clause 44 (see section 9.1.8). On the other hand, *all* claims for *cost* are submitted and dealt with in accordance with Clause 52(4) even though the cost element of a claim will almost certainly be inter-related with the time element.

The procedure for submitting a claim under Clause 52(4) is covered by parts (a) and (b) of the clause and depends upon the nature of the claim. In this respect, claims may be categorised as follows:

(1)   Claims in connection with variations ordered pursuant to Clause 51:

Failing agreement between the Engineer and the Contractor over the valuation of an ordered variation, Clause 52(1) (see section 9.1.9) authorises the former to determine a fair valuation for such and to notify the Contractor accordingly. If, by reason of the nature or amount of any ordered variation, certain Bill rates are, in the opinion of *either* the Engineer or the Contractor, rendered *unreasonable* or *inapplicable*, the Engineer is empowered by Clause 52(2) (see section 9.1.9) to revise such, to the extent that he thinks reasonable and proper, and to notify the Contractor accordingly. In the event that the Contractor wishes to claim a higher rate or price than that notified him by the Engineer in either of the above instances, Clause 58(4)(a) requires him, '... within 28 days after such notification ...', to give the Engineer written notice of his intention.

(2)   Claims in connection with changes of measured quantities:

If the Engineer is of the opinion that the *actual* quantities of work executed differ from the *Billed* quantities to an extent which renders the relevant Bill rates unreasonable or inapplicable, Clause 56(2) (see section 9.1.10) empowers him, after consultation with the Contractor, to revise the rates as he thinks fit and to notify the Contractor accordingly. If the Contractor wishes to claim a higher rate that that notified him, the appropriate course of action is as detailed in (1) above.

If, on the other hand, the *Contractor* is of the opinion that a rate revision is in order, by reason of changes in item quantities, he is required (by implication from Clauses 52(4)(b) and 56(2)) to notify the Engineer accordingly, *as soon as reasonably possible after the relevant work has been executed*, and to keep such contemporary records as may reasonably be necessary to support a possible subsequent claim. The procedure thereafter is as detailed in the preceding paragraph.

(3)   Claims submitted for reasons other than (1) and (2) above:

If the Contractor wishes to claim any additional payment under any clause in the Conditions *other* than those mentioned in (1) and (2) above, Clause 58(4)(b) requires him to give the Engineer notice of his intention '... as soon as reasonably possible after the happening of the events giving rise to the claim ...' and to keep such contemporary records as may be necessary to support his claim.

The procedure which follows the submission of the appropriate 'notice of intent' by the Contractor is detailed in Clauses 52(4)(c) and 52(4)(d).

Clause 52(4)(c) provides the Engineer with the *discretionary* authority to instruct the Contractor to keep any *further* records which are adjudged necessary. *All* records pertaining to the claim are to be made available for inspection by the Engineer at any time and should be copied to the Engineer if he so requests.

Clause 52(4)(d) requires that the Contractor submit, as soon as reasonably possible after he has notified the Engineer of his intention to claim, full and detailed particulars (to the extent possible in the circumstances) of the claim and the grounds on which it is based. In the event of the claim being of the 'protracted' variety, the Contractor is required to provide the Engineer with the appropriate 'updates' as and when the latter requires.

Clause 52(4)(e) refers to possible limitations on the Contractor's entitlement to payment in the event of his non-compliance with the provisions of Clause 52(4).

Finally, Clause 52(4)(f) clarifies the Contractor's entitlement to have certified the *whole* or *any part* of a claim for which he has supplied sufficient particulars to enable the Engineer to determine the amount due. By way of avoiding any misunderstanding over this clause, it should be emphasised that the clause is intended to cover *full and final payment* of either the *whole of a claim* or a *part of it for which full and detailed particulars have been supplied*. It should *not* be interpreted as providing the Contractor with the right to part-payment 'on-account' pending the supply of full and detailed particulars.

6th Edition Changes

The previous Clause 12(1) has now been split into three separate clauses. The content is substantially the same except that: Clause 12(1) now calls for the Engineer to be given the earliest possible written notification of a 'Clause 12' situation; Clause 12(2) now requires separate (or simultaneous) notification, pursuant to the appropriate clauses, of the Contractor's intention to claim additional payment or an extension of time. Clause 12(4)(a) replaces the original Clause 12(2)(a) (which empowered the Engineer, post—notification, to require a cost estimate from the Contractor regarding the latter's proposals) in that the Engineer, following his receipt of notice of a 'Clause 12' situation, is now authorised to (inter alia) require the Contractor to investigate and report upon the practicability, cost and timing of the measures he is proposing to take.

Clause 13(3) is (regrettably) substantially unchanged except for a reservation that the Contractor will not now be entitled to claim for any cost or delay associated with an instruction issued as a consequence of his default (that, at least, removes one 'bone of contention'!) and that, in respect of any additional Permanent or Temporary Works, profit may be added to cost.

Clause 31(2) is substantially unchanged save that profit may now be added to cost in respect of any additional Permanent or Temporary Works.

9.1.12 Disputes

Clauses: 66;

Where a dispute exists between the Employer and the Contractor, Clause 66 sets out the procedure to be followed.

Before outlining the contents of this clause, however, it should particularly be noted that a dispute does not exist simply because the Engineer has modified or rejected a claim submitted by the Contractor. It is only when the Contractor questions the Engineer's actions, and states that he is not satisfied with such, that a dispute exits. Stated simply, the sequence is:

(Claim) + (Rejection of Claim) + (Rejection of Rejection) = DISPUTE

Where a dispute is adjudged to exist, Clause 66 requires that the matter be referred to, and settled by, the Engineer. The Engineer is required to state his decision in writing, and to notify the Employer and Contractor accordingly, whereupon such decision becomes binding on *both* parties unless *either* of them requires the matter to proceed to arbitration.

Where, at the time the dispute arises, a Certificate of Completion has *not* been issued for the *whole* of the Works, the Engineer is required to make such decision *within one calendar month of the matter being referred to him*. In the event of *either* the Employer or the Contractor being dissatisfied with the Engineer's decision, *either* party may refer the dispute to arbitration but must do so *within three calendar months of receiving such decision*. If the Engineer *fails* to give a decision within one month, either the Employer or the Contractor may refer the dispute to arbitration *within three calendar months of the expiry of the said period of one month*.

Where a Certificate of Completion *has* been issued for the *whole* of the Works, the procedure is identical to that detailed above except that the Engineer is required to give his decision *within three calendar months of the matter being referred to him* – thus extending the overall period to six months.

Upon *either* party serving *written* notice of arbitration on the other party, *both* parties must agree upon the appointment of an Arbitrator (or *Arbiter* in Scotland) to conduct the proceedings. Failing agreement being reached *within one calendar month of such notice being served*, the matter will be referred to a person to be appointed, on the application of *either* party thereto, by the current President of the Institution of Civil Engineers.

The procedure following the appointment of the Arbitrator can vary and will depend upon the parties involved in the arbitration and the circumstances surrounding the dispute. In this respect, a suitable procedure is set out in the Institution of Civil Engineers' *Arbitration Procedure (1983)* which accompanies, in loose-leaf form, the I.C.E. Conditions of Contract. If the Works are situated within Scotland, the equivalent procedure is set out in the I.C.E. *Arbitration Procedure (Scotland) (1983)* which takes account of minor differences between Scots and English Law.

Notwithstanding the nature and form of the proceedings, the following points are worthy of particular note:

(a) Unless the parties agree otherwise in writing, disputes may be referred to arbitration at any time during the execution of the Contract;

(b) The Arbitrator has full power, in the course of the proceedings, to open up, review and revise *any* decision, opinion, instruction, direction, certificate or valuation of the Engineer;

(c) The Engineer may be called as a witness during the proceedings – at which time (by implication) he may be required to account for, *and* justify, any decision he has made in respect of the dispute;

(d) Neither party to the dispute is limited as to the evidence or arguments they may advance in support of their respective cases even though such may not have been put before the Engineer for the purpose of obtaining his original decision (the cause of the dispute!).

(e) The award of the Arbitrator shall be *binding on all parties*;

6th Edition Changes

Clause 66 has been re-written to include the intermediate option of referring the dispute (following either *party's rejection of the Engineer's decision) to a* Conciliator *prior to requesting arbitration. In the event that this option is chosen by the parties to the dispute, a suitable procedure is set down in the* I.C.E. Conciliation Procedure (1988). *The only other change to the clause is that, when a dispute is adjudged to exist, the Engineer should be notified by a* Notice of Dispute *which states the nature of the dispute.*

9.1.13 Default of the Contractor

Clauses: 3, 33, 41, 53–54, 63;

In the aforegoing discussion, coverage has been given to a variety of sanctions which are available to the Engineer and/or the Employer in the event of the Contractor's failure to comply with the terms of the Contract. In the main, however, the sanctions covered have been those which are available in the event of comparatively minor transgressions – i.e. those which do not *fundamentally* affect the Contract – on the part of the Contractor.

In the event of a *serious* default on the part of the Contractor – i.e. one which goes to the *root* of the Contract, such as his persistent and flagrant failure to discharge his contractual obligations – Clauses 53, 54 and 63 are of particular relevance.

The essence of Clause 53 is that, although the Contractor is expressly entitled to exclusive use of such for the purpose of executing the Works, all plant, goods and materials brought on to the site, and owned by the Contractor (or by a company in which the Contractor has a controlling interest), shall be deemed thereafter to be the property of the Employer. The ownership of such will revert to the Contractor in the event of its *authorised* removal from site or upon completion of the Works, whichever comes first – that is, of course, providing the forfeiture clause (Clause 63 – see below) has not been invoked.

In this respect, it should be particularly noted that, since the operation of this clause is intended to provide the Employer with a measure of protection in the event of the Contractor's serious default, such plant, goods or materials cannot be removed from site without the Engineer's prior *written* consent. Where such plant etc. is not immediately required for the purpose of completing the Works, however, the clause provides that the Engineer's consent for its removal should not be unreasonably withheld.

By way of safeguarding the Employer against the possibility of the Works being seriously affected by the repossession of items of *hired* plant in the event of the Contractor failing or being in dispute with the owners thereof, the clause contains special provisions relating to the Contractor's use of such plant:

Broadly speaking, the use of hired plant is prohibited unless the relevant hire–agreement contains provisions which allow the Employer, following his giving 7 days written notice of such, to take over the hire of the plant on the same terms as were applicable to the Contractor. Furthermore, the agreement must contain provisions which permit the Employer to allow a replacement Contractor to make use of the plant for the continued execution of the Works. By way of enabling the Engineer to ascertain the Contractor's compliance with this requirement, the latter is obliged to provide the former with details of the ownership of all items of hired plant as and when requested so to do.

In the event that the above requirement is facilitated by the Employer being included in any agreement for the hire of plant by the Contractor, any additional costs associated therewith are payable by the Employer. In the event that the forfeiture clause (see below) is invoked, however, such costs will be considered as part of the cost, to the Employer, of completing the Works subsequent to the forfeiture. Such cost will be deducted from any monies to which the defaulting Contractor is entitled in respect of work completed up to the date of forfeiture.

Finally, it should be noted that, upon completion of the Works, Clause 33 (see section 9.1.5) requires the Contractor to remove all plant and equipment from the site, along with any surplus material, within such reasonable time as the Engineer requires. In the event of the Contractor's failure to remove any such items, the Employer is entitled to sell those which belong to the Contractor and to return to the owners, at the Contractor's expense, those which are not his property. Whilst the Contractor is (not unreasonably) entitled to the proceeds of any such sale of his property, the Employer is entitled to deduct from such any expenses which he has incurred in connection with the sale and/or return of the relevant items. In the event, however, that the proceeds of the sale are insufficient to cover the Employer's expenses, the Contractor will be liable to the Employer (either under the Contract or, failing that, at law) for the outstanding balance.

For the purpose of securing payment for certain approved goods and materials prior to their delivery to the site, Clause 54 enables the Contractor to transfer the ownership of such to the Employer *providing*:

(a) the goods and materials are listed, for this purpose, in the Appendix to the Form of Tender (see section 6.2);

(b) the goods and materials have been manufactured or prepared and are *substantially* ready for incorporation in the Works;

(c) the Engineer has been provided with documentary evidence that the goods or materials are the property of the Contractor, or will become his property upon delivery to the site;

(d) the goods or materials have been stored (on the manufacturer's premises), and suitably marked or identified, to the satisfaction of the Engineer;

(e) the Engineer has been provided with a schedule listing and giving the value of each relevant item;

(f) the Engineer has been given the opportunity to inspect the goods and materials;

It should be noted that, notwithstanding the ownership of such goods and materials having been transferred to the Employer, the Contractor is responsible for any loss or damage to such and must insure himself accordingly. The Contractor is similarly responsible for any storage, handling or transportation costs.

It should be further noted that the Engineer's approval of such goods or materials, for the purpose of this clause, does not prejudice his right to subsequently disapprove such in the event of their non-compliance with the Contract requirements.

It should be mentioned that, whilst the procedure set out in this clause works, with proper legal handling, in England, it is fraught with legal pitfalls in Scotland.

Clause 63 deals with the forfeiture of the Contract, as a consequence of the actions or omissions of the Contractor, and has the most far-reaching implications of any of the clauses contained within the Conditions of Contract.

The clause states that the *Employer* (note: *not* the Engineer!) may, after giving the Contractor 7 days *written* notice of his intention, expel the Contractor from the Site and the Works if the Contractor:

(a)   has been adjudged by a court to be bankrupt, or:

(b)   has had a receiving order made against him, or:

(c)   has presented his petition in bankruptcy, or:

(d)   has made an arrangement with, or assignment in favour of, his creditors, or:

(e)   has agreed to carry out the Contract under a Committee of Inspection of his creditors, or:

(f)   has gone into liquidation (other than voluntarily for the purpose of amalgamation or re-structuring), or:

(g)   has assigned the Contract without the Employer's written consent (contrary to the provisions of Clause 3), or:

(h)   has had an execution levied on his goods;

The foregoing will *also* apply if the Engineer certifies, to the Employer, that, in his opinion, the Contractor:

(a)   has abandoned the Contract, or:

(b)   has, without reasonable excuse, failed to commence the Works in accordance with Clause 41 (see section 9.1.8), or:

(c)   has, without reasonable excuse, suspended the progress of the Works for 14 days after having received the Engineer's written notice to proceed, or:

(d)   has failed to remove condemned or rejected goods or materials from the Site having been given written instructions by the Engineer so to do, or:

(e)   has failed to pull down and replace faulty work having been given written instructions to that effect, or:

(f)   is failing to proceed with the Works with due diligence despite previous warnings from the Engineer, or:

(g)   is *persistently* or *fundamentally* in breach of his contractual obligations, or:

(h)   has sub-let any part of the Contract *either* to the detriment of good workmanship *or* in defiance of the Engineer's instructions to the contrary;

In expelling the Contractor from the site for any of the above reasons, it should be noted that the Employer neither avoids the Contract nor releases the Contractor from any of his obligations or liabilities under the Contract. Furthermore, the rights and powers of the Engineer and the Employer, under the Contract, remain unaffected.

Having expelled the defaulting Contractor, the Employer has two courses of action open to him: he may complete the Works himself (i.e. with his own workforce) or, alternatively, he may employ a replacement Contractor. In either event, use may be made of the defaulter's plant, Temporary Works, goods or materials which are deemed, under Clauses 53 and 54 (see above), to belong to the Employer. Furthermore, and by way of offsetting any losses incurred in connection with the forfeiture, the Employer may sell any of such items at any time.

By the aforementioned 7 days notice, or by further notice given within 14 days following such, the Contractor may be required, by the

Engineer, to assign to the Employer the benefit of any agreement which he (the Contractor) has entered into for the supply of goods or materials and/or for the execution of work for the Contract.

As soon as is practicable following the Employer's entry upon the Site and his expulsion of the Contractor, the Engineer must fix, determine and certify the amount earned by, or reasonably accruing to, the Contractor up to the date of his expulsion. He must also ascertain the value of any plant, temporary works or unused, or partially used, goods or materials which are the property of the Employer under Clauses 53 and 54.

In the event of entry and expulsion under Clause 63, the Employer is not liable to pay the defaulting Contractor any monies due under the Contract until the expiry of the Period of Maintenance and, then, *only* the amount due after the Employer has deducted *all* the costs and expenses incurred by him as a consequence of the forfeiture *and* any liquidated damages due as a result of late completion. If this results in a 'negative balance' – i.e. if the Employer's expenses exceed the amount which would otherwise be due the Contractor upon completion – such balance is regarded as a debt due by the Contractor to the Employer and is recoverable from the Contractor at law.

It should be noted that the liquidated damages element in the above calculation will cover the period from the *due* date of completion to the *actual* date of completion even though, for part of the time, the progress of the Works will be outwith the Contractor's control. That said, however, the Contractor will be entitled to an extension of time for any delay on the part of the Employer in arranging for the remainder of the work to be completed (whether by his own workforce or by a replacement Contractor) and for any unnecessary delay in the subsequent completion operations.

It should also be noted that, irrespective of the course of action chosen by the Employer *vis-a-vis* completion of the remainder of the Works, the 'loss-mitigation' requirement of contract law (see section 3.6.1) obliges him to be *reasonably* thrifty in the matter.

6th Edition Changes

Clause 53 is substantially unchanged except that vesting is now restricted solely to equipment, goods or materials owned by the Contractor and does not, therefore, apply to hired plant.

Clause 54 is substantially unchanged except that a provision has now been included to the effect that vesting shall take place if the Engineer so directs.

The grounds for forfeiture of the Contract under Clause 63 now include the Contractor having had an administration order made against him.

9.2 Exercises

The following exercises have been compiled with a view to providing the reader with an opportunity to gain a certain amount of 'hands on' experience regarding the use of the I.C.E. Conditions of Contract.

Whilst the exercises have been specifically formulated with the 5th Edition of the Conditions in mind, considerable value may be derived by considering the effects (if any) which the 6th Edition may have upon each situation.

Exercise 1.

The Bill of Quantities for a drainage contract indicated a total requirement for 20,000 metres of reinforced-concrete pipes of various diameters. During the course of the contract, without any variation orders being issued or requiring to be issued, the total length of pipework has increased to 30,000 metres resulting in substantial disruption to the Contractor's operations and a six-month overrun in contract duration. The Employer has declared his intention to deduct liquidated damages to the extent specified in the Appendix to the Form of Tender.

What are the Contractor's rights and how should he go about securing them?

What should the Engineer do?

Is the Employer entitled to deduct liquidated damages?

Exercise 2.

In the course of a Contract for the construction of a sewage works, the Contractor was issued with a drawing showing the pump–house layout in relation to two ground stations, A and B, set–out by the Engineer's staff during the original site survey.

A further drawing was subsequently issued showing the layout of the settlement tanks. The drawing did not show the pump–house but it did show stations A and B along with two other ground stations, C and D, on a line parallel to AB and 70m from it. Points C and D were set–out by the Engineer's staff at the same time as A and B.

Prior to the pump–house being set–out, one of the Resident Engineer's assistants told the Contractor's setting–out engineer that point C was 0.28m below point A.

The Contractor's setting–out engineer, for his own convenience, used points C and D for setting–out the pump–house. The setting–out was checked by a member of the Resident Engineer's site staff, using the same two points, and was found to be correct.

Following the completion of the pump–house foundations, the pump–house superstructure was set–out from points C and D. This time, however, the setting–out was checked using points A and B and was found to be incorrect although it related correctly to the completed foundations. Further investigation revealed that AB was *not* parallel to CD and that C was, in fact, 0.28m *above* point A.

What is the contractual position?

What would the position be if it were subsequently found that the foundation concrete did not meet the specified strength?

Exercise 3.

The 'general arrangement' drawing of a bridge shows 76 Type 1 wing–wall copes numbered consecutively. A separate 'typical detail' drawing gives the dimensions of a Type 1 cope.

The Bill of Quantities item for Type 1 copes gives the quantity as 72 No.

On the basis of the 'general layout' drawing, the Contractor orders 76 copes. He subsequently finds that he has 4 surplus copes. On relating the total length of the wing–walls to the dimensions of the

Type 1 cope given in the 'typical detail' drawing, and taking into account the specified gap between copes, he finds that only 72 copes were required.

Who is responsible for the surplus copes?

Would your answer be affected if:

(a)   The Bill item had shown the quantity of copes to be 74 No.?

(b)   The units on the 'general arrangement' drawing had not been numbered?

Exercise 4.

The Contract completion time for a four–storey office block, with basement, is two years with a Maintenance Period of one year.

The basement and ground floor are completed in the first year and, as the Employer is anxious to take possession as soon as possible, the Engineer issues a Certificate of Completion for this part of the Works.

The whole job is completed in the specified time, the Engineer certifies accordingly and the Employer takes possession of the entire building.

Six months later, severe distortions become apparent in the false ceilings on the second and third floors. At the same time, a leak appears in the bituminous tanking to the basement, resulting in wet patches on the basement walls.

What is the contractual position?

Following satisfactory resolution of the above problem, further defects become apparent five years into the Employer's occupancy of the building. What is the position now?

Exercise 5.

A Contract has a Tender Total of £2,500,000 and a completion time of 80 weeks.

The Bill of Quantities includes a Provisional Sum of £300,000 for contingencies if ordered.

During the course of the contract, the Engineer orders work to the value of £200,000 to be carried out under the Provisional Sum item.

Is the Contractor entitled to an extension of time and, if so, is it possible to say approximately how much time is due?

Exercise 6.

In the course of a reservoir Contract in the North West of Scotland, the Contractor ran into two difficulties in the course of excavating for the stilling basin:

(a) Following ten successive days of heavy rainfall, the excavation became flooded and the Contractor was forced to install powerful pumps to get rid of the water.

(b) The excavation disclosed a series of old foundations which required to be broken up and removed. This resulted in the excavation taking an extra three days. There was no indication of the presence of these foundations either on the drawings or on the surface.

The Bill of Quantities provides a rate for breaking out old foundations in the vicinity of the control house but not in the vicinity of the stilling basin.

The Schedule of Daywork includes a rate for the use of pumps.

The Contractor claims an extension of time and additional payment for coping with both of these difficulties. Is he entitled to such?

If the principles of the claims are accepted, to what extension of time would the Contractor be entitled and to how much additional payment?

Would your answer be affected if the contract had been in the South-East of England?

Exercise 7.

The Contractor submitted the design of a temporary retaining wall to support the side of a deep excavation immediately adjacent to the site boundary. The Engineer approved the design and authorised the Contractor to proceed.

The retaining wall subsequently collapsed, damaging a building on neighbouring land belonging to a chemical company. One of the Contractor's employees was injured, as was one of the employees of the chemical company.

Both employees sued their respective employers. The Employer was sued by the chemical company.

What is the extent of liability of the Contractor, the Engineer and the Employer:

(a)   Under the contract?

(b)   At law?

PART 3

SAFETY

CHAPTER 10

ACCIDENTS AND ACCIDENT PREVENTION

In the middle of the nineteenth century, at the start of the so-called 'great railway boom', seven years were taken to construct the Woodhead Tunnel carrying the Sheffield to Manchester railway under the Pennines. In the course of construction, 32 men were killed and 140 were injured – a casualty rate to which the 'management' were, by all accounts, supremely indifferent despite the fact that it was proportionally heavier than that suffered by the British Army at Waterloo.

At the turn of the century, a 22-arch viaduct was constructed to carry the Fort William to Mallaig railway across the glen at the head of Loch Shiel in the western highlands of Scotland. It is part of local folk-lore (not to mention part of the 'stock in trade' of the local tourist industry!) that a man died for each arch of the viaduct.

Whilst the situation has improved substantially since those days, the safety record of the construction industry *still* leaves much to be desired in that, *each week*, several hundred construction-related accidents are reported somewhere in the United Kingdom. Although most of these result in minor injuries, such as strains, cuts and abrasions, around a third of them result in serious injuries – major fractures, amputations and the like – and two or three of them are fatal.

Although the figures are sufficient in themselves to give cause for considerable concern, the seriousness of the situation is compounded by the general recognition that, with the implementation of elementary safety precautions – and the application of a modicum of that most elusive of commodities: common sense – the vast majority of such accidents could be prevented!

10.1 Root Causes and Preventative Measures

In the context of accident *statistics*, an accident is defined as an unforeseen or unexpected occurrence which results in injury or damage. In the context of accident *prevention*, however, such an occurrence is termed an accident irrespective of the end result. For example, the dropping of a hammer from a high scaffold is no less an accident because the hammer has the good fortune to land harmlessly on the ground and not on somebody's head. The point here is that it could have landed on somebody's head and, in the event of a repeat performance, might well do so!

The practice of accident analysis – i.e. the statistical examination of accident records *vis–a–vis* location, cause and effect – is useful for the purpose of identifying particularly risk–fraught work processes, thereby enabling appropriate 'process–related' safety precautions to be devised, but is of limited effectiveness in regard of accident prevention in *general* terms. For example, a particular accident record will provide the information that Joe Bloggs suffered a fractured skull because he fell off a ladder and landed on his head. It might also provide the information that the fall occurred because the ladder was insecurely founded. What it will *not* provide, however, is information as to *why* the unfortunate Mr Bloggs climbed an insecurely–founded ladder in the first place.

It is often suggested that the *root* cause of an accident is that, somewhere along the line, someone has either done something which he ought *not* to have done (e.g. dropping a hammer from a scaffold) or has *not* done something that he *ought* to have done (e.g. ommitting a scaffold safety–rail). Quite simply, someone, somewhere, somehow, has made a mistake.

Whilst the general principle of the above suggestion is correct – namely, that accidents are *primarily* caused by *people* and not *processes* – many safety experts are of the opinion that the *real* root cause of an accident lies in the reason for the mistake having been made.

The following root causes of accidents may be cited as being the most common:

(a) Ignorance and lack of training

The majority of plant- or equipment-related accidents stem directly from the operator's misuse of such. Whilst there are numerous reasons for the incorrect usage of plant and equipment, by far the most common is the operator's ignorance of the *correct* method of usage which, in almost all cases, is not discovered until it is too late!

The solution is obvious. If an operative is required, for example, to use a jackhammer, it is advisable to *ensure* (i.e. don't just take his word for it!) that he knows how to use it *correctly* and is aware of the potential dangers. If there is any doubt, he must be shown how to use it by a suitably experienced person. It is similarly advisable to ensure, by means of adequate supervision, that he *continues* to use it correctly.

(b) Complacency and cavalier attitudes

The '"it can't happen to me!"' attitude, complacency is generally regarded as the biggest single root cause of construction-related (or any other type, for that matter) accidents.

Whilst cavalier attitudes are very often a direct consequence of complacency, it has to be said that, to a certain extent, they derive from the traditional 'macho' image which many male construction workers (workforce and management alike!) have of themselves - "a *man* working in a *man's* industry" etc. etc. - and which engenders the feeling that safety precautions, in some obstruse way, constitute an affront to their perceived masculinity.

The only way to combat such attitudes is to constantly emphasise the need for safety-consciousness *at all times* and to 'drum home' the message to all concerned that 'it' not only *can* happen to them but that, given half a chance, *will* happen to them - 42-inch biceps notwithstanding!

That said, however, it should be realistically appreciated that, as a means of engendering safety-consciousness, there are limits to the effectiveness of 'preaching the gospel'. It is often suggested, in this respect, that, regrettable though it may be, nothing concentrates the mind quite like seeing the results of a serious accident or, failing that, having to break the news of such to an unsuspecting spouse or parent!

(c) Carelessness

Whilst by no means the cause of as many accidents as might be imagined, carelessness will always be a prime contender for the role of 'root cause' in any accident involving a young (i.e. fresh from school, college or university) person. Apart from the fact that such persons tend to be full of enthusiasm (for the first few months, at least!) and, consequently, tear into what they are doing without a moment's thought, they will (more importantly) be largely unaware of the potential dangers which face them.

A substantial number of potential accidents can be avoided by instilling in such people, at the earliest possible opportunity, the need to take care in whatever they are doing and, subject to the limitations discussed in (b) above, by apprising them of what can happen if they don't.

It should be *particularly* noted, that since carelessness is a natural by-product of complacency, the problem will not be limited to the young alone!

(d) Lack of Discipline

This is another common cause of accidents involving the young. Whilst nobody would reasonably expect military-style discipline on a construction site - indeed, any attempt to impose such would probably be totally counter-productive in today's social climate - there is a definite need to encourage a proper respect for authority.

Whilst a reasoned approach will suffice in most cases, instances of flagrant or persistent ill-discipline should be dealt with in such a way as to set the necessary example to others. This is of particular importance when horseplay occurs on site - as, almost inevitably, it will from time to time - since more than a few accidents have resulted from horseplay which has got out of hand.

(e) Distraction

Accidents resulting from distraction are numerous and range from the comparatively minor variety - such as a carpenter hammering his thumb, instead of the nail, whilst concentrating more on a workmate's conversation than on the job in hand (no pun intended) - through to serious, and possibly fatal, instances – such as an excavator cutting through a buried 10kV electricity cable as a result of the operator's attention being momentarily diverted.

There would, therefore, appear to be the twin–fold requirement, firstly, to instil the need for job concentration and, secondly, to discourage workers from distracting their fellow workers.

(f) Poor communication

There is no finer way of illustrating the problem of poor communication than by quoting the words of a well–known politician (no names, no pack–drill!) who was attempting to play down one of his celebrated gaffes by accusing the media of misrepresenting him:

> "I know that you believe you understand what you think I said – but I'm not sure that you realise that what you heard was not what I meant."

In order to avoid the possibility of having to come out with something similarly devastating – for example, to some unfortunate operative who has just cut off his toes with a concrete–breaker because he wasn't clear as to what *exactly* was required of him – there are two 'golden rules' regarding communication:

1. Ensure that all instructions are *clear* and *unambiguous*. If possible, illustrate the point in some way.

2. By way of ensuring that what was heard was what was meant, check that the instructions have been fully understood by the recipient.

In applying these rules, however, extreme care should be taken not to make the entire exercise counter–productive by engendering, within the recipient, the impression that he is being treated like a somewhat unreliable and backward infant!

(g) Management indifference

As a root cause of accidents, management indifference to safety is becoming less and less common in the case of the large well–known construction companies, most of whom take safety very seriously indeed. Regrettably, however, the reverse is becoming increasingly the case with the smaller firms who, largely existing on the very limits of commercial viability at the best of times, often tend to place a very low priority on the safety of their employees and, on account of the costs involved, even less priority (i.e. none whatsoever!) on the training of such. This is particularly the case with the 'one man and his dog' type outfits who, for a variety of reasons (most of which centre around their avoidance of the responsibilities and obligations imposed by employment legislation!), prefer to employ the majority of their workforce on an *ad hoc* or 'semi-casual' basis.

It is unfortunately the case that little, if anything, can (realistically) be done to remedy the situation except to rely on the combined efforts of the Health and Safety Executive (see section 11.3.2) and the appropriate legislation (see Chapter 11). (At the time of writing, the author is aware that some unions provide a safety 'hotline' facility for members to 'blow the whistle' if they consider that their personal safety is being compromised by their employer's disregard of his fundamental safety-related obligations. Whilst this is obviously a step in the right direction, judgement should be reserved as to its effectiveness as a potential solution to the problem since such a facility has, in effect, always been provided by the Health and Safety Executive.)

10.2 Process-related Causes and Preventative Measures

Various statistical analyses of accidents, carried out over the years, have identified that about two thirds of all accidents which occur on construction sites in any given year fall into one or other of the following main categories:

(a) Persons falling;

(b) Falls of Materials;

(c) By agency of plant, machinery or transport;

These accident categories are considered in turn, together with a 'miscellaneous' category which, in effect, consists of accidents which do not fall into any of the three main categories. For each category, typical accident causes are noted and are illustrated by specific examples either drawn from accident records or from the author's personal experience. A 'checklist' of appropriate preventative measures is presented in each case.

10.2.1 Persons Falling

About 25% of all accidents fall into this category and may be further categorised as follows:

(a) From scaffolds etc.

Most falls from scaffolds occur through personnel overbalancing, tripping or slipping. The rest may be attributable to general defects in the scaffolding such as boards slipping or breaking, the absence of guard rails, insecure foundations etc.

A welder was dragging burning gear along a gantry surrounding the top of a power−station chimney. The gear snagged. In the course of the welder's attempts to jerk the gear free, a grating was dislodged and the welder fell 160m (fatal).

A labourer tripped on a piece of loose timber on a scaffold walkway and fell 5m (multiple abrasions and bruised spine).

Safety Precautions

1.   Erection and dismantling should be carried out *efficiently* and by *experienced* personnel. The *dismantling* sequence, wherever possible, should be the exact reverse of the *erection* sequence.

2.   *Proper* stagings and work platforms should be erected.

3.   A *thorough* inspection should be made before first use. Weekly inspections (more often in bad weather) should be made thereafter.

4.   Temporary alterations to scaffolding must be authorised and made good immediately the relevant work (i.e. that which necessitated the alteration) is completed.

5.   Where suspended scaffolds or bosuns chairs are to be used, a *thorough* inspection of the scaffold/chair, ropes, pulleys and hoisting gear should be made before first use. Weekly inspections should be made thereafter.

6.   Where *system* scaffolding is used, the manufacturer's instructions should be *strictly* adhered to.

7.   Keep walkways free of obstructions.

8. Do not overload the scaffold.

9. Provide *proper* access to scaffolding (i.e. suitably lashed ladders) to avoid personnel slipping off whilst emulating Tarzan.

10. Ensure that the foundations are *secure* and *level* (if necessary, excavate rather than pack-up!) and that the standard bases are seated on *level* ground.

11. Ensure that the *proper* couplers are used (i.e. that a putlog coupler is not used where a load-bearing coupler is required).

12. Ensure that the supports (i.e. tie-ins, braces, lashing) are adequate and secure.

13. Ensure that walkways are of a *sufficient* width (a minimum of *three* boards width), a *constant* width and *securely seated* (i.e. no 'cantilever' boards). Defective scaffold boards should be scrapped.

14. Provide toe boards (to avoid tools or materials being inadvertently kicked off) and guard-rails (to stop personnel falling off).

15. Incomplete or unsafe scaffolding must not be used and a notice to that effect must be posted.

(b) From ladders

Whilst most falls are caused by the person slipping or overbalancing and falling from the ladder, a considerable number are caused by movement of the ladder (either the bottom slips outward or the top slips sideways). A number of accidents also arise out of defects (e.g. missing or broken rungs) in the ladder or, in some cases, by the ladder itself breaking. This particular type of accident is almost invariably associated with badly repaired or 'home-made' ladders.

The ladder on which a joiner was working was resting on a concrete floor and was not secured. The ladder slipped at the base causing the man to fall 3.5m (head injuries and concussion).

A labourer, working from a 3m ladder adjacent to a 15m lift shaft, was removing a nail from a wall. The exertion of the job caused the ladder to move and the labourer fell off the ladder into the liftshaft (fatal).

Safety Precautions

1. Stand ladders on a *firm, level* base. If necessary, either level the ground or bury one foot. Never pack–up one side.

2. Never stand a ladder which is too short on something (e.g. an oil–drum) to gain extra height – use a longer ladder.

3. Never support a ladder on its rungs (e.g. across a plank).

4. Ladders should be erected with any rung– or stile–reinforcement on the *underside*.

5. Ensure a correct angle of repose – 1m out for every 4m of height – and at least 1m top projection.

6. Ensure that there is sufficient space behind a ladder for proper footholds on the rungs.

7. Ensure the correct overlap of adjacent sections on an extending ladder – length less than 5m: $1\frac{1}{2}$ rungs; length 5 – 6m: $2\frac{1}{2}$ rungs; length over 6m: $3\frac{1}{2}$ rungs;

8. *Always* lash the *top* of the ladder to prevent sideways slippage. *If possible*, lash the bottom as well. Failing that, a person should hold the base of the ladder when in use but this is generally only effective with short ladders.

9. Beware of wet, icy or greasy rungs. If necessary, clean the rungs before mounting. Always clean mud from boots before climbing.

10. Face the ladder when climbing or descending. Grasping the rungs is preferable to grasping the uprights (the *stiles*). If necessary, wear gloves!

11. Never over–reach from a ladder – move the ladder. Remember that it takes *two* people to carry a ladder!

12. Avoid carrying materials when using a ladder. Tools should either be carried in pockets or in holsters designed for that purpose.

13. For high work, consider staging (no further than 9m apart) rather than a single ladder.

14. Check ladders regularly for defects, wear and damage. If a ladder cannot be repaired *properly*, it should be scrapped.

15. *Never* paint a ladder – it hides defects and is illegal. If necessary, *clear* varnish and/or preservative may be used.

16. *Never* use a 'home–made' ladder.

17. To avoid them warping, do not leave timber ladders lying on wet ground or exposed to the weather.

18. Store ladders correctly to avoid damage – e.g. never hang them horizontally from brackets as this tends to pull out the rungs.

(c) From structures and other heights (except plant)

Such falls occur in a variety of ways – e.g. falls from shuttering (often as a result of a collapse of the shuttering on which persons are standing), falls from barrow runs whilst transporting materials, falls through unprotected openings and fragile roofs – and are almost always the result of a lack of care on the part of the individuals concerned.

Whilst fixing shuttering on a bridge abutment, a carpenter slipped and fell 15m on to the wing–wall starter bars (fatal and very messy).

The supports for a temporary plywood covering to a lift–shaft were removed. Due to an oversight, the plywood covering was not removed at the same time. A labourer subsequently stepped on the plywood (thinking it was still supported) which collapsed. He fell 30m down the lift–shaft (inter alia fractured spine and resultant permanent paralysis).

Safety Precautions

1. Ensure that all roof and floor edges are fitted with barriers.

2. Erect fences around openings. Failing that, openings should be covered with heavy material, securely fastened down, with appropriate warning notices.

3. Ensure that warning notices are displayed on fragile roofs. Provide crawling boards where access is required.

4.   Never walk along, or across, pipes.

(d) From plant

Falls from plant are surprisingly frequent on construction sites. The majority of such falls occur when drivers are climbing on to, or descending from, their machines or when 'unauthorised' passengers are riding machines not adapted for such. The rest tend to occur when machines are being serviced or loaded. The falls are often associated with wet conditions and/or muddy boots.

Whilst crawling along the jib of his machine to inspect a jammed pulley wheel, a crane driver slipped and fell 13m into a concrete based trench (fatal).

Whilst checking the mix, a junior engineer slipped off the operating platform of a concrete batcher and fell 1.5m (two bones cracked in spine).

Safety Precautions

1.   Provide ladders from ground level to working platforms on cranes and other high plant.

2.   Prohibit riding on tow−bars, running boards, steps etc.

3.   Prohibit the carrying of passengers unless there is a *safe* spare seat.

4.   Prohibit the carriage of passengers by plant which is not intended for such (e.g. dumpers, crane buckets etc.)

5.   Instruct drivers accordingly and ensure that appropriate warning notices are displayed wherever necessary.

6.   *'Pour encourager les autres'*, crack down hard upon transgressors.

(e) Into excavations and manholes

In the main, falls into excavations usually occur when personnel slip whilst trying to jump them or whilst trying to climb out of them by means other than a ladder. A surprising number of personnel fall into uncovered manholes (particularly those which have become obscured by vegetation) whilst walking across the site.

A labourer slipped on wet greasy ground whilst trying to jump a 0.75m wide trench and fell 2m into the trench (sprained knee).

Whilst assisting with setting-out, a chainman fell into an open manhole on an overgrown site (fractured leg).

Safety Precautions

1. Warn personnel to look where they are going.

2. Provide proper walkways across trenches and thereby remove the temptation for personnel to jump across.

3. Provide ladders for access to, or exit from, excavations.

4. Erect substantial (and brightly coloured!) barriers around excavations. Display appropriate (and legible!) warning notices if the public are likely to be affected.

5. Do not leave manholes uncovered or unfenced – particularly if their presence is likely to be subsequently obscured by vegetation.

(f) On the level

Although rarely, if ever, fatal, falls on the level account for a substantial proportion of injuries suffered by construction workers. Such accidents generally comprise simple falls or slips which are often associated with slippery conditions and/or the carrying of objects which are either heavy, awkward or partially obscure forward vision. In cases where the injured person has either tripped on something, or has stepped down onto some obstruction, the root cause is almost inevitably that the person concerned has not been looking where he has been going. Although this category mostly comprises falls at ground level, it also includes falls on scaffolding etc. where the person has not actually fallen off.

A plant driver stepped down off his machine into a small heap of debris (fractured ankle).

Whilst carrying a shutter panel, a charge-hand carpenter slipped on frosty ground and sat on a nail projecting from some previously stripped shuttering (punctured backside and severely dented pride). (It might be mentioned that several onlookers – alerted by the positive explosion of expletives which emanated from the carpenter - very nearly died from laughter-induced heart-attacks. The carpenter, perhaps not surprisingly, did not see the funny side of his plight at the time!)

Safety Precautions

1. Warn personnel to look where they are going.

2. Keep a tidy site - remove the inevitable rubbish at regular intervals.

3. Ensure that walkways are kept free of materials and other obstructions which might cause falls.

10.2.2 Falls of Materials

Material objects, whether heavy or comparatively light, gather tremendous energy in falling. For example, a brick dropped by a man on his own foot can fracture a bone. Something very much lighter - a 20mm nut, for example - will, fairly obviously, not have the same effect. Dropped from a height of 20m, however, that nut can penetrate a man's skull and kill him.

Falls of material are responsible for about 15% of injuries suffered on construction sites and may be classified as follows:

(a) From heights or into excavations

All sorts of things - ranging from small tools to wheelbarrows, scaffold members and girders - fall from heights. Mostly, the falls occur from working positions such as scaffolds, or from structures where personnel are working. Often, persons are hit by spoil falling from the working faces of excavations, or from lorries and dumpers. Additionally, accidents occur when materials or other objects fall, or get knocked, on to personnel working in excavations. Demolition workers are at particular risk in that around 50% of all injuries suffered by such are the result of falling debris.

An engineer was setting-out near the base of a 30m scaffold which was in the process of being erected. He had temporarily removed his safety helmet whilst sighting through his theodolite and was struck on the head by a spanner dropped by a scaffolder near the top of the scaffold (fatal).

Whilst his mates went to find some props, a labourer continued to remove brickwork supporting a reinforced concrete roof slab which collapsed and crushed him (fatal).

Concrete tipped into a foundation fell on a labourer who was spreading the concrete (bruises and shock).

Safety Precautions

1. Provide toe–boards to scaffolds to prevent tools or materials being inadvertently kicked off.

2. Provide tool boxes for tools.

3. Place materials where they are unlikely to fall.

4. Position personnel so that they are unlikely to be struck by falling material. Provide some form of overhead protection where necessary.

5. Consider the use of facade netting, particularly where there is a potential risk to the general public. Ensure that the netting does not become overloaded with debris.

6. Do not place materials, plant or spoil too near the edges of excavations. Stack such materials securely.

7. Wherever possible, batter the working faces of excavations. If not possible, take particular care to avoid temporary instability as a result of (say) undermining. Unstable faces must be stabilised – even if this results in overbreak – before personnel are allowed to work near the face.

8. Lower materials and other objects *properly*. Do not throw them down. Use a chute if necessary.

9. Take particular care with demolition. Such operations should always be *planned* and *never* rushed. Demolition of a structure should *always* be carried out in the reverse sequence to the erection thereof – i.e. start at the top and work downwards! Beware of any temporary instability.

10. Encourage the wearing of safety helmets. Crack down hard on defaulters.

(b) On the level

Most accidents result from personnel dropping materials on their own (or other peoples') feet, often during loading or unloading operations. A substantial number of accidents also arise from the collapse of unstable stacks of materials such as pipes. 'Pinched finger' type accidents occur with monotonous regularity when materials are being stacked or moved.

In the course of loading spoil on to a lorry, a mechanical excavator hit a number of sheet piles leaning up against a nearby retaining wall. The piles collapsed on a ganger (fatal).

Whilst a labourer was moving a large piece of old asphalt by hand, a portion broke off and fell on to his foot (broken big toe).

Safety Precautions

1. Institute a safe working procedure when materials are being moved or stacked. Particular care should be taken with materials – e.g. pipes – which can roll.

2. Ensure that stacks of material are stable. Use chocks with materials such as pipes. Stack materials on level ground wherever possible.

3. To facilitate the safe withdrawal of material, ensure that material stacks are not too high.

4. The removal of material from stacks should be in the reverse sequence to that used to stack the material.

5. Do not stack material where it causes obstruction and might, therefore, be hit by moving plant. Take particular care to avoid unnecessary protrusions. Protrusions which cannot be avoided should have distinctive markers attached to them.

6. Do not stack material near overhead power lines.

7. Do not stack materials against a wall. Apart from the fact that such stacks are inherently unstable, the wall may not be able to take the consequent lateral thrust.

8.   When stacking materials on upper floors of a building, ensure that the floor is able to take the weight of the stack. Check with the designer if necessary.

9.   Instruct personnel in safe methods of lifting and carrying – not only to avoid strains, hernias and the like but, equally important, to prevent droppages of heavy or awkward materials.

10. Encourage the wearing of safety boots.

10.2.3 By agency of Plant, Machinery and Transport

About 25% of accidents occurring on construction site are associated with plant, machinery and, to a lesser extent, transport. For fairly obvious reasons, such accidents are usually serious and often fatal.

According to statistics, such accidents may be classified as falling into five distinct sub-categories:

(a) Striking personnel, collisions and overturning

The majority of plant- or transport-related accidents fall within this category. Typical examples include colliding with personnel, or other machines, due to the restricted vision of the driver or operator, or as a result of the inadvertent starting of the plant (particularly those items which have automatic transmission – such as scrapers – and tend to move if started) or as an indirect consequence of the plant or transport being driven too fast.

> *A technician was taking a compaction test on the sub-base of a major road contract. The driver of a grader working nearby was, due to his vision being restricted, unaware of the technician's presence and backed the machine over him (severe internal injuries – eventually fatal).*

> *The driver of a road-roller missed his gears whilst descending a steep hill. The driver was thrown off whilst attempting to negotiate (at, it was later estimated, around 40mph!) a bend at the bottom of the hill (general cuts and contusions).*

Safety Precautions

1.   Route moving plant in accordance with a definite traffic pattern. Use traffic lights (or a man with a 'stop/go' board) if necessary.

This latter measure is mandatory in the case of haul roads which impinge on the public highway.

2. Position personnel and plant to avoid the former being struck by the latter (the converse is not, generally, a problem!).

3. Take care when reversing plant, particularly when personnel are working in the vicinity or where material stacks may cause a problem. If necessary, use an assistant to guide the driver.

4. Erect barriers around workmen if they are in the vicinity of moving plant or transport. Encourage the use of 'day-glo' type clothing, particularly where work is to be carried out on the public highway.

5. Place chocks near the edges of excavations to prevent tipper trucks and suchlike from over-riding and falling in.

6. Do not track moving plant too near the edges of excavations.

7. Instruct plant operators in the correct and safe use of their machines, particularly in cases of unfamiliarity (new operators and/or new machines).

8. Prohibit racing or other dangerous driving. This is particularly relevant with scrapers and dump trucks. Crack down hard on offenders.

(b) Lifting appliances (except hoists) and associated tackle

The majority of such accidents consist either of personnel being struck by swinging loads or of the 'pinched finger' variety sustained in slinging operations or whilst using pulley blocks. Additionally, loads sometimes fall due to faulty slinging or, exceptionally, due to the slings themselves breaking. Accidents also occur occasionally due to the safe working load of a crane being exceeded – either the jib collapses or, in the case of mobile cranes, the crane ends up on its side. Incorrect signalling, or the misunderstanding of signals is also the cause of a number of accidents.

A labourer's thumb was caught between the crane hook and a concrete skip when the crane driver lifted prematurely (first joint of thumb severed).

A tracked excavator was moving sheets of reinforcing mesh, three at a time, from a stack to a floor slab 10m away. A single-leg chain sling, with an open, distorted hook, was being used. One load fouled the ground, was displaced and fell on to a labourer (fatal).

Safety Precautions

1. Ensure that the driver or operator can operate the appliance efficiently and correctly.

2. Never use a crane when the wind speed, at jib height, exceeds 30mph.

3. Always site the appliance on firm level ground. Shore up the base if necessary.

4. Never exceed the safe load of the appliance *or* the lifting tackle. Ensure that the necessary warning devices (if supplied) are operating correctly and are obeyed. If not supplied, determine the weight of the load before lifting. Remember that the total load will include the weight of the tackle.

5. Never attempt to raise a load unless the rope is vertical (or as near vertical as makes no difference).

6. Ensure that the load is free before lifting – never attempt to jerk it free with the appliance.

7. Instruct personnel in the correct slinging procedure and ensure that the correct slings are used. In the event of the correct slings not being available, do not lift. In particular, do not attempt a compromise merely to get the job done.

8. Inspect lifting appliances *thoroughly* before first use and at weekly intervals thereafter.

9. Inspect lifting tackle (chains, ropes, slings, hooks etc) *thoroughly* before first use and at *minimum* six-monthly intervals thereafter. Damaged tackle should be repaired as soon as a defect is spotted. Never attempt to repair damaged hooks, rings and eye bolts – they should be scrapped. Damaged chains should be cut up to prevent subsequent use.

10.   Position personnel to avoid them being struck by swinging loads.

11.   Institute a proper signalling system between banksmen and drivers. Remember that *only one* banksman is required per crane (i.e. *no one* other than the 'official' banksman should make signals to the driver).

(c)   Hoists

Hoist have special hazards all of their own — hence the individual category. Personnel fall down shafts, or stick their heads into shafts and get struck by the hoist as it goes up or down. They may actually go into the shaft — for example, to clear up rubbish — and the hoist comes down on top of them. Comparatively common—place accidents involve the hoist striking projections into the shaft or, conversely, projecting loads catching on the shaft. Wheelbarrow handles are a regular culprit in this respect. Occasionally, an accident will occur as a result of the hoist ropes breaking.

A labourer was hit by a descending hoist whilst clearing rubbish from the base of the hoist shaft (fatal).

A labourer was bending down placing scaffold frames ready for loading on to a hoist. The rear part of his body, which was projecting into the hoist shaft area, was struck by the descending hoist (deep bruising to side of chest and leg).

Safety Precautions

1.   Inspect hoists weekly.

2.   Differentiate, by means of the appropriate notices, between *passenger* hoists and *goods* hoists. Prohibit personnel from riding on goods hoists and *vice—versa*. Crack down on offenders.

3.   Install appropriate safety gates at *all* stopping points. *Interlocking gates* should be used for *passenger hoists*.

4.   Fence the hoist shaft adequately to prevent personnel falling in and to prevent objects projecting into the shaft.

5.   Load materials *stably*, *securely* and with *no projections*.

6.   Institute a *proper* signalling system.

7.   Control should be effected from the *highest* point of delivery and by *one person only*. The controller and the signaller should be intervisible *at all times*.

(d) Pneumatic and power tools

Most accidents in this category occur either through the misuse of tools or a lack of concentration on the part of the user. Notable, and fairly common, examples include injuries to the feet of operators of pneumatic breakers and suchlike – either the steel slips, breaks or comes out – and injuries arising out of the absence of guards on the moving parts of mechanical saws, planers, grinders and the like. Accidents also occur through the blowout of compressed air hoses (usually damaged or defective ones) when personnel are clearing blockages.

A labourer was struck by a steel released from a pneumatic breaker as a result of a faulty retaining clip (three broken toes).

A labourer inverted a pneumatic drill to replace the point, causing the trigger to be depressed on the ground. The point struck him in the chest (bruised ribs).

Safety Precautions

1.   Ensure the operator knows what he is doing. Instruct him if necessary.

2.   Provide the appropriate safety clothing and/or equipment (i.e. safety goggles, safety boots etc.) where necessary.

3.   Ensure that there is sufficient obstruction-free working space round each tool.

4.   Ensure that jackhammer steels are kept sharp – a sharp steel behaves more predictably than a blunt one.

5.   Protect compressed air hoses from damage (e.g. from being driven over).

6.   Inspect compressed air hoses regularly for any damage. Discard damaged hoses or shorten (if practicable) to remove the damaged section. Never attempt a repair.

7.   Never uncouple a hose when the pressure is on.

8.   Always release the pressure when a compressor has been stopped, when tools are being changed or when steels are being replaced.

9.   Construction workers can be extremely inventive when it comes to horseplay involving compressed air. They can be similarly inventive with poker vibrators. Such antics can be very dangerous indeed. Horseplay, therefore, should be 'strongly discouraged' and exemplary measures should be taken with offenders.

10.   Ensure that guards are fitted, where appropriate, to the moving parts of power tools. To avoid accidents resulting from the 'stroboscope effect', never use grinders, power saws and the like in locations lit *solely* by fluorescent lighting.

11.   Instill the need for job concentration in operators.

(e)   General

This section comprises a myriad of plant— and machinery—related accident types which do not fit into any of the previous four sub—categories. Typical examples include hair, or unsuitable clothing, getting caught in moving parts of machinery, damaged fingers sustained whilst effecting repairs to a running engine, injuries incurred through an engine backfiring whilst being hand cranked etc. etc. The list of potential accidents is virtually endless as is the list of possible safety precautions. Most precautions, however, can be covered by the simple statement: use a bit of common sense!

The engine backfired when a labourer was hand cranking a dumper into life. The handle struck him on the hand (dislocated thumb and two broken fingers).

A fitter was sent to replace the bottom hydraulic hose on a fork—lift truck. He was instructed to lower the forks before starting work. He didn't, and was found with one of the forks across the back of his neck (fatal).

Safety Precautions

1.   Check the ability of machine operators to operate their machines *correctly* and *safely*. Provide instruction if necessary.

2.   Provide guards to moving parts of machinery.

3.   Never repair plant with the engine running **unless** absolutely necessary (as, for example, with engine–tuning adjustments).

4.   Instruct personnel in the correct method of cranking an **engine** – i.e. thumb and fingers on the *same* side of the handle.

5.   Never repair hydraulic machinery (e.g. excavators, tipper **trucks** etc.) without propping or grounding the hydraulically–operated units.

6.   Inspect *all* plant and machinery (including hired plant) **before use.**

7.   Maintain all plant and machinery at regular intervals to ensure that it remains in good and safe working condition.

8.   Ensure that the test certificate for any lifting appliance or item of lifting tackle, together with the last statutory inspection report (see section 11.7), is held before first use.

10.2.4 Miscellaneous causes

In essence, this category comprises accident causes and types which cannot be classified under any of the three heads covered previously. The overall category is subdivided into 6 *specific* sub–categories and 1 *general* sub–category which, although not *individually* of sufficient incidence levels to merit their being classified as so–called 'principal' causes of accidents, account *cumulatively* for around a third of all accidents occurring on construction sites.

(a) Stepping on, or striking against, objects

Around 50% of the injuries from accidents in this category result from people inadvertently stepping on nails protruding from (mostly gash) timber. The remainder result from a variety of causes such as striking against protruding scaffold members or reinforcing bars, or handling broken pipes, glass or similar sharp–edged material without gloves.

A steel fixer's radial artery was punctured when his wrist struck against the protruding end of a reinforcement tie (shock induced by severe loss of blood).

A labourer stepped on a nail protruding from a piece of timber partially buried in mud (punctured foot).

A carpenter turned suddenly in reponse to a call from a workmate and his head came into sharp contact with the end of a length of reinforcing bar used to replace a missing shear—pin on an adjustable prop (severely damaged eye, resulting in partial loss of sight).

Safety Precautions

1. Keep the site tidy and free from unnecessary rubbish. Institute regular clearing—up operations and either bury the rubbish (if practicable) or remove it from site.

2. Remove the nails from gash timber.

3. Avoid protrusions from scaffolding, reinforcement cages and suchlike where people walk or work. Tie distinctive and brightly coloured markers on unavoidable protrusions.

4. Never use a length of reinforcement to replace a missing shear—pin on an adjustable prop. Apart from it posing an immediate hazard to personnel (see above), the shear—resistance of any reinforcing bar small enough to fit in the hole is unlikely to be sufficient for the intended purpose.

5. Ensure that workers have sufficient workspace.

6. Provide protective clothing (e.g. safety boots, goggles and gloves) where necessary.

(b) Hand tools

The majority of such accidents are caused by personnel striking themselves, or others, with picks, shovels and suchlike. Many accidents also result from defective tools or from misuse of tools (either using them incorrectly or for purposes for which they were not intended — e.g. a wood chisel being used as a screwdriver).

A ganger was using a pick to cut up timber. The pick glanced off the timber and struck his ankle (severe wound to ankle).

A labourer was struck on the head by a workmate wielding a

*pick in a somewhat cavalier manner (head injuries to struck man
- broken jaw to pick wielder).*

*A fitter had used a piece of scaffold tube to 'lengthen' a wrench
to tighten a difficult nut. The bolt thread stripped suddenly and
the tube hit the fitter in the face (broken cheekbone and facial
injuries).*

Safety Precautions

1. Use the right tool for the job and instruct personnel in the
    correct usage thereof.

2. Position personnel to avoid them being struck by picks, shovels
    and suchlike wielded by others.

3. Inspect hand tools regularly and take defective tools out of use
    until repaired (if, indeed, repair is possible - it may not be!).
    Particular attention should be paid to hammers and picks with
    loose heads or split handles, splay-jawed spanners and
    mushroom-headed chisels.

4. Do not 'lengthen' a spanner with a piece of tube to exert more
    force - the length of any given size of spanner is appropriate to
    the force required.

5. Do not use a screwdriver or other wedge-shaped object to 'adjust'
    a non-adjustable spanner - use the correct size of spanner.

6. The 'pull' on an adjustable wrench should be in the direction
    which forces the jaws further on to the nut - generally *with* the
    bend in the wrench.

7. Load wheelbarrows *evenly* and do not *overload* - make another
    trip or use an alternative method.

(c) Collapse of excavations

Such accidents, which are almost exclusively due to inadequate
support-work, are usually serious and often fatal. Other factors which
contribute to such accidents are the movement of plant, or stacking of
materials, too close to the edges of excavations. Erosion of the sides of
excavations, usually due to prolonged severe rainfall, is another factor
which should not be overlooked.

An excavator, engaged in the removal of excavation spoil, was tracking close to the edge of a drainage trench in which men were working. The trench side, which was otherwise adequately supported, collapsed, trapping a pipelayer (no physical injuries of any substance — however, owing to the fact that it took around 30 minutes to free the trapped man, during most of which the excavator was jammed (but gradually slipping downwards) in the trench immediately above him, the psychological side-effects were substantial and long-lasting).

Safety Precautions

1. Support work will generally be required in all excavations except those in sound rock. Tables of requirements may be found in the appropriate B.S. Code of Practice and in most safety handbooks.

2. Personnel erecting support work should be in a protected position.

3. Personnel (other than those erecting supports) should not work in unsupported sections of excavations and should be prohibited from so doing.

4. To preclude the need for subsequent *ad—hoc* adjustments, ensure in advance that the support work will not impinge upon the permanent work and will not unduly interfere with the construction thereof.

5. Prohibit personnel from clambering on support work as this weakens it.

6. Take note of any 'creaking' in the support work as this may be the first indication of incipient collapse.

7. Do not stack materials, place excavated spoil, or track plant too close to the edges of excavations. If unavoidable, additional support work should be erected.

8. Beware of 'piping' in saturated ground, particularly in sandy soils. If this is a possibility, either de-water the excavation area or isolate the excavation from the surrounding waterlogged soil by sheet-piling to a comparatively impermeable stratum.

9. Large open excavations should (if possible) have battered sides. Safe batter angles depend on the soil type and are available from

the appropriate Code of Practice.

10. Prior to commencing work, divert any surface–water streams which will be interrupted by the excavation.

11. Provide a means of rapid access or exit for personnel.

12. Check excavations regularly for problems, particularly during adverse weather. Plan the work in order to avoid excavations remaining open any longer than absolutely necessary.

(d) Electricity

A source of enormous potential danger – largely because it is an 'unknown quantity' to most people – electricity regularly accounts for between 5% and 10% of fatalities in the construction industry. Although a number of electricity–related accidents result from crane–jibs and suchlike coming into contact with overhead lines, or from excavators or personnel cutting through buried cables, the majority of such accidents result from the use of defective or badly maintained electrical hand–tools.

Two steel erectors were manhandling into position a steel joist suspended from a mobile crane. The hoist rope of the crane made contact with an 11kV overhead line (both erectors were killed).

In connecting a 3–pin plug to a portable electric tool, a labourer connected the earth wire to the positive pin (electric shock).

Safety Precautions

1. *A 110V single–phase supply, suitably centre–tapped to earth*, should be used for *all* power tools. The tools should, of course, be suitable for such.

2. All power tools should be effectively earthed. Despite popular belief to the contrary, rubber boots are *not* an effective measure of preventing electricity 'earthing' through the human body!

3. Ensure that electrically operated tools and other apparatus are inspected *regularly*. Defective tools or apparatus *must* be taken out of service immediately and *must* be repaired *by an electrician*.

4. Wherever possible, use *insulated* cables to carry power. To prevent

damage to them by traffic and suchlike, sling them overhead (suitably drawing attention to them by means of appropriate notices) or bury them and mark their position.

5.   Always use splashproof plugs and sockets for electrical tools.

6.   Avoid working high plant near overhead lines. Where unavoidable, install audible warning devices which sound when crane jibs and suchlike come too close to an overhead line. Remember that, since electricity in high–tension lines can arc several feet, accidents can arise without contact actually being made.

7.   Prohibit plant from passing beneath overhead lines except at pre–determined points where 'goal posts' have been erected to ensure jibs and suchlike are properly lowered.

8.   Beware of buried cables when using excavating plant or when 'bottoming–out' trenches by hand. Check with the local electricity board if necessary.

9.   Remember that electricity and water do not mix!

(e)   Burns and scalds

Although the majority of such injuries are sustained in connection with hot bitumen, steam piling and welding or burning, a surprising number arise from paraffin stoves in rest rooms and from (believe it or not!) making tea.

A plant man, on an asphalt mixer, was cleaning a choked bitumen pipe when the hot bitumen blew out on his face and hands (severe burns).

A plant driver, who was not wearing welding goggles, was assisting a welder when the electric arc welder flashed (flash burns to eyes – 'arc–eyes').

Safety Precautions

1.   Ensure that the working system is as safe as possible.

2.   Provide protective clothing (i.e. gloves and goggles) wherever necessary and ensure that it is used.

3.   Prohibit the use of petrol or paraffin to light fires.

4.   Never apply heat to empty, or partially empty, drums until the contents have been identified and the appropriate safety precautions taken.

5.   Institute suitable fire precautions in offices and accomodation facilities. Ensure that they are strictly adhered to.

(f)   Strains and twists

Most of these accidents result from personnel handling heavy or awkwardly-shaped objects. The injury is often the cumulative effect of strain occuring over a period.

Safety Precautions

1.   Instruct personnel in safe methods of lifting or carrying. Display the 'standard' illustrations of such.

(g)   Other causes

Notable example are accidents associated with work over, or in, water, accidents sustained in connection with the use of explosives and those which are connected with the handling of toxic or caustic substances.

A labourer was hit by flying debris following blasting. He had taken cover 100m away from the blast but had left his leg exposed (broken bone in foot).

A labourer contracted dermatitis from handling cement.

A diver was working on a damaged valve on a submerged pipeline. The valve came off and the diver was sucked into the pipe by the inrush of water (fatal).

A dumper driver backed his machine off the end of a jetty and into the sea (fatal).

Safety Precautions

1.   Explosives should be stored *safely* and *securely* in a purpose-designed magazine. They should be used *only* by *qualified*

personnel. When blasting, effective warning systems should be instituted and *strictly* adhered to.

2. Provide lifejackets for personnel working on, or over, water. Ensure that they are worn at all times.

3. Provide a means of rescue when working on, over or near water.

4. Diving work should *only* be carried out by *qualified personnel* under the supervision of an *appointed Diving Contractor*.

5. Provide means of protection against toxic or caustic material.

6. When using toxic or caustic material, ensure that the manufacturer's instructions are *strictly* adhered to and the appropriate precautions taken.

7. Ensure cleanliness after working in contact with sewage or other contaminated materials.

10.3 Accident Costs

In an ideal world, a general concern for the health, safety and general well-being of the work force would, in itself, provide sufficient motivation for employers and management to pay the necessary attention to safety and the prevention of accidents. We do not, however, live in an ideal world and, whilst such altruistic concern may be a 'prime mover' in many instances, in other instances it will not.

That being the case, it is relevant to consider site safety from the more clinical standpoint of its cost-effectiveness in the hope that *this*, if nothing else, will provide the necessary motivation *vis-a-vis* accident prevention and general safety consciousness.

Some idea of the tremendous cost of accidents may be inferred from the fact that accident insurance premiums currently cost the construction industry in excess of £20 million *per annum* – i.e. around £17 to £20 per employee per year! The costs, however, do not stop there in that accidents involve a considerable number of *additional* costs which are *not* covered by insurance and which will, therefore, have to be met from profits.

These costs, which may be *direct* or *indirect*, will comprise some, or all, of the following:

1.   'Unproductive' wages paid to:

    (a)   the injured party(s). For a comparatively minor accident (such as a cut finger), this will amount, at most, to an hour's wages. In more serious cases, where the accident results in the injured party taking time off work, the cost may be substantial since the employer's recovery of 'unproductive' wages under the statutory sick–pay scheme will be limited to those for the first three days only. The remainder may, or may not, be recoverable from insurance.

    (b)   supervisors involved in reporting, or investigating, the accident and arranging for continuance of work. Obviously, the extent of time spent on such will depend on the severity of the accident.

    (c)   operatives who stop work from curiosity, sympathy or to give assistance. Depending upon the severity of the accident, and upon their perception of the cause thereof, operatives may well refuse to continue working until the situation has been remedied – particularly if the accident resulted from an inherently unsafe condition which they consider still exists.

2.   Administration expenses such as First Aid, phone calls (for ambulances, to relatives, to the HSE, to head office etc. etc.) and the inevitable mountain of paperwork which is generated by any accident of substance.

3.   Possible compensation to the injured party(s). Depending upon the extent of culpability of the injured party's employers, this may, or may *not*, be recoverable from insurance.

4.   Possible fines in the event of the accident being adjudged to have resulted from the employer's breach of statutory duty (see section 11.3.1).

5.   Costs of repairing damage to plant or equipment, or to the permanent or temporary works.

6.   Cost of plant or labour standing idle whilst waiting for the above repairs to be effected.

7. Interruption to the planned sequence of work. The costs may be substantial if the contract completion time is delayed as a result.

8. Initial lower productivity of the injured party(s) upon returning to work.

9. Possible site—wide reduction in productivity as a consequence of site morale being affected by the accident.

10. Taking on, and instructing, replacement labour if necessary.

11. Head office administrative costs.

It is, of course, possible that, in many instances, a number of the above components will not apply. On the other hand, it is equally possible that some accidents will involve certain cost components over and above those listed. Whichever is the case, however, it is quite likely that the *overall* cost of any given accident will be significantly greater than that of providing such safety measures as would probably have prevented it occurring in the first place.

CHAPTER 11

SAFETY AND THE LAW

As discussed in Chapter 10, there are two basic reasons why employers and supervisory personnel *should* pay due regard to the safety of their workforce, namely that it is their humanitarian duty to do so and, if *that* fails to impress them sufficiently, that to do so makes sound commercial sense. In respect of the latter, it might also be mentioned that, in certain circumstances, a contractor's safety record may play a significant role in influencing the Employer's decision *vis-a-vis* the award of a contract (see section 12.3).

There is, however, one overriding reason why they *must* pay due regard to the safety of their workforce, namely that to do otherwise will bring them into conflict with the law.

This chapter examines in detail the safety-related obligations and responsibilities imposed by the law on employers, and others involved in the work process, and also considers the penalties which may be levied against them in the event of their failure to satisfactorily discharge such.

11.1 Civil Law and Statute Law

The *common law* duties of care (see section 2.2) imposed upon an *employer* are that he should provide:

(a) a *reasonably* safe place of work;

(b) a *reasonably* safe system of work;

(c)   *reasonably* safe plant and equipment;

(d)   a *reasonably* safe means of access to, and egress from, the place
of work;

The common law duties of care imposed upon *employees* are,
broadly speaking, that they should exercise *reasonable* skill and diligence
in the performance of their work and, in so doing, should pay
reasonable regard to the safety of *themselves* and *others who may be
affected thereby*.

A breach of any such duty of care, *where such breach causes
injury or damage*, may give rise to a *civil* action, instigated by the
injured party, usually for *negligence* (see section 2.2). If the alleged
breach is proven *on the balance of probabilities*, the injured party is
entitled to *compensatory damages* from the party who committed the
breach.

A breach of a *statutory* duty of care (i.e. one *expressly* imposed by
an Act of Parliament or item of delegated legislation), however, may
result in a *criminal* prosecution being brought by the *state*, against the
offending party, *irrespective of whether or not the breach results in
injury or damage*. In the event of the offender's guilt being proven
beyond reasonable doubt, the appropriate *punishment* (i.e. a fine or
imprisonment) is levied against him.

It should particularly be noted (see section 2.2) that the two
possibilities are not mutually exclusive in that a breach of a *statutory*
duty, which results in injury or damage, may give rise to a criminal
action (for the breach) *and* a civil action (for the injury). If such is
the case, the two actions are, in the eyes of the law, regarded as
separate issues *independent* of one another with the outcome of the
criminal action having no bearing on the potential outcome of the civil
action and *vice versa*.

11.2 The Factories Act 1961

It is generally recognised that the first *serious* attempt at industrial
safety legislation was the *Explosives Act 1875*. In the course of the
next eighty years or so, a myriad of Acts of Parliament and
Regulations came into existence in a rather piecemeal fashion and,
during that period, the construction industry became subject to such
diverse legislation as the *Electricity Regulations 1908*, the *Boiler
Explosions Act 1922* and the *Employment of Women, Young Persons*

and Children Act 1920.

This rather fragmentary process culminated with the enactment of the *Factories Act 1961* (preceded by the 1937 and 1948 versions thereof) which, in essence, laid down a general framework of statutory regulation for specific types of workplaces including construction sites. In particular, the Act authorised the issue of *specific* provisions to cover *particular* operations and their associated hazards. Those specifically relating to the construction industry were the *Construction Regulations* which were drawn up by the (then) Minister of Labour (*now* the Secretary of State for Employment) under authority invested in him by the Act.

11.2.1 The Construction Regulations

The four sets of Regulations generated by the Factories Act are detailed hereunder, along with a *further* set of Regulations spawned subsequently by the Health and Safety at Work Act 1974.

(a)   The Construction (General Provisions) Regulations 1961

The regulations cover the following:

1.   supervision of the safe conduct of work;

2.   work in excavations, shafts and tunnels;

3.   work involving the use of cofferdams or caissons;

4.   storage and use of explosives;

5.   work in dangerous or unhealthy atmospheres;

6.   work on, or adjacent to, water;

7.   the use of plant or transport;

8.   demolition;

9.   miscellaneous operations not covered by the above list;

(b)   The Construction (Lifting Operations) Regulations 1961

These regulations cover the construction, erection and inspection of lifting appliances, lifting gear/tackle and hoists used on construction sites.

(c) The Construction (Working Places) Regulations 1966

As the title would imply, these regulations cover the safety of working places and the safe access to, and egress from, such. In addition to regulations covering construction sites in general, there are regulations relating *specifically* to the use of scaffolds, ladders, walkways and suchlike and also to work on fragile roofs.

(d) The Construction (Health and Welfare) Regulations 1966

These regulations, as amended by the *Construction (Health and Welfare) (Amendment) Regulations 1974*, cover the following:

1. the provision of facilities for shelter during inclement weather;

2. the provision of facilities for clothing and for the taking of meals;

3. the provision of washing and sanitary facilities;

4. the provision of safe access to the above facilities;

5. the provision of protective clothing for personnel required to continue their work in inclement weather;

Although the original version of the regulations covered the provision of first–aid facilities, this matter is now covered by the *Health and Safety (First–Aid) Regulations 1981*.

(e) The Construction (Head Protection) Regulations 1989

These regulations, which came into force in March 1990, cover employers' responsibilities *vis–a–vis* the provision and maintenance of suitable means of head protection (i.e. hard hats) and their rights to formulate and enforce rules governing the wearing of such on sites under their control.

Two particular points should be noted in respect of the above sets of Regulations, the first of which is that *at least one* copy of *each* relevant set of regulations *must* be kept on site for reference purposes.

The second point worth noting is that exemption from *individual* regulations may, in the cases of the *General Provisions*, the *Lifting Operations*, the *Working Places* and the *Health and Welfare* Regulations, be granted by *H.M. Chief Inspector of Factories* if compliance with such is *not necessary to ensure and maintain the safety of employees*, or is *not reasonably practicable* (see section 11.3.1 for what is considered to be 'reasonably practicable'). In the case of the *Head Protection* Regulations, exemption may be granted by the *Health and Safety Executive* (see section 11.3.2) *only if the health and safety of employees will not be prejudiced thereby.*

11.3 The Health and Safety at Work Act 1974

It is generally recognised that the Factories Act had two *principal* defects, the first of which was that the use of regulations to exercise *detailed* control over *specific* work processes tended to produce a complicated and poorly understood system which, as a consequence of the rapid development of improved construction techniques during the 1960s and early 1970s, was often out of date. The second major defect, again related to the ever-changing face of the construction industry, was that the self-employed were not covered.

In order to remedy these and other defects, a committee of investigation (the Robens Committee on Safety and Health at Work) was set up in 1972. Amongst other things, the committee criticised the *negative* approach (i.e. *don't* do this, *don't* do that) of existing legislation and proposed a radical change of emphasis towards a *positive self-regulatory* system. The result was the *Health and Safety at Work Act 1974* which, in contrast to the Factories Act and other similar legislation, was aimed *principally* at *people and their activities* rather than at *premises and processes*.

Although much of the previous legislation is still in existence, the Health and Safety at Work Act *supplements* such by imposing the general principal that *all* persons involved in the work process have a responsibility for safety.

The Act is subdivided into four parts as follows:

Part I: contains the provisions for health, safety and welfare at work;

Part II: concerns the Employment Medical Advisory Service (EMAS) and is, in effect, a re-enactment of the *Employment Medical Advisory Service Act 1972* with certain minor changes;

Part III: modifies the law relating to building regulations under the *Public Health Act 1936*;

Part IV: contains minor amendments to the *Radiological Protection Act 1970*, the *Fire Precautions Act 1971* and the *Companies Act 1967*;

In addition to the above, the Act also contains ten schedules, the most important of which (for site safety purposes) is Schedule 1 which lists existing enactments containing relevant statutory provisions which continue to remain in force until progressively replaced by health and safety regulations supported by approved codes of practice.

From the point of view of the Act's applicability, it is of interest to note that the Act applies only to a very limited extent in Northern Ireland and that the majority of Part III is inapplicable to Scotland.

11.3.1 General Duties imposed by the Act

In the context of maintaining the health and safety of those involved in the work process, Part I of the Act is the most important part thereof and, by progressively replacing existing statutory provisions (see above) with a system of regulations and codes of practice designed to maintain or improve existing standards of health, safety and welfare, aims to:

1.   secure the health, safety and welfare of *people at work*;

2.   protect *other people* from risks arising out of work activities;

3.   control the keeping and use of explosive, highly flammable and dangerous substances;

4.   control the emission of noxious or offensive substances into the atmosphere;

With a view to securing these aims, the following general duties are imposed on the various categories of people involved in the work process:

(a) Employers (in respect of their employeees) must:

1. provide and maintain, *so far as is reasonably practicable*, machinery, equipment etc. and systems of work which are safe and without risk to health;

2. arrange, *so far as is reasonably practicable*, for safe and healthy systems of use, handling, storage and transportation of machinery, equipment, appliances and materials;

3. provide, *so far as is reasonably practicable*, whatever information, instruction, training and supervision is necessary to ensure health and safety at work;

4. provide and maintain, *so far as is reasonably practicable*, any workplace under their control in an environmentally safe and healthy condition, with means of access to, and egress from, that are without risk to health or safety and with adequate welfare facilities for employees whilst at work;

5. prepare, and bring to the notice of *all* employees, a written statement of company safety policy (note: this does not apply to employers with less than *five* employees);

It is worth noting that the first four duties cited make *statutory* requirements of what were, prior to the Act coming into being, an employer's *common law* duties of care. As a consequence, the penalties for a breach of such are more stringent than previously.

(b) Employers (in general) must:

1. carry out their work in such a way that persons not in their employment, who might be affected thereby, are not exposed to risk to their health and safety;

(c) Employees must:

1. take *reasonable* care for the safety of themselves *and* others who may be affected by their activities whilst at work;

2. cooperate with their employers, *or anyone else*, where it is necessary for them to discharge their statutory duties and obligations;

(d) Self–employed persons must:

1. carry out their work, *so far as is reasonably practicable*, in such a way as not to create health and safety risks to *themselves or anyone else*;

(e) Controllers or Owners of Premises must:

1. ensure, *so far as is reasonably practicable*, that the premises, and means of access to, and egress from, are safe and without risk to health;

2. use the *best practicable means* of preventing noxious or offensive emissions therefrom into the atmosphere, and of rendering harmless and inoffensive any that *do* escape into the atmosphere;

In respect of the former duty, it is worth noting that this confirms a similar duty imposed on owners and controllers of premises by the *Occupiers Liability Act 1984*.

In respect of the latter duty, it should be noted that control *also* exists under other legislation such as the *Control of Pollution Act 1974*, the *Clean Air Act 1966* etc. etc.

(f) Designers, Manufacturers or Suppliers of Plant, Machinery or Equipment must:

1. ensure, *so far as is reasonably practicable*, that the articles are without risk to health or safety whilst being used, cleaned or maintained;

2. make arrangements for any testing or examination which may be necessary to ensure their compliance with the above requirement;

3. supply users with adequate information on the correct usage of their products, together with any conditions which must be observed to ensure minimal risk during use, cleaning, maintenance, dismantling or disposal;

4. ensure, *so far as is reasonably practicable*, that such information is updated in the light of new knowledge or experience and that the updated information is passed on to existing users;

(g) Manufacturers, Importers or Suppliers of Materials or Substances must:

1. ensure, *so far as is reasonably practicable*, that the materials or substances are safe to use and without risk to the health of the user;

2. ensure, *so far as is reasonably practicable*, that sufficient reliable testing or research has been carried out to ensure compliance with the above requirement;

A number of points are worthy of particular note in respect of the above-listed statutory duties, the first of which is that such duties are *mandatory* upon the relevant parties in that they cannot relieve themselves of such by delegating the duties to others, irrespective of the competence thereof.

Secondly, the phrase: 'so far as is reasonably practicable' is generally taken to mean that a person, or organisation, is considered discharged from a statutory duty if it can be shown that the risk was *remote* and could *only* have been averted at *unreasonably high cost* in terms of time, effort or money. In a criminal action for an alleged breach of a statutory duty, the onus rests with the *defence* to prove the case for *non-compliance* with the duty and *not*, as might be imagined from the general tenets of criminal law, with the *prosecution* to prove the case for *compliance*.

Thirdly, *everyone* involved in the work process is placed under a duty not to misuse *anything* provided under a statutory requirement *vis-a-vis* health and safety.

Finally (and this will be of *particular* interest some employees!), an employer is not allowed to charge his employees for *anything* done or provided under a health and safety statutory requirement. This may be taken to include *anything* provided under *any statute or regulation* in the interests of health or safety.

11.3.2 Enforcement

The national authority and policy-making body for all matters relating to health and safety at work is the *Health and Safety Commission* which is responsible, to the Secretary of State for Employment, for *inter alia* making arrangements to implement the enabling measures contained within the Health and Safety at Work Act.

The executive wing of the Commission is the *Health and Safety Executive* (the HSE) whose principal duty is to act on behalf of, and in accordance with the wishes of, the Commission and to arrange for the enforcement, in conjunction with other enforcing authorities, of the statutory regulations.

The actual enforcing (except where the Local Authority is the *sole* enforcing authority – as will be the case with, for example, office work or catering services where they are the *main* activities carried out on the premises in question) is carried out by the *HSE Inspectorate* appointed by the Executive for that very purpose.

It has sometimes been suggested, somewhat irreverently, that HSE Inspectors have powers with which God, ships' captains and public school headmasters have yet to be invested! Whilst that is a subject for debate (except, of course, for sailors and inmates of public schools for whom the issue is a 'no contest'!), the fact remains that the powers of the Inspectorate are, to say the very least, extensive.

In particular, HSE Inspectors are empowered to:

(a)   enter premises (usually unannounced, for fairly obvious reasons), including construction sites, and carry out examinations or inspections;

(b)   take a police officer with them if obstruction is envisaged;

(c)   instruct that the premises, or specified parts thereof, remain undisturbed until such times as any investigations have been completed;

(d)   take such measurements, photographs, recordings or samples as may be considered necessary;

(e)   take possession of plant or equipment, or order it to be dismantled, or order materials or substances to be tested, if they consider them to be a source of *potential* danger to health or safety;

(f)   seize plant, equipment or substances, and destroy them or cause them to be rendered harmless, if they consider them to be a source of *immediate* danger to health or safety;

(g)   demand, for inspection or copying purposes, the production of any records, registers or documents which are required under statutory regulations;

(h)   request relevant information from *any* person and require that person to sign a declaration of truth regarding the information supplied (note: the information *cannot* be used in evidence against the supplier thereof but *can* be used in evidence against third parties);

(i)   issue *Improvement* and/or *Prohibition Notices* (see below);

(j)   institute *criminal* proceedings against *employers*, *employees* or *self-employed persons* as appropriate;

An *Improvement Notice* is issued for *each* contravention of a statutory regulation which, in the opinion of an Inspector, *may* cause, or result in, *subsequent* personal injury. Such notice, which must state the regulation contravened and the reason for issuance of the notice, requires remedial action to be taken within a *specified* time limit.

A typical example of a situation warranting the issue of an Improvement Notice is where a scaffold has a few minor defects (such as a lack of toe-boards or guard-rails) which *may* result in an accident if not rectified.

A *Prohibition Notice*, on the other hand, is much more serious than an Improvement Notice in that it is issued for *each* contravention of a statutory regulation which, in the opinion of an Inspector, poses a *grave* risk of *immediate* and *serious* personal injury. The essential difference between such a notice and an Improvement Notice is that a Prohibition Notice generally (but not always) requires *immediate cessation* of the relevant activities until such time as the contravention has been rectified and the risk of danger averted.

In the case of the abovementioned scaffolding example, a Prohibition Notice would undoubtedly be issued if the scaffold was blatantly unsafe and was in imminent danger of collapsing. In such a case, any personnel working on the scaffold would be instructed to 'dismount' and further use of the scaffold would be prohibited until the necessary remedial work had been effected.

In both cases, the Inspector *may* (but is not obliged to) specify the remedial action required.

It should be noted that an appeal against the terms of *either* notice may be made to an Industrial Tribunal but *must* be lodged *within 21 days* of the date of issue of the notice. In the case of an Improvement Notice, the lodging of the appeal *automatically* suspends the notice until the Tribunal decision has been reached. The notice is, of course, cancelled if the appeal is upheld. In the case of a Prohibition Notice, the prohibition remains in force until the Tribunal decision has been reached whereupon it is either cancelled or confirmed.

It should be further noted that, for purposes of the *Environment and Safety Information Act 1988*, either notice would be regarded as a 'relevant notice' (and annotated to that effect) if an Inspector was of the opinion that persons other than those 'at work' (e.g. members of the public) were, or were likely to be, affected by the unsafe condition referred to in the notice. In such a case, the notice would be entered into a public register and kept there for a period of *at least* three years. Following compliance with the terms of the notice, the relevant register-entry would be annotated to that effect.

Finally, it should be noted that, for the purpose of enabling employees to be adequately informed about matters relating to their health, safety or welfare, an Inspector has a *duty* to supply them, or their Representatives (see section 11.6), with relevant factual information concerning their workplace and full details of any action he has taken or is proposing to take. For fairly obvious reasons, the same information *must* be given to their employer.

11.3.3 Offences and Sanctions

In accordance with Section 33 of the Act, *any* contravention of the statutory provisions of the Act (or of *any other* relevant health and safety Regulations or Acts), or a failure to discharge any statutory duty imposed thereby, constitutes a *criminal offence* and renders the person(s) or organisation(s) concerned liable to prosecution.

By way of illustrating the sort of offences which are likely to be committed, the main possibilities are listed hereunder:

(a) contravention of any part of, or failure to discharge any statutory duty imposed by, Sections 2 to 9 (see section 11.3.1) of the Act;

(b) contravention of any of the Construction Regulations;

(c) contravention of the terms of an Improvement or Prohibition

Notice;

(d) *intentional* obstruction of an HSE Inspector in the performance of his duty;

(e) preventing another person from appearing before an HSE Inspector to answer questions;

(f) *intentionally* making a false statement to an HSE Inspector;

(g) *intentionally* falsifying an entry in a statutory register or document;

As far as the appropriate punishment is concerned, it depends upon the seriousness of the offence and, perhaps more importantly, the attitude of the court in which the case is tried. There are, however, statutory guidelines in this respect as follows:

Summary offences (i.e. *comparatively* minor transgressions), tried in a Magistrate's Court (or a Sheriff Court in Scotland), are punishable by a *fine not exceeding £2000 for each offence.*

Indictable offences (i.e. more serious transgressions), tried in a higher court before a jury, are punishable by *an unlimited fine* or *a maximum of two years imprisonment* or *both.* It should be noted that the imprisonment sanction only applies to certain serious offences which include the carrying out of an activity without an HSE licence if such is a statutory requirement, the contravention of any terms of such a licence, the contravention of any statutory provision relating to the acquisition, possession or use of explosives and, perhaps most importantly from the standpoint of the average construction site, contravention of the terms of a Prohibition Notice.

The following points should also be noted:

(a) *Continued* contravention of the terms of an Improvement or Prohibition Notice following conviction for such, or failure to comply with a court order following an offence, constitutes a *further* offence which renders the offending person(s) or organisation(s) liable for a *maximum fine of £100 for each day the contravention or failure continues.*

(b) If an offence committed by a given party can be proved to be the fault of another party, the other party can be prosecuted and convicted of the offence irrespective of whether or not the *prima facie* offender is prosecuted.

(c) If it can be proved that an offence committed by an organisation was committed with the connivance of, or due to negligence on the part of, an official of that organisation, the official *and* the organisation can be prosecuted. In this respect, it should *particularly* be noted that, whilst an organisation *per se* cannot be imprisoned (see section 1.5.2), officials from such *can* if individual culpability is proved and the offence merits a custodial sentence.

(d) The guilt or innocence of an offender *vis–a–vis* a breach of statutory duty does not prejudice the outcome of a subsequent *civil* action brought against the offender by the injured party(s).

11.4 The Written Statement of Safety Policy

In accordance with Section 2(3) of the Health and Safety at Work Act, it is the *statutory duty* of *every* employer (except those having less than *five* employees) to prepare, and bring to the notice of *all* employees, a *written* statement of his policy regarding the health and safety of his workforce.

As might be imagined, it is virtually impossible to lay down hard and fast rules concerning the compilation of such statements since they will vary from company to company and will depend upon such factors as the size of the company, its organisational structure and the type of work carried out by it.   That notwithstanding, however, compliance with the wording of Section 2(3) of the Act requires the statement to contain:

(a) a *general* statement of safety policy;

(b) details of the *organisation* within the company with responsibility for implementing the policy;

(c) details of the company's *arrangements* for ensuring health and safety in its workplace(s);

11.4.1 The General Statement

By way of emphasising the importance which the company places upon matters of health and safety, the general statement should start with a *clear* and *unequivocal* declaration of the company's intention to ensure and maintain the health and safety of its workforce and all others who might be affected by its work operations.

This should be followed by a brief description of the means by which this will be achieved. In this respect, it is generally sufficient (and, indeed, common practice) to list the general duties imposed on the company, as an employer, by the Health and Safety at Work Act (see section 11.3.1), together with any other duties (either self-imposed or imposed by other relevant legislation) which are considered appropriate.

By way of demonstrating the company's compliance with the requirements of Section 2(3) of the Act, it is generally recommended that this section of the safety policy contain details of the procedure by which the policy will be brought to the notice of *all* employees. In this respect, posting copies of such on notice boards, or distribution in book form to all employees, will constitute the necessary compliance.

Other important points worthy of inclusion in this section of the policy are:

(a) the degree of support required from *all* personnel to achieve the objectives of the safety policy;

(b) the degree of support which will be given to any employee taking health- or safety-related action likely to result in increased costs or diminished profits − i.e. an indication of just how seriously the company really takes safety when it comes to the 'crunch'!

(c) sanctions which will be imposed by the company (as opposed to the law) in the event of breaches of the safety regulations by *anybody*; in this respect, it should go without saying that sanctions should only be specified if the company is in a legal position to levy them and is prepared to do so if necessary;

(d) steps which have been, or might need to be, taken by the company to obtain any specialist advice necessary to determine risks to the health or safety of employees and the relevant precautions;

(e) the provision of relevant information, and/or training, to employees in respect of risks to their health or safety which may arise out of their work or at their workplace;

Finally, and by way of giving the policy document the necessary authority, the general statement should be signed *and* dated (thereby implying that the policy will be kept up to date and revised, if necessary, in the light of new knowledge or changes in work practices) by the director (or partner, if appropriate) having *ultimate* responsibility for safety within the company or organisation.

11.4.2 The Organisation

This should be a statement of the personnel structure of the company, ideally in 'family tree' form and covering *everyone* from those at the 'sharp end' right up to the director or partner with whom the 'buck' stops, showing a *clear* and *logical* apportioning of individual responsibility *vis-a-vis* health and safety matters.

Particular mention should be made of responsibilities in respect of:

(a)   training;

(b)   the performance of statutory duties in connection with the Construction Regulations etc. (see section 11.7);

(c)   monitoring compliance with the safety policy;

(d)   maintaining contact with sources of specialist advice (the HSE etc.);

(e)   consulting with employees' Safety Representatives (see section 11.6);

Most importantly, the organisational statement should name, and define the duties and responsibilities of, the company Safety Officer and Supervisor(s) (see sections 11.5.1 and 11.5.2), giving a *clear* definition of the extent of authority of such, concerning health and safety matters, in relation to other management.

Since the organisational statement is essentially about responsibilities, it should also contain evidence that the company has ensured that those with individual responsibilities *understand* them and are sufficiently competent and qualified to discharge them satisfactorily.

11.4.3 The Arrangements

This section details how the above responsibilities are carried out and should cover the full range of the company's activities.

It should include the arrangements for:

(a) the identification of *general* and *specific* risks to health and safety and the devising and implementation of appropriate precautions for dealing therewith;

(b) ensuring the safety of working systems and methods;

(c) ensuring the safety of the workplace environment;

(d) carrying out any statutory inspections or examinations (see section 11.7);

(e) the provision and use of protective clothing or equipment;

(f) the identification of training needs and the procedures for securing adequate training;

(g) ensuring adequate communication between management and workforce in respect of health and safety matters;

(h) accident reporting and investigation (see sections 11.8.1 and 11.8.2);

(i) dealing with emergencies – such as fires or explosions;

(j) carrying out health and safety inspections or audits – with a view to monitoring the effectiveness of the above arrangements – and the keeping of records therefrom;

11.5 Safety Supervisors

In accordance with Regulation 5 of the *Construction (General Provisions) Regulations* (see section 11.2.1), every contractor who employs a total of more than 20 employees is required to appoint, *in writing*, one or more Safety Supervisors experienced in the type of work undertaken by the contractor, and suitably qualified to:

(a) advise the employer on the requirements of all relevant health and safety legislation and on health and safety matters generally;

(b) supervise and ensure the observance of the appropriate legal requirements;

(c)   promote the safe conduct of work;

In large and medium–sized construction companies, safety supervision is usually effected at two levels: that of the *Safety Officer*, who looks after the safety interests of the *company*, and that of the *Site Safety Supervisor* who looks after the safety interests of *individual sites*.

Small construction companies, on the other hand, may operate a so–called '*group safety scheme*' whereby a number of such companies, operating in the same area, jointly form an independent organisation employing a Safety Officer who fulfils the requirements of Regulation 5 (see above) on behalf of each member company. Site Safety Supervisors may, or may not, be employed by the individual companies.

11.5.1 Safety Officers

If a Safety Officer is employed to look after the health and safety interests of a company, he must be *specifically* employed *full–time* as such and should be properly trained and experienced in the construction industry. Although the requirement is not *mandatory*, it is generally *recommended* that he be a corporate member of the *Institution of Occupational Safety and Health (Construction Division)*.

His (or her!) main responsibilities may be summarised as follows:

(a)   to advise management on:

1.   preventing injury to personnel and damage to plant and equipment arising from the use of unsafe working practices;

2.   preventing hazards leading to occupational ill–health;

3.   the determination of safe systems and methods of working, and on the preparation of work method statements (from a health and safety standpoint);

4.   *further* improvements in existing sound working practices;

5.   the preparation and drafting of site safety regulations;

6.   legal requirements concerning health, safety and welfare;

7.   the adequacy of facilities provided under statutory regulation;

8.   the provision, use and adequacy of protective clothing and equipment;

9.   the suitability, from a safety standpoint, of new and hired plant and equipment and the validity of all relevant test certificates;

10.   potential hazards on new contracts before work starts, and on the site safety organisation and fire precautions required;

11.   methods of safe working arising from new developments;

12.   changes in health- and safety-related legislation;

(b)   to carry out regular site inspections with the site management to ensure that:

1.   the methods of working are safe and without risk to health;

2.   work method-statements are *strictly* adhered to;

3.   *all* statutory and site regulations are being observed;

4.   *all* statutory examinations and inspections have been carried out *and* that the appropriate records have been kept up to date;

5.   *all* statutory notices are posted;

6.   *all* statutory welfare facilities have been provided *and* are being adequately maintained;

7.   first-aid facilities *and* a *qualified* 'first-aider' are provided if more than 50 personnel are employed on the site; for a workforce of 5–50 persons, a suitably equipped first-aid box (at least!) must be provided; note that these are *all* requirements of the *Health and Safety (First Aid) Regulations 1981*;

8.   the location of the first-aid facilities is known to *all* persons on the site;

(c)   to be available during tendering, planning and other pre-contract activities to advise on all health, safety and welfare matters and on any pre–contract training requirements;

(d) to investigate accidents and dangerous occurrences (see sections 11.8.1 and 11.8.2) and recommend means of preventing their re–occurrence;

(e) to assist with the identification of training requirements and with subsequent training for *all* levels of employees;

(f) to maintain contact with official (i.e. regulatory) and professional bodies involved with health and safety at work;

(g) to keep up to date with safety legislation, recommended codes of practice, health and safety literature etc. and to circulate any relevant information gained therefrom to each level of employee;

(h) to liaise with, and act on any recommendations of, the HSE or Factories Inspectorate following site visits;

(i) to foster, within the company, an understanding that injury prevention and damage control are integral parts of business and operational efficiency;

(j) to set a personal example;

11.5.2 Site Safety Supervisors

Unlike a Safety Officer, who must be employed *full–time* as such, a Site Safety Supervisor may be engaged on other tasks *provided* they do not interfere with, or hinder, the efficient performance of his health and safety supervisory duties. In other words, if an engineer is appointed as Site Safety Supervisor, his health– and safety–related duties must come first – legally!

A typical set of duties and responsibilities for a Site Safety Supervisor might be as follows:

(a) to *read*, *understand* and *implement* the company Safety Policy in so far as it affects individual sites;

(b) to accompany the Safety Officer, or any member of his department, during a site visit;

(c) to act upon any instructions or recommendations given by the

Safety Officer;

(d)   to accompany any member of the Health and Safety Executive during a site visit;

(e)   to ensure that all statutory inspections or examinations (see section 11.7) are carried out and to complete and sign (if appropriate – see also section 11.7) the site register of such;

(f)   to ensure that work method–statements are circulated to all those affected by them and to ensure that a copy of each is kept in the site register;

(g)   to ensure that work method–statements are *strictly* adhered to *and* to report any deviations therefrom;

(h)   to report, at internal site meetings (see section 12.5), on all matters relating to health or safety;

(i)   to set a personal example;

11.6 Safety Representatives and Committees

In accordance with Section 2(4) of the *Health and Safety at Work Act 1974*, and Regulation 3(1) of the *Safety Representatives and Safety Committees Regulations 1977*, a *recognised* trade union (i.e. one defined by sections 29(1) and 30(1) of the *Trade Union and Labour Relations Act 1974*) having membership at the workplace may appoint *Safety Representatives* from amongst the employees.

If a union considers that its members require representation on health and safety matters, a full–time official of such will normally discuss, with the employer, the number of Representatives required and the appropriate qualifications of such. Under Regulation 3(4) of the above Regulations, the *minimum* qualification is adjudged to be *two years employment by the employer* (immediately preceding the appointment) or, failing that, *two years similar experience with another employer*. Dependent upon circumstances, however, and subject to the agreement of the employer, this period may be reduced to one of *twelve months*.

Following the appointment of the appropriate number of Representatives, each Representative is issued with the appropriate credentials by the union and the employer is accordingly notified *in writing*, as soon as possible, of the appointment(s) and the group(s) of employees represented.

The following points should be noted:

(a)   Where Safety Representatives have been appointed, Section 2(6) of the Health and Safety at Work Act places a statutory duty upon employers to consult with them with a view to making and maintaining arrangements for co-operation in promoting, developing and checking the effectiveness of measures to ensure the health and safety of *all* employees whilst at work.

(b)   If a Safety Representative requires training in order to satisfactorily fulfil his functions (see below), Regulation 4(2) of the above Regulations obliges the employer to allow him *paid* time off to complete such. A similar obligation applies in respect of any other time off required by the Representative in connection with the performance of his functions.

(c)   Regulations 7(1) and 7(2) of the above Regulations oblige the employer, upon receipt of reasonable *written* notice, to provide Representatives with, or allow them access to, such information or documentation as is required for them to fulfil their functions. This is subject to the exceptions of information relating specifically to individuals (such as medical records etc.), unless the individual concerned has given the appropriate permission, and any information which the employer is legally prohibited from disclosing or which might compromise national security or due legal process.

(d)   When an employer is requested, *in writing by at least two Representatives*, so to do, he is obliged, by Regulation (2) of the above Regulations, to set up a Safety Committee, within three months of receiving the request, with a composition and brief to be ascertained by negotiation between himself and *all* recognised trades unions having membership at the workplace, and to notify his employees accordingly.

The functions (which, in accordance with the Regulations, may not be construed as constituting statutory *duties*) of a Safety Representative are set out in Regulation (4) of the above Regulations and may be summarised as follows:

(a)   to investigate potential hazards and dangerous occurrences (see section 11.8.1) at the workplace;

(b)   to examine the causes of accidents at the workplace;

(c)   to investigate health− and/or safety−related complaints from the employees he represents;

(d)   to make representations, normally confirmed in writing, to the employer in repect of:

    1.   matters arising from (a), (b) or (c) above;

    2.   health and safety matters generally;

(e)   to carry out workplace inspections − subject to the employer being given *prior written notice* of such;

(f)   to represent employees in discussions with the HSE and other enforcing authorities;

(g)   to receive information, from the HSE or Factories Inspectorate, concerning the health or safety of employees (see section 11.3.2);

(h)   to encourage co−operation between the workforce and the employer in health and safety matters;

(i)   to attend meetings of Safety Committees;

11.7 Statutory Examinations, Inspections and Reports

Each of the examinations listed below must be carried out by a so−called 'competent' person. Whilst there is no legal definition of a 'competent' person, it is advisable that such a person have such theoretical and practical knowledge, and actual experience of the type of machinery and plant to be examined, as will enable him to detect relevant defects or weaknesses and to assess their importance.

The relevent legislation is indicated in each case as in the following examples:

GP 17(2)   −   The Construction (General Provisions) Regulations: Regulation 17(2);

LO 28(1)   −   The Construction (Lifting Operations) Regulations: Regulation 28(1);

WP 8   –   The Construction (Working Places) Regulations:
Regulation 8;

1. Scaffolding, suspended scaffolding, bosuns chairs etc. [WP 22]:

    (a)   weekly *inspection*, more often in bad weather;

2. Excavations, earthworks, trenches, shafts, tunnels, cofferdams and caissons [GP 9]:

    (a)   *thorough examination* before personnel are allowed to work therein;

    (b)   *thorough examination* if damaged or affected by explosives;

    (c)   weekly *thorough examination*, and daily *inspection*, whilst in use;

3. Materials or timber used to construct scaffolding, or to construct or support trenches, excavations, cofferdams or caissons [WP 8; GP 10(1) and 17(2)]:

    (a)   *inspection* on each occasion before use;

4. Cranes, crabs and winches [LO 10(1)(c), 28(1), 28(2) and 28(3)]:

    (a)   *test and thorough examination* once every four years, and after substantial alteration or repair affecting strength or stability;

    (b)   *thorough examination* every fourteen months, and after substantial alteration or repair;

    (c)   weekly *inspection*;

5. Pulley blocks, gin wheels and sheer legs [Regulations as for 4. above]:

    (a)   *test and thorough examination* before first use, and after substantial alteration or repair, unless used only for loads under 1 ton;

    (b)   *thorough examination* every fourteen months, and after substantial alteration or repair;

(c)   weekly *inspection*;

6.   Cranes: anchorage or ballasting arrangements [LO 19(3), 19(4) and 7]:

(a)   *test* before crane is taken into use, after each erection or re-erection on a site, or whenever anchorage or ballasting arrangements have been changed;

(b)   *examination* after exposure of crane to weather conditions likely to have affected its stability – a re-test might be necessary – and after each erection or re-erection;

7.   Cranes (non-mobile): automatic safe load indicator [LO 30(1)]:

(a)   *test* after erection or installation of the crane, and before it is taken into use;

(b)   weekly *inspection*;

8.   Cranes (mobile): automatic safe load indicator [LO 30(2)]:

(a)   *test* before crane is taken into use, after it has been dismantled or after anything has been done which is likely to affect the proper operation of the indicator;

(b)   weekly *inspection*;

9.   Other lifting appliances (i.e. excavator, dragline, piling frame, aerial cableway or ropeway, overhead runway) [LO 28(3) and 10(1)(c)]:

(a)   *thorough examination* every fourteen months, and after substantial alteration or repair;

(b)   weekly *inspection*;

10.   Goods hoists [LO 46 and 10(1)(c)]:

(a)   *test and thorough examination* before use, and after substantial alteration or repair;

(b)   *thorough examination* every six months;

(c)   weekly *inspection*;

11. Passenger hoists [regulations as for 10. above]:

(a)   as with goods hoists, except that an additional *test and thorough examination* is required after re–erection or alteration in the height of travel;

12. Chains and lifting gear except rope and rope slings (LO 34(1), 40 and 41]:

(a)   *test and examination* before first use, and after substantial alteration or repair;

(b)   *thorough examination* every six months, except when used only occasionally;

13. Wire rope and wire rope slings [LO 34(1) and 40]:

(a)   *test and examination* before first use;

(b)   *thorough examination* every six months, except when used only occasionally;

14. Fibre rope and fibre rope slings [LO 40]:

(a)   *thorough examination* every six months, except when used only occasionally;

15. Chains, rings etc. which have been altered or repaired by welding [LO 35]:

(a)   *test and thorough examination* before re–use;

16. Dangerous atmospheres [GP 21(2)]:

(a)   *thorough examination* before personnel are employed therein, and as frequently thereafter as considered necessary;

17. Air receivers on compressors [Factories Act: Section 36]

(a)   *test and thorough examination* (by manufacturer) to determine safe working pressure;

(b)   *thorough examination* every 26 months;

The following should be noted:

(a)   The specified frequencies of examinations and/or inspections stated are the *minimum* required by law. They may be increased (but *not* decreased!) if circumstances so warrant.

(b)   The records and results of the above statutory examinations and inspections must be entered on the appropriate forms (see the individual Regulations for details) which must be retained either on site or at the company's head office, and must be made available for inspection by the enforcing authority *at all reasonable times.*

(c)   Lifting appliances, lifting gear and hoists must be *tested* and *thoroughly examined* by the manufacturer prior to being supplied to potential users. The records and results thereof should be supplied to the user. The validity of any test certificate supplied must be checked by the Safety Officer or Site Safety Supervisor before first use.

11.8 Reporting of Injuries, Diseases and Dangerous Occurrences

In accordance with the *Reporting of Injuries, Diseases and Dangerous Occurrences Regulations 1985 (RIDDOR)*, certain *injuries* and *dangerous occurrences* are required to be reported directly to the appropriate enforcing authority along with *diseases* relating to specific types of work.

11.8.1 Reportable Events

For the purposes of the Regulations, 'reportable events' are classified as followed:

(a)   a death;

(b)   a *specified* major injury or condition (e.g. skull fracture, unconsciousness resulting from the inhalation of any substance);

(c)   an injury which results in more than three *consecutive* days incapacity for work;

(d)   an injury of type (b) or (c) which results in death within one year of incurrence;

(e) a *scheduled* disease diagnosed by a doctor and contracted, or exacerbated, as a *direct* result of work practices (e.g. mesothelioma from contact with asbestos);

(f) a *dangerous occurrence* which either resulted in injury or, if it did not, *might* have resulted in injury (e.g. collapse of scaffolding over 5m high);

Further details, regarding *specified* major injuries and *scheduled* diseases, may be found in Schedule 1 and Parts I and III of RIDDOR.

In the event of *any* of the above occurring, a report must be submitted to the appropriate enforcing authority (which, in the case of a construction site, will be HM Factories Inspectorate) as shown in Figure 11.1.

REPORTABLE EVENT	REPORT REQUIRED
(a), (b), (f)	Immediately by telephone; confirmed in writing, on Form 2508, within 7 days of occurrence;
(c)	In writing, on Form 2508, within 7 day of occurrence;
(d)	In addition to any reports required as above, a further report is required following death;
(e)	In writing, on Form 2508A, following diagnosis;

Figure 11.1 Reporting of 'Reportable Events'

The duty to report rests with the 'responsible person' as shown in Figure 11.2.

The 'responsible person' must take the following *minimum* records and retain them for a *minimum* period of *three years* during which the records must be made available, *at all reasonable times*, to the

REPORTABLE EVENT	SUFFERED BY	RESPONSIBLE PERSON
(a), (b), (c), (e)	An employee	The employer
(a), (b), (c), (e)	A non-employee under training	The person responsible for providing the training
(a), (b), (c)	A self-employed person working in premises under the control of somebody else	The person in control of the premises
(b), (c)	A self-employed person working in premises under his own control	The self-employed person or his representative
(e)	A self-employed person	The self-employed person
(a), (b)	A person not at work but affected by the work of someone else	The person in control of the premises
(f)	—	As above

Figure 11.2 Persons Responsible for Reporting 'Reportable Events'

appropriate enforcing authority:

1.   A death, injury or condition:

   (a)   date and time;

   (b)   the full name of the injured or deceased person;

   (c)   the nature of the injury or condition (if applicable);

   (d)   the place where it occurred;

   (e)   a brief description of the circumstances;

2.   A dangerous occurrence:

   (a)   date and time;

   (b)   the nature of the dangerous occurrence;

(c)   the place where it occurred;

(d)   a brief description of the circumstances;

3.   A scheduled disease:

(a)   the name and occupation of the diseased person;

(b)   the name and nature of the disease;

(c)   the date of its diagnosis;

(d)   any circumstances which might be relevant;

No specified form is required for the keeping of these records. Whilst many of the larger construction companies have their own specific forms for this purpose, the retention of photocopies of Forms F2508 and F2508A or, in the case of an injury or condition, an entry in the site Accident Book, is adjudged sufficient for purposes of compliance with the requirements of RIDDOR.

11.8.2 Accident Investigation

Whilst the law requires certain *basic* information (see section 11.8.1) to be recorded in the event of an accident on site, it should be appreciated that such details are essentially the *minimum* required for legal purposes – if there is the possibility of subsequent legal proceedings being brought – and for the purposes of accident analysis (see section 10.1) by the Health and Safety Executive.

For the principal purpose of determining and dealing with the fundamental (i.e. *root*) cause of the accident, however, a *thorough* investigation is required. In the case of comparatively minor incidents, this will generally be carried out by the Site Safety Supervisor (see section 11.5.2). In the case of *major* incidents, however, the investigation will probably be carried out by the company Safety Officer (see section 11.5.1) in conjunction with the Site Safety Supervisor and the employees' Safety Representative(s) (see section 11.6).

The main objectives of such an investigation are generally recognised to be:

(a)   to determine the cause(s) of the accident with a view to preventing its recurrence;

(b)   to gather information for use in any subsequent civil or criminal proceedings;

(c)   to confirm or refute a subsequent claim, by, or on behalf of, the injured party(s), for industrial injury benefit;

(d)   to prepare notifications to be made to the appropriate enforcing authority (see section 11.8.1);

Whilst it is regrettably common practice (and a peculiarly *British* trait, to judge from the attitudes of those involved in, and reporting upon, the public enquiries into a number of major disasters which have occurred over the years) for such investigations to become obsessed with the need to blame someone, this approach should be *rigorously* avoided – for the simple reason that it tends to obscure the *principal* objective of the exercise which is to determine the underlying cause of the accident and the means by which a recurrence could be prevented.

That said, however, there will be occasions on which an investigation will unearth a blatant disregard of safety procedures on the part of someone (not always the injured party!) and will, as a consequence, lead to disciplinary action being taken or, in the case of a breach of a *statutory* duty, legal proceedings being brought. Such instances, however, are comparatively few and far between.

With the cited objectives in mind, the investigator should firstly establish the following basic facts regarding the *injured party* or, if more than one, *each injured party*:

(a)   name, sex and age;

(b)   address, marital status and occupation;

(c)   whether employed by the Main Contractor, employed by a sub-contractor, self-employed or employed under training;

(d)   time and date of accident;

(e)   nature of injuries received;

(f)   time and date of ceasing work as a result of the accident (note that these may not always be the same as the time and date of the accident itself);

(g)   whether or not the party was disabled in any way prior to sustaining the injuries which resulted from the accident in question;

Having established these facts, the investigator should then proceed to establish the basic facts regarding the *accident*:

(a)   the *exact* location of the accident;

(b)   the nature of the accident (e.g. the injured party fell 10m from scaffolding);

(c)   the *prima facie* cause of the accident (e.g. the injured party slipped);

(d)   the *exact* operation which was being carried out, and which led to the accident;

With a view to determining effective precautionary procedures which might be instituted to reduce the chances of a recurrence, consideration should then be given to establishing whether:

(a)   any *specified* system of work had been established (i.e. whether or not a method statement had been prepared) and, if so, whether it was safe and adhered to;

(b)   the work was undertaken in accordance with good established practice (failing the provision of a specific method statement);

(c)   adequate instructions (or training) had been given to the party responsible for doing the work which resulted in the accident, and whether these instructions had been complied with;

(d)   safe access to the location of the work had been provided;

(e)   any plant or equipment used:

   1.   was suitable for the work being carried out;

   2.   was in good condition, was safe to use and had the appropriate guards fitted;

   3.   had such instructions supplied with it (if appropriate) as to enable it to be used safely and correctly;

4. was being safely and correctly used under adequate supervision;

(f) any tests – e.g. atmospheric tests – were required and, if so, whether they were effected by a suitably experienced person using correct equipment in good working order;

(g) any personal protective equipment or clothing was required and, if so, whether it was provided and used correctly;

(h) any environmental conditions (e.g. wind, rain, untidiness of the work area) may have contributed to the accident;

(i) any *statutory regulation* or *duty of care* had been disregarded and, if so, whether the offender(s) had been aware of this fact;

Finally, it should be mentioned that *all* accidents should be investigated *as soon as possible after their occurrence* – the longer the delay, the less likely it is that the *true* facts will be established.

CHAPTER 12

MANAGING HEALTH AND SAFETY IN CONSTRUCTION

For the purpose of considering health and safety matters throughout the 'life' of a construction project (i.e. from the initial 'bright idea' stage through to final completion), five *key* stages in the project may be identified:

1. the *Design* stage – during which the preliminary and final designs of the project are completed by the Engineer's design staff;

2. the *Contract Documents* stage – at which the Contract Documents are drawn up by the Engineer;

3. the *Selection of a Contractor* – the stage at which the submitted tenders are appraised by the Engineer with a view to the Employer's selection of a suitable Contractor to carry out the construction of the project;

4. the *Construction Planning* stage – the stage, following the award of the Contract, at which the Contractor formulates his plan of action and draws up the construction programme;

5. the *Construction* stage – which is self explanatory and, from the standpoint of health and safety, extends from the Date of Commencement of the Contract to the date of issue of the Maintenance Certificate;

Whilst the coverage hereunder relates *specifically* to the 'conventional' contractual arrangement (i.e. Employer – Engineer –

Main Contractor – Domestic and Nominated Subcontractors), the majority of the principles will be applicable to other types of contract such as management–type contracts (see section 5.4.8) and the like.

12.1 The Design Stage

The 'golden rule' here is that, when designing a project, the design team should always take account of the health and safety of those involved in the construction process by anticipating and minimising potential construction hazards.

For example, much can be done to reduce the risk to construction personnel by minimising the complexity of a steel– or precast concrete–framed structure and thereby reducing the number of occasions on which men are required to work at high level to make connections between structural units. A similar reduction in the level of risk may be achieved by the design allowing for the early installation of floor slabs or permanent staircases to reduce the need for less safe temporary means of access such as scaffolding or ladders. In the latter case, the 'early installation' facility should, of course, be drawn to the Contractor's attention via the Contract Documents (see section 12.2) to enable him to take advantage of such if he so desires.

A further factor to consider, particularly regarding the anticipation of potential hazards, is the possible use of complex, highly specialised or 'frontiers of knowledge' type construction techniques. If such is likely to be the case, expert advice should be sought particularly if the designer is not entirely familiar with the technique(s) in question or is unaware of particular hazards which might be associated with its use.

As if the construction process was not enough to contend with, the design team should *also* give consideration to potential hazards associated with the *maintenance* of the finished product and, in view of the fact that, with the possible exception of Westminster Abbey and other similar structures, everything that goes up will have to come down some day, its eventual *demolition*.

In respect of the maintenance of the finished product, an obvious example which comes to mind is the provision of windows, in high–rise buildings, which may be cleaned from the *inside*. Where this is not possible, suitable equipment for safety of access – e.g. non–ferrous (to avoid corrosion) fixings cast into the roof slab to hold down outriggers for suspended access equipment – should be considered at the design stage. Similarly, the building–in of provisions for fixing scaffolds, which may be required during the life of the building, would prevent a large number of accidents occurring in the course of the subsequent maintenance of the building. In both cases, any relevant data – such as maximum permissable loadings on fixings – should be included in the 'as–built' drawings presented to the eventual owner (who may not be the Employer) of the building.

As far as the eventual demolition of the structure goes, it is sufficient to state that the designer should include, in the 'as–built' drawings or accompanying data, any details which he feels might be of value to those responsible for the demolition.

Under most Conditions of Contract, the *primary* responsibility for matters concerning health and safety on site rests with the Contractor. In the process of carrying out the design, however, the design team might well discover factors which have a bearing on the health and safety of those involved in the construction process. In this event, the Contractor should be informed by the appropriate inclusions in the Contract Documents.

Examples of such factors are:

(a) potentially hazardous ground conditions such as contamination, subsidence (usually in areas where there are old mine workings) or gross instability;

(b) the location of existing underground services such as electricity, water, gas and sewerage;

(c) Permanent Works design factors relevant to the safe design of temporary works in cases where the Permanent Works will be, or are likely to be, supporting the temporary works (e.g. scaffolding lashed to the side of the structure etc.);

(d) special or unusual health or safety hazards associated with the use of specified construction materials – i.e. anything out of the ordinary which the Contractor might overlook or be unaware of;

(e) any feature of the design which might make the partially completed structure temporarily weak or unstable; in this respect, it would be unwise to assume that a possibility of temporary instability or weakness which is obvious to the designer will be equally obvious to the Contractor;

(f)   plans of existing structures on the site which require to be demolished or incorporated into the Permanent Works – particularly if such structures have any features which are likely to make them unstable or dangerous under certain circumstances;

In respect of the general implications of construction upon design, it is vital for the design team to make a note, for further reference, of any construction method–related assumptions made in the course of the design.

For example, the lifting into position of, say, a precast concrete bridge beam may induce hogging stresses which may not be present when the beam is in place. If a particular lifting method has been assumed by the designer in order to quantify those stresses and include the appropriate reinforcement, a note should be made of it with a view to informing the Contractor accordingly. Better still, the design of the beam should include the casting–in of lifting eyes, in the appropriate positions, to ensure that the assumed lifting method is used.

Finally, a general point should be noted in respect of the consideration of safety–related factors which might affect the design and that is that, with the best will in the world, the design team will almost inevitably have overlooked *something*. With this possibility in mind, they should arrange for, and be prepared to deal with, any feedback from the Contractor or the Engineer's site staff regarding hazards identified in the course of construction.

12.2 The Contract Documents Stage

The inclusion of *specific* safety provisions in the Contract Documents is, and always has been, a contentious issue.

One school of thought contends that, if *specific* safety provisions are *not* included, reputable and safety–conscious Contractors will be at a distinct disadvantage when competitively tendering against unscrupulous Contractors who are prepared to cut corners safety–wise in an attempt to win jobs and maximise profits.

The other school of thought contends that the inclusion of specific safety provisions might well improve safety during the construction phase of the project but will also place certain restrictions on the Contractor's freedom to devise his own working methods. Furthermore, the specific provisions might tempt some Contractors into thinking that they are the only safety provisions to consider and that construction safety is automatically ensured thereby (in much the same way as a 30mph sign induces the feeling, in some drivers, that it is safe to travel at 30mph all the time *and* irrespective of prevailing road conditions).

Whilst there is obviously no simple solution, one oft-mooted suggestion which goes some way to ensuring that contracts are not won at the expense of safety is that *priceable* items, which can *clearly* be identified as being *essential* for health and safety, should be included in the Bill of Quantities. Whilst this approach will not necessarily *ensure* subsequent safety, it will at least ensure that the items are not overlooked or ignored by Contractors and will provide the Engineer with a *contractual* remedy in the event of non-performance of the items.

By way of reinforcing their importance, reference to such provisions should be made in the Instructions to Tenderers (see section 6.1) and the Preamble to the Bill of Quantities (see section 8.2.2).

If it is felt that the particular circumstances of the project merit consideration of *specific* safety aspects, special clauses, in addition to the existing 'blanket' safety clauses contained in most standard Conditions of Contract, may be included to cover:

(a) *specific* safety considerations and precautions relating to *specific* operations – without, of course, implying that specific *methods* must be used;

(b) *specific* site rules – dependent upon the circumstances of the project;

(c) compliance with specific Standards or Codes of Practice;

Although it is generally accepted that the Contractor should be free to decide upon his working methods, any *specific* construction methods or sequences required to avoid temporary structural instability, or the generation of excessive or unnecessary 'construction' stresses which might lead to structural failure, must be specified in the Contract Documents advisedly in the form of a 'method' (or 'recipe') Specification (see section 6.4).

In this respect, the designer should not assume that a particular method of construction, or particular sequence of operations, which is blindingly obvious to *him* will be equally obvious to the Contractor (see the 'Water Main' case study in section 3.3).

In the case of a Contract executed in accordance with the I.C.E. Conditions of Contract, these matters are, to a certain extent, covered by Clause 14(1) (see section 9.1.8) which requires the Contractor to submit his construction programme, along with a general description of his proposed methods of construction, for the Engineer's approval. Furthermore, and by way of enabling the Engineer to satisfy himself that the Contractor's proposed methods may be adhered to without detriment to the Permanent Works, Clause 14(3) (see section 9.1.5) requires the Contractor, upon being requested so to do, to submit further details of his proposed construction methods (including those pertaining to Temporary Works and the use of Constructional Plant), along with details of their anticipated effects on the Permanent Works, for the Engineer's approval.

Clearly, any methods or sequences which are unsatisfactory from a safety standpoint can be rejected at this point. However, if this requires the Contractor to possibly re-plan or re-sequence the operations in question, there could be serious financial ramifications – particularly if the Contractor was unaware, at the time of tender, of any method- or sequence-related constraints. Consequently, if any specific methods or sequences *are* required, it is best to include full details in the Tender Documents to let everybody know where they stand before tenders are compiled.

12.3 Selection of Contractors

Whilst the selection of a suitable Contractor to undertake the construction will, in most cases, be *primarily* effected on the basis of a *financial* comparison of the submitted tenders, it is *also* important to assess tenders from a safety standpoint.

If *selective* tendering (see section 7.2) is proposed, health- and safety-related factors should be considered when drawing up the shortlist of prospective tenderers.

Such tenderers should be assessed, for inclusion on the shortlist, on the basis of, firstly, competence in managing health and safety. In the absence of any personal knowlege of such on the part of the Engineer, or the Employer, this will usually be evidenced by the submission of an acceptable written statement of company safety policy (see section 11.4) compiled in accordance with the recommendations of the Health and Safety Executive. If the types of project for which the tenderers are being shortlisted require *specific* safety provisions (which should, of course, be detailed in the pre–qualification invitation) not covered by the company safety policy, the policy should be amended accordingly.

The second assessment criterion should be the past performance of the prospective tenderers from a safety standpoint. There are a number of ways in which a given tenderer's record may be adjudged in this respect, the most obvious of which is to consider the previous incidence of serious accidents or dangerous occurrences on sites under his control and the level of enforcement action (i.e. enforcement notices or prosecutions) which has been previously taken against his company. Similarly, consideration may also be given to the number of *justified* health– or safety–related complaints previously levied against the tenderer both from internal (i.e. employees or their representatives) and external (i.e. the public) sources.

The above information can either be obtained from the prospective tenderers themselves (in which case, a further assessment criterion may be applied, namely: the willingness of a tenderer to supply such information) or from the Health and Safety Executive.

In respect of all of the above criteria, and principally to avoid the 'give a dog a bad name' syndrome, any assessment should be based on the latest available information and, for the purpose of updating an Employer's list of potential tenderers, should be regularly reviewed for indications of possible improvement. For example, a substantial improvement in a Contractor's previously bad safety record might indicate a change in attitude or management, thus making him a more worthwhile proposition for inclusion on a future select list than, say, a Contractor whose previously good safety record showed distinct signs of deteriorating to an unacceptable level.

If, as might possibly be the case with a public sector Employer, *open* tendering (see section 7.1) is proposed, the above assessment should be made following the receipt of tenders and should form part and parcel of the overall appraisal of tenders.

A final aspect which merits consideration is that pertaining to the selection of *sub–contractors* who, advisedly, should be even more stringently appraised, from a safety standpoint, than should *Main Contractors*.

This is largely due to the fact that, whilst many sub–contractors will be large, well–organised firms who, under different circumstances, might well be operating as Main Contractors, many will be small, less well-organised outfits who, for a variety of reasons (mostly financial), tend to place a low priority on safety.

According to the I.C.E. Conditions of Contract (and, indeed, many other equivalent standard Conditions), the Main Contractor is responsible for *all* the activities of *all* sub-contractors working on the site and can, therefore, be held accountable under the Contract (but not *tortiously*, since sub-contractors are, in *this* context, regarded by the law as *independent contractors*) for any accident incurred or caused by a sub-contractor. This being the case, the health- and safety-related appraisal of prospective sub-contractors by the Main Contractor should be based on similar criteria to those detailed above, with the *additional* requirements that, as pre-requisites for their selection, sub-contractors should *formally* accept the Main Contractor's right to manage the site *and* should *formally* undertake to co-operate with the Main Contractor and other sub-contractors in matters of health and safety.

The situation is, of course, slightly different in the case of *nominated* sub-contractors since their selection is effected by the Engineer and not by the Main Contractor. In such cases, therefore, the health and safety appraisal of nominated sub-contractors should advisedly be carried out by the Engineer prior to their appointment. In the event that, following the award of the *main* contract, the Main Contractor has reservations about the safety competence of a nominated sub-contractor, or if a nominated sub-contractor refuses to sign either of the abovementioned undertakings, the Main Contractor should not hesitate to exercise his power of veto over the appointment of the sub-contractor (see section 9.1.6).

12.4 The Construction Planning Stage

This is the stage at which, having been awarded the Contract, the Main Contractor assumes *full* responsibility for the construction of the Works and develops his *outline* plan (drawn up solely for the purpose of compiling his tender) into the more detailed form of a *construction* plan.

For fairly obvious reasons, it is impossible to lay down hard and fast rules relating to the consideration of safety matters at this stage. As a general guide, however, the following list provides a flavour of the sort of health and safety points which should be considered by the Main Contractor when drawing up the construction plan:

(a)　the identification of specific hazards associated with particular operations (see section 10.2);

(b)　the selection and provision of suitable plant and equipment in order to minimise potential hazards;

(c)　the use of safe work practices (see section 10.2);

(d)　the provision of safe working places and safe means of access to, and egress from, such (see section 10.2);

(e)　the sequencing and coordination of construction operations to minimise hazards – e.g. since water and electricity do not generally mix, the possibility of plumbers and electricians being scheduled to work in the same vicinity at the same time should be avoided;

(f)　the need to provide training or instruction on *general* site safety and regarding hazards *specific* to the site and/or certain working practices;

(g)　emergency procedures – e.g. fire, rescue, medical aid;

(h)　the need to carry out environmental monitoring – e.g. for smoke, dust, noise etc. – or health surveillance;

(i)　risks to adjacent properties, buildings or plant – e.g. the method of supporting the sides of a deep excavation adjacent to a neighbouring property; note *also* the need, in this respect, to consider the possibility of committing a *private* or *public nuisance* (see section 2.3) or a breach of *Strict Liability* (see section 2.5);

(j)　safe access to the site – i.e. that the access is positioned such that hazards to site personnel and the general public are minimised; pay special regard to this aspect if, say, a school or old-peoples' home is in the vicinity;

(k)　site layout – to minimise the level of potential risk to site personnel;

(l)　risks to the general public – such as the possibility of passers-by being hit by falling rubble or falling into trenches immediately adjacent to public footpaths and the like;

(m) adequate (i.e. compliant with the requirements of the Construction Regulations – see section 11.2.1) accomodation and welfare facilities for site personnel;

(n) the allocation of responsibilities for health and safety on the site;

12.5 The Construction Stage

In order to get everything off on the right footing, and to let everyone involved in the construction process know where they stand, and what is required of them, a project meeting should be convened, following provision of the basic facilities (site offices etc.) and immediately prior to the start of construction, with the following parties in attendance:

(a) The Site Agent (Chairman);

(b) Senior site supervisory staff – Sub–Agents, General Foreman, Section Foremen;

(c) Company Safety Officer;

(d) Site Safety Supervisor(s);

(e) Senior management and safety advisors from such sub–contract organisations as have been selected at this stage in the proceedings;

(f) Safety Representatives from the workforce;

(g) Representatives from the design team;

In order to 'set the tone' regarding general management of the site, and particularly health and safety management *on* the site, the following health- and safety-related points should be covered:

1. The Construction Programme.

   In this respect, points worthy of particular coverage include:

   (a) the start and finish dates of significant phases of the Works;

   (b) any special measures necessary to prevent hazards arising from *concurrent* work operations;

In the absence of any *overall* management and coordination of their activities, sub-contractors tend to have a distressing habit of concentrating exclusively on their own tasks without paying too much attention as to who they might be affecting in the process. Consequently, particular attention should be paid to the above points where sub-contract organisations are involved.

2.   Planned procedures.

By and large, any complex and/or potentially hazardous operations will have been identified by the Main Contractor at the construction planning stage (see section 12.4) and the appropriate method statements drawn up. If, for any reason, this has *not* been done, it should be dealt with at *this* stage before any work commences. Again, particular attention should be paid to such operations where they are to be performed by, or will involve, sub-contractors. In this respect, *written* method statements should be insisted upon.

In the case of *particularly* hazardous operations – such as those which require to be performed in flammable, explosive or toxic atmospheres – where safety can only be ensured by compliance with *every* aspect of a *detailed* system of work, formal 'permit-to-work' systems should be instituted.

3.   Arrangements for coordination, liaison and communication.

The principal point to emphasise at this stage is that, by way of ensuring *effective* management of the site, the *overall* responsibility for the coordination of site operations should rest fairly and squarely on the shoulders of the Main Contractor's senior manager on site – generally the Site Agent – irrespective of whether specialist coordinators have been appointed for particular sections of the Works.

Since such coordination will involve a considerable amount of liaison between the Main Contractor and the various sub-contract organisations involved, it is vitally important to agree and define clear lines of communication between organisations by way of avoiding the all-too-common confusion which results from not knowing who to deal with.

With particular regard to the communication of health- and safety-related matters, safety supervisors for *each* organisation should be identified. Even though some of the sub-contract organisations may well not have safety supervisors *per se*, someone from each organisation should be nominated as having responsibility for safety. That way, the Main Contractor's Agent (or anybody else, for that matter!) knows who to contact about safety matters. It should also be mentioned that such an arrangement has the additional benefit of precluding (or, at least, *reducing*) the inevitable 'buck-passing' when something goes wrong!

Finally, procedures for liaising with the design team (see section 12.1) should be agreed and defined. In this respect, it should be emphasised (for fairly obvious reasons) that any liaison between sub-contractors and the design team should *always* involve the Main Contractor.

4. Arrangements for monitoring health and safety performance on site.

Firstly, and most importantly, the duties, responsibilities and extent of authority of the Main Contractor's Safety Supervisor should be defined. In other words, if he is responsible for *all* safety on the site, such should be made clear at this point.

Secondly, the arrangements for carrying out regular safety inspections and audits should be defined. If it is intended that the sub-contractors' safety nominees are to be involved in such, in addition to the Main Contractor's Safety Supervisor, it should be stated at this point.

Finally, if a safety committee (see section 11.6) is considered appropriate under the circumstances, its composition and functions should be defined.

5. Safety Representatives.

Where recognised trade unions, having membership on the site, have considered it appropriate to appoint Safety Representatives (see section 11.6) from amongst the workforce, such parties should be identified and their duties and functions defined. Obviously, this should go hand-in-hand with a definition of the arrangements for joint employer/employees consultation on matters of site safety in the course of the construction.

6.   Statutory examinations/inspections and record keeping.

Full details of statutory examinations or inspections (see section 11.7) should be given to all interested parties and the persons responsible for carrying them out should be identified.

The statutory requirement that all records from such should be kept by the Main Contractor in a central site register, to be available for inspection, at any reasonable time, by the appropriate enforcing authority, should be emphasised.

A particular point to note, in this respect, is that, whilst the Main Contractor is not *statutorily* obliged to carry out examinations and inspections of plant or equipment supplied or hired by sub-contractors for their own use, that being *their* responsibility, he is strongly advised to monitor the performance of such and the keeping of the relevant records. If such is to be the case, the appropriate arrangements should be agreed and defined at the outset.

7.   Use of common plant, equipment and facilities.

To avoid the perennial arguments arising as to who is entitled to use what and, more importantly, as to who is responsible if things go wrong, items for common use by the Main Contractor and the various sub-contractors on site should be identified at the outset and the rules and conditions for such usage should be defined. Generally speaking, this will be something to the effect that sub-contractors are entitled to use specified facilities, or items of plant or equipment, which have been provided by the Main Contractor principally for his own use, on condition that they do so in a responsible fashion and, most importantly, that they do so at their own risk.

8.   External liaison.

In this respect, contact points for liaison with external bodies − such as the HSE or Factories Inspectorate, the emergency services, the local authorities and suchlike − should be provided. Most importantly, the point should be made that external liaison should *always* be carried out *via* the Main Contractor except where statutorily required otherwise (such as in the case of an injury to a sub-contractor's employee where the duty to report lies with the sub-contractor − see section 11.8.1).

9. Arrangements for training, instruction and the dissemination of health and safety information.

    As discussed previously (see section 12.4), the Main Contractor should assess, at the planning stage, any requirement regarding workforce training or instruction in respect of *general* site safety, or in respect of the avoidance of hazards *specific* to the site and/or certain working practices, and should make the appropriate arrangements for ensuring that such is provided. By way of encouraging them to discharge *their* duties in this respect (which, it should be mentioned, are of the *statutory* variety in their capacities as employers), the sub-contractors' policies in this respect must be clarified.

    Additionally, any arrangements proposed by way of ensuring an adequate supply of health and safety information to the workforce – which, again, is a *statutory* requirement – should be agreed and defined.

10. Responsibilities of sub-contractors.

    First and foremost, sub-contractors' *statutory* responsibilities as employers (see sections 11.3.1 and 11.8.1) should be emphasised.

    Additionally, since the Main Contractor is *contractually* responsible for *all* the activities (including the safety thereof) of *all* sub-contractors, such parties should confirm their pre-tender agreements (see section 12.3) to co-operate with the Main Contractor, particularly in matters of health and safety. Furthermore, in view of the Main Contractor's *overall* responsibility for the coordination of *all* site operations (see section 12.4), the sub-contractors should formally undertake not to deviate from planned programmes and/or agreed procedures without the consent of the Main Contractor.

    Finally, in the event that particular sub-contractors propose to use sub-contractors of their own, the appropriate arrangements should be made to ensure that such 'sub-sub-contractors' (for want of a better term!) adhere to safety standards equivalent to those required of the sub-contractors themselves.

11. Responsibilities of individuals.

    In this particular context, it is sufficient to state that the Main

Contractor should ensure that the appropriate arrangements are made by way of enabling each individual employed on the site to be fully apprised of his/her statutory responsibilities as an employee and of any other *specific* responsibilities accorded them by virtue of their position or appointment.

Although there appears from the aforegoing that there is a considerable amount to cover at this initial project meeting, particularly since safety matters will constitute only a *part* of the overall agenda, it should be re-emphasised that the whole point of the meeting is to establish the 'ground rules' and to leave nobody in any doubt as to where they stand. In this light, therefore, it is of considerable importance to cover each and every one of the aforementioned points.

As might be imagined, things do not (or, at least, *should* not) stop with the initial project meeting. In order to continue in the same vein, *further* project meetings (with the same participants as detailed above) should be held at regular intervals (usually at the end of each month) throughout the construction period at which health and safety should be given *at least equal priority* with other matters (and *not*, as is regrettably the case in a number of instances, something which is cursorily covered as the last item on the agenda before adjourning for lunch at the nearest pub!).

Whilst the format of each meeting will depend upon prevailing circumstances, the following points should be covered under the general heading of 'health and safety':

1. Safety performance.

   Principally, an appraisal should be made of the effectiveness of the existing arrangements for ensuring the maintenance of health and safety on the site and, if necessary, improvements thereto should be considered for possible implementation. By way of ensuring the effectiveness of any safety committee which has been set up, and in order to emphasise the importance of 'worker-participation' (which is the whole point of having such a committee in the first place!), any recommendations therefrom should be *fully* considered and the appropriate action taken.

   Additionally, since nothing concentrates the mind quite like the prospect of being publicly 'pilloried', poor health and safety performance, *particularly* by sub-contractors, should be identified and *firm* commitments sought to improve matters.

2. Particular safety problems.

   Such problems will generally be found to emanate from one or other (or both!) of two principal sources: the design of the structure or the work practices used in its construction.

   If, on the one hand, particular design features are considered to be the source of abnormal, and largely unforeseen, health or safety hazards, the possibility of reducing the hazards by design modification should be investigated (obviously in conjunction with the design team as mentioned above). On the other hand, abnormal hazards arising through particular work practices should be alleviated either by modification of the relevant procedures or, if not possible, by the introduction of additional safety precautions.

   In both cases, it should be noted that a certain amount of re-scheduling may be required to allow sufficient time for modifications to be made or the appropriate safety precautions to be devised.

3. The construction programme.

   It is generally the case that the bulk of any given monthly project meeting is devoted to the construction programme – or, more precisely, to why it has not been adhered to and to what can be done about it! From a safety standpoint, however, the discussion of the programme should be carried out with a view to taking the appropriate corrective action in the event that departures from the planned work schedule have given rise to possible sequence-related hazards.

   As an example of the sort of hazards which may arise, and of the corrective action which might be taken in such an event, consider the construction of a steel-framed multi-storey building. Whilst the project network may indicate the possibility of an overlap between the steelwork erection and floor-construction operations, the latter may well have been sensibly re-scheduled to avoid floor-workers being at risk from falling materials. In the event of steel erection being delayed by a few days inclement weather, and a consequent re-scheduling of the floor construction operations resulting in an unavoidable delay to the project completion, the only solution would be to take advantage of the aforementioned overlap. Since this would revive the risk of materials falling on workers, the obvious corrective action would be to introduce some sort of overhead protection for the floor workers.

INDEX